Constructive Negations and Paraconsistency

TRENDS IN LOGIC
Studia Logica Library

VOLUME 26

SCOPE OF THE SERIES

Trends in Logic is a bookseries covering essentially the same area as the journal *Studia Logica* – that is, contemporary formal logic and its applications and relations to other disciplines. These include artificial intelligence, informatics, cognitive science, philosophy of science, and the philosophy of language. However, this list is not exhaustive, moreover, the range of applications, comparisons and sources of inspiration is open and evolves over time.

Volume Editor
Heinrich Wansing

The titles published in this series are listed at the end of this volume.

Sergei P. Odintsov

Constructive Negations and Paraconsistency

Springer

Sergei P. Odintsov
Russian Academy of Sciences
Siberian Branch
Sobolev Institute of Mathematics
Koptyug Ave. 4
Novosibirsk
Russia

ISBN 978-1-4020-6866-9 e-ISBN 978-1-4020-6867-6

Library of Congress Control Number: 2007940855

Printed on acid-free paper.

9 8 7 6 5 4 3 2 1

springer.com

Contents

Chapter 1

Introduction

The title of this book mentions the concepts of paraconsistency and constructive logic. However, the presented material belongs to the field of paraconsistency, not to constructive logic. At the level of metatheory, the classical methods are used. We will consider two concepts of negation: the negation as reduction to absurdity and the strong negation. Both concepts were developed in the setting of constrictive logic, which explains our choice of the title of the book. The *paraconsistent* logics are those, which admit inconsistent but non-trivial theories, i.e., the logics which allow one to make inferences in a non-trivial fashion from an inconsistent set of hypotheses. Logics in which all inconsistent theories are trivial are called *explosive*. The indicated property of paraconsistent logics yields the possibility to apply them in different situations, where we encounter phenomena relevant (to some extent) to the logical notion of inconsistency. Examples of these situations are (see [86]): information in a computer data base; various scientific theories; constitutions and other legal documents; descriptions of fictional (and other non-existent) objects; descriptions of counterfactual situations; etc. The mentioned survey by G. Priest [86] may also be recommended for a first acquaintance with paraconsistent logic. The study of the paraconsistency phenomenon may be based on different philosophical presuppositions (see, e.g., [87]). At this point, we emphasize only one fundamental aspect of investigations in the field of paraconsistency. It was noted by D. Nelson in [65, p. 209]: "In both the intuitionistic and the classical logic all contradictions are equivalent. This makes it impossible to consider such entities at all in mathematics. It is not clear to me that such a radical position regarding contradiction is necessary." Rejecting the principle "a contradiction implies everything" (*ex contradictione quodlibet*) the paraconsistent logic allows one to study the phenomenon of contradiction itself. Namely this formal logical aspect of paraconsistency will be at the centre of attention in this book.

We now turn to constructive logic. Constructive logic is the logic of constructive mathematics, logic oriented on dealing with the universe of constructive mathematical objects. The common feature of different variants of constructive mathematics is the rejection of the concept of actual infinity and admitting only the existence of objects constructed on the base of the concept of potential infinity. In any case, passing to constructive logic from the classical one changes the sense of logical connectives. For example, Markov [60] defines the constructive disjunction as follows: "The constructive understanding of the existence of a mathematical object corresponds to the constructive understanding of the disjunction of sentences of the form "P or Q". Such a sentence is considered as accepted if at least one of the sentences P, Q was accepted as true." Of course, this understanding of disjunction does not allow one to accept the law of excluded middle and leads to the rejection of classical logic. In the setting of constructive logic, there are two basic approaches to the concept of negation and they are considered in our investigation.

Since the Brouwer works, the negation of statement P, $\neg P$, is understood as an abbreviation of the statement "assumption P leads to a contradiction". Note that this concept agrees well with paraconsistency. The above understanding of negation does not assume the principle "contradiction implies everything" (*ex contradictione quodlibet*) responsible for the trivialization of inconsistent theories. The first formalization of intuitionistic logic suggested by A.N. Kolmogorov [44] in 1925 was paraconsistent. In this work, A.N. Kolmogorov reasonably noted that *ex contradictione quodlibet* (in the form $\neg p \to (p \to q)$) has appeared only in the formal presentation of classical logic and does not occur in practical mathematical reasoning. However, A. Heyting was sure that using *ex contradictione quodlibet* is admissible in intuitionistic reasoning and he added the axiom $\neg p \to (p \to q)$ to his variant **Li** of intuitionistic logic [35]. Note that adding *ex contradictione quodlibet* creates some problems with interpretation of **Li** as calculus of problems [45]. One cannot consider the implication $P \to Q$ as the problem of reducing the problem Q to the problem P. In **Li**, the implication $P \to Q$ means that the problem Q can be reduced to the problem P or the problem P is meaningless. This difficulty was known to A. Heyting, but he did not considered this as a serious problem. According to A. Heyting [36, p. 106], "... it (*ex contradictione quodlibet* — S.O.) adds to the precision of the definition of implication" and "I shall interpret implication in this wider sense."

Only in 1937 I. Johansson [41] questioned the using of *ex contradictione quodlibet* in constructive reasoning and suggested the system, which we denote by **Lj**. Axiomatics for **Lj** can be obtained by deleting *ex contradictione quodlibet* from the standard list of axioms for intuitionistic logic, more exactly, $\mathbf{Li} = \mathbf{Lj} + \{\neg p \to (p \to q)\}$. In [41], Johansson proved that many important properties of negation provable in the Heyting logic **Li** can be proved also in the system **Lj**. Since that the logic has the name "Johansson's logic" or "minimal logic"(see the title of Johansson's article). Note that, in fact, Johansson came back to the Kolmogorov's variant of intuitionistic logic. More exactly, the implication-negation fragment of **Lj** coincides with the propositional fragment of the system from [44]. Kolmogorov considered the first-order logic, but in the language with only two propositional connectives, implication and negation.

Unfortunately, the logic **Lj** was for a long time on the borderline of studies in the field of paraconsistency, which was traditionally motivated by the following "paraconsistent paradox" of **Lj**. Although **Lj** is not explosive, admits non-trivial inconsistent theories, we can prove in **Lj** for any formulas φ and ψ that

$$\varphi, \neg\varphi \vdash_{\mathbf{Lj}} \neg\psi.$$

This means that the negation makes no sense in inconsistent **Lj**-theories, because all negated formulas are provable in them. In this way, inconsistent **Lj**-theories are positive. It should be noted that studies in the field of paraconsistency were directed during a long period to searching for "the most natural system" of paraconsistent logic, which is maximally close to classical logic (cf. [39, p. 147]). The above paradox obviously shows that **Lj** cannot play the role of such logic. However, recently more attention has been paid to the study of paraconsistent analogs of well-known logical systems. In this respect, Johansson's logic **Lj** is worthy of attention as a paraconsistent analog of intuitionistic logic **Li**.

Turning to the second main approach to negation in constructive logic, the concept of strong negation. Note that the strong negation is namely a proper constructive negation.

As happens with most fundamental logical concepts, the concept of strong negation was developed independently by many authors and with different motivations. Constructive logic with strong negation was suggested for the first time by D. Nelson in 1949 [64]. The truth of a negation of statement in intuitionistic and minimal logic can be stated only indirectly, via reducing a negated sentence to an absurdity. As a consequence of this, the negation in these logics has the following feature, unsatisfiable from the constructive point of view. When the negation of a conjunction $\neg(\varphi \wedge \psi)$ is provable, it does not follow in the general case that either $\neg\varphi$, or $\neg\psi$ is provable. In the mentioned work, D. Nelson suggested a new constructive interpretation of the negation connective based on the idea that the falseness of atomic formulas can be seen directly, which leads to parallel constructive procedures reducing the truth and falseness of complex statements to the truth and falseness of their components. As a result, D. Nelson obtained a logical system possessing the property:

$$\text{if } \vdash\sim(\varphi \wedge \psi), \quad \text{then } \vdash\sim\varphi \text{ or } \vdash\sim\psi,$$

where $\sim$ denotes the negation connective and $\vdash$ the derivability in Nelson's system. Now, the above property is traditionally considered as a characteristic property of constructive negation, and Nelson-type negations are called *strong*. One year later constructive logic with strong negation was considered by A.A. Markov [59]. The propositional variant of Nelson's logic was studied by N.N. Vorobiev [114, 115, 116]. Independently, Gentzen-style calculus equivalent to Nelson's system was developed by F. von Kutschera [49]. A system closely related to strong negation systems also arose in the work by J.P. Cleave [18], who constructed the predicate calculus adequate for the algebra of inexact sets by S. Körner [46]. The paraconsistent variant of Nelson's system was studied independently by R. Routley (later R. Sylvan) in the propositional case in [96], by Lopez-Escobar in [51] and by Nelson himself [1], both in the first-order case. It should be noted that the term "strong negation" is connected not with the idea of direct falsification, but with comparing strong and intuitionistic negations in the explosive variant of Nelson's logic [64]. In this logic, one can define an intuitionistic negation $\neg$ via a strong negation as follows $\neg\varphi := \varphi \rightarrow\sim\varphi$, and prove the implication $\sim\varphi \rightarrow \neg\varphi$ showing that the negation $\sim$ is stronger than the intuitionistic one. In the paraconsistent version of Nelson's logic, one cannot define the intuitionistic negation and the above comparison loses its meaning, but traditionally the name "strong negation" is used also in this case.

We now say a few words about denotation of logics under consideration. There is no generally accepted convention. Nelson used the denotation N and N^- for his system of strong negation and for its paraconsistent variant (see [64, 1]), respectively. In Dunn's systematization [26], these systems receive the denotation **N** and $\mathbf{BN}_1$, respectively. We will follow another tradition (see, e.g., [120]) and denote explosive Nelson's logic by **N3** and paraconsistent Nelson's logic by **N4**. This choice is motivated by the Kripke-style semantics for these logics. Kripke semantics for **N3** was developed by R. Thomason [107] and R. Routley [96]. As in the case of intuitionistic logics,

N3-frames are partial orderings. But since verification and falsification are treated in **N3** independently, **N3**-models have two valuations, v^+ for verification and v^- for falsification, with the additional restriction that $v^+(p) \cap v^-(p) = \varnothing$, i.e., no atomic statement can be true and false in the same world simultaneously. Omitting the latter restriction we obtain a semantics for **N4**. It is not hard to check (see [93]) that from the pair (v^+, v^-) one can pass to one many-valued valuation, which is three-valued (true, false, neither) in case of **N3** and four-valued (true, false, neither, both) in case of **N4**.

Of course, the logic **N4** is more attractive for applications, because it allows one to work with inconsistent information. A view of **N4** as a logic convenient for information representation and processing is reflected in a series of books (see [40, 117, 118]). Also, **N4** has proved useful for solving some well-known philosophical logic paradoxes [119, 121]. At the same time, the attention paid to this logic is incomparable with that for **N3**. In particular, semantic investigations of **N4** were restricted mainly to the case of Kripke-style semantics. There was no specific information about the class of **N4**-extension, except for the information about its proper subclass, the class of **N3**-extensions. It should be noted that the latter was thoroughly studied (see [33, 47, 99, 100, 101]).

Thus, we have two explosive logics **Li** and **N3**, and their paraconsistent analogs **Lj** and **N4**. It will be shown that **Li** can be faithfully embedded into **Lj**, whereas **N3** is faithfully embedded into **N4**. In this way, refusing the explosion axiom does not lead to a decrease in the expressive power of a logic. Here arises the question: which new expressive possibilities have the logics **Lj** and **N4** as compared to the explosive logics **Li** and respectively **N3**, and how regularly this family of new possibilities is structured? In this book we give answers to these question by studying the lattices of extensions of the logics **Lj** and **N4**.

Studying the lattices of extensions of different logics such as, e.g., the intuitionistic logic **Li** (see, e.g., [16]), the normal modal logic **K4** [30, 31], etc., plays an extremely important part in the development of modern non-classical logic. In the first part of the book (Chapters 2–7) we concentrate on the study of the class of extensions of Johansson's logic. This was the first attempt to systematically study the lattice of extensions for a paraconsistent logic. We will see that there is one important feature, which distinguishes the class of **Lj**-extensions from the classes of extensions of the explosive logics **Li** and **K4**. The class Jhn of non-trivial extensions of minimal logic has a non-trivial and interesting global structure (it is three-dimensional in some sense), which allows one to reduce its description (to some extent) to the well-studied classes of intermediate and positive logics.

More exactly, the class Jhn is the disjunctive union of three classes: the class Int of intermediate logics, which are explosive; the class Neg of negative logics, i.e., logics with degenerate negation containing the scheme $\neg p$; and the class Par of properly paraconsistent extensions of **Lj** containing logics which do not belong to the first two classes. So we have

$$\mathsf{Jhn} = \mathsf{Int} \cup \mathsf{Neg} \cup \mathsf{Par}.$$

Note that negative logics are definition equivalent to positive ones.

For any $L \in \mathsf{Par}$, one can define its intuitionistic counterpart L_{int} (negative counterpart L_{neg}) as the least logic from the class Int (respectively, from the class Neg) containing L. There are strong translations (i.e., translations preserving the consequence relation) of logics L_{int} and L_{neg} to the original paraconsistent L. The logic L_{int} may also be obtained by adding *ex contradictione quodlibet* to L. In this way, the above-mentioned translation of L_{int} shows that the usual explosive reasoning can be modelled in a paraconsistent logic. On the other hand, as was noted above, the important advantage of paraconsistent logics is that they allow one to distinguish contradictions: different contradictions are not equivalent in them. In case of **Lj**-extensions, the structure of contradictions in the paraconsistent logic L can be presented as a formal system, and namely the logic L_{neg} plays this part. The strong translation of L_{neg} in L can be done via the contradiction operator $C(\varphi) := \varphi \wedge \neg\varphi$. Due to this fact, the logic L_{neg} can really be treated as the logic of contradictions of the logic L.

We conclude our study of the class Jhn with an effort to describe the structure of Jhn up to the *negative equivalence*. Two logics $L_1, L_2 \in \mathsf{Jhn}$ are said to be *negatively equivalent* if they have the same negative consequence relation, i.e., $X \vdash_{L_1} \neg\varphi$ iff $X \vdash_{L_2} \neg\varphi$ for an arbitrary set of formulas X and any formula φ. The negative equivalence of logics from **Lj** is equivalent to the fact that they have the same family of inconsistent sets of formulas. From the constructive point of view, these facts mean that negatively equivalent logics have essentially the same concepts of negation and of contradiction.

Concluding the first part of the book, we suggest a way to overcome the above mentioned paradox of minimal logic. It can be done via introducing the unary operator of absurdity $A(\varphi)$ instead of the constant $\bot$ and defining the negation as the reduction to this generalized absurdity:

$$\neg\varphi := \varphi \to A(\varphi).$$

The idea of such a definition arose from comparing the contradiction operator in the logic **Le** of classical refutability [22] with the necessity operator in Łukasiewicz's modal logic Ł [52, 53]. For the first time, a similar interconnection between **Le** and Ł was noted by Porte [84, 85]. We prove that one of the modal paradoxes of Ł exactly corresponds to the fact that the absurdity operator is constant, i.e., is like in **Le**. Moreover, it turns out that negation in several well-known paraconsistent logics can be defined in this way. For example, in the logic **CLuN** of Batens [5, 6] and in Sette's maximal paraconsistent logic P^1 [102, 88], the negation can be presented as the reduction to a unary absurdity operator.

In the second part of the book we study the lattice of extensions of paraconsistent Nelson's logic. This investigation was motivated not only by the interest in Nelson's logic as an alternative formalization of intuitionistic logic, but also by the desire to prove whether is it possible to apply to this new object the approach developed in the first part of our work? The answer to this question is positive, although we discovered essential differences in the structures of lattices of extensions of minimal logic and paraconsistent Nelson's logic.

In connection with the paraconsistent Nelson logic there also arises a question: in which language should this logic be considered? The explosive **N3** is usually considered in the language $\langle \vee, \wedge, \rightarrow, \sim, \neg \rangle$ with symbols for two negations, strong $\sim$ and intuitionistic $\neg$. As was noted above, the intuitionistic negation is superfluous in this case, because it can be defined via the strong one. If we pass to the paraconsistent **N4**, the interpretation of $\neg$ is not clear and it looks natural to consider the language with only the negation symbol $\sim$. This variant of the paraconsistent Nelson logic will be denoted **N4**. However, it turns out that the presence of intuitionistic negation is natural and desirable. The conservative extension of **N4** in the language $\langle \vee, \wedge, \rightarrow, \sim, \bot \rangle$ obtained by spreading **N4**-axioms to the new language and adding axioms $\bot \rightarrow p$ and $p \rightarrow \sim \bot$ for the new constant is denoted $\mathbf{N4}^\bot$. The intuitionistic negation is defined in $\mathbf{N4}^\bot$ in the usual way, $\neg\varphi := \varphi \rightarrow \bot$.

To study the class $\mathcal{E}\mathbf{N4}$ ($\mathcal{E}\mathbf{N4}^\bot$) of extensions of Nelson's logic **N4** ($\mathbf{N4}^\bot$) we need adequate algebraic semantics. This means that we have to describe the variety of algebras determining **N4** ($\mathbf{N4}^\bot$) such that there is a dual isomorphism between the lattice of subvarieties of this variety and the lattice of **N4**($\mathbf{N4}^\bot$)-extensions. For explosive **N3**, the algebraic semantics is provided by N-lattices, which are well studied [90, 28, 29, 33, 99, 100, 110]. The **N4**-lattices introduced in [72] provide this kind of semantics for **N4**. The algebraic semantics for $\mathbf{N4}^\bot$ is provided by $\mathbf{N4}^\bot$-lattices, a natural modification of **N4**-lattices. An interesting peculiarity of **N4**(and $\mathbf{N4}^\bot$)-lattices is that they have a non-trivial filter of distinguished values.

The advantage of the language with intuitionistic negation becomes obvious, when we start the investigation of the class of $\mathbf{N4}^{\perp}$-extensions. Its structure differs essentially from that of Jhn. First of all, unlike Jhn containing the subclass Neg of contradictory logics, $\mathbf{N4}^{\perp}$ does not admit contradictory extensions. Despite its paraconsistency the logic $\mathbf{N4}^{\perp}$ admits only local contradictions, adding any contradiction as a scheme to $\mathbf{N4}^{\perp}$ results in a trivial logic. However, the class $\mathcal{E}\mathbf{N4}^{\perp}$ decomposes into subclasses of explosive logics, normal logics, and logics of general form. This decomposition reflects the local structure of contradictions inside $\mathbf{N4}^{\perp}$-models and is very similar to the decomposition of Jhn into subclasses of intermediate, negative and properly paraconsistent logics.

Note that the negative equivalence relation, which played an important role in the study of extensions of minimal logic, degenerates if we pass to $\mathbf{N4}(\mathbf{N4}^{\perp})$-extensions. Two extensions of $\mathbf{N4}$ ($\mathbf{N4}^{\perp}$) are negatively equivalent if and only if they are equal.

We shall now describe more precisely the structure of the book. Chapter 2 contains definitions of the most important logics from the class Jhn and necessary information concerning algebraic and Kripke-style semantics for **Lj**-extensions. Chapter 3 is devoted to the logic of classical refutability, the maximal paraconsistent extension of **Lj** playing the key role in the studying the class of **Lj**-extensions.

In Chapter 4, we investigate the logic $\mathbf{Le}' = \mathbf{Lj} + \{\perp \vee (\perp \to p)\}$ and prove that the class of its extensions coincides with the class of all possible intersections of intermediate and negative logics. Moreover, any logic L extending $\mathbf{Le}'$ has a unique presentation as an intersection of intermediate logic L_1 and negative logic L_2. The logic L_1 (resp., L_2) will be taken as *intuitionistic* (resp., *negative*) *counterpart* of L. The notions of intuitionistic and negative counterparts allow a generalization to the class of all **Lj**-extensions and it turns out that the class Par of properly paraconsistent **Lj**-extensions decomposes into a disjoint union of classes $Spec(L_1, L_2)$ consisting of all logics having L_1 and L_2 as its intuitionistic and negative counterparts, respectively. Each of the classes $Spec(L_1, L_2)$ forms an interval in the lattice Par with the upper point $L_1 \cap L_2$. In this way, studying the structure of Jhn reduces to the investigation of intervals of the form $Spec(L_1, L_2)$.

The next chapter will be devoted to constructing an adequate algebraic semantics, in fact, a suitable presentation of j-algebras, which is convenient to determine the location of different logics inside the intervals $Spec(L_1, L_2)$. The effectiveness of the obtained presentation will be demonstrated via its application to numerous extensions of **Lj** considered by K. Segerberg [98]. We also provide several facts concerning Kripke semantics for **Lj**-extensions.

In Chapter 6, we introduce the negative equivalence of logics (see above), which we denote as $\equiv_{neg}$, and by modifying the technique of Jankov's formulas prove that the quotient lattice $Spec(L_1, L_2)/\equiv_{neg}$ is isomorphic to the interval $Spec(\mathbf{Lk}, L_2)$. We also prove that every interval $Spec(L_1, L_2)$ contains infinitely many classes of negative equivalence and that there is a continuum of negative equivalence classes in Jhn.

The last chapter of the first part of the book, Chapter 7, will be devoted to studying absurdity as a unary operator.

Chapter 8 starts the second part of the book, devoted to strong negation. In the first section, we define two variants of paraconsistent Nelson's logic. The logic **N4** is determined in the language $\langle \vee, \wedge, \rightarrow, \sim \rangle$, where $\sim$ is a symbol for strong negation, whereas the logic $\mathbf{N4}^{\perp}$ is a logic in the language $\langle \vee, \wedge, \rightarrow, \sim, \perp \rangle$ with an additional constant $\perp$. Moreover, $\mathbf{N4}^{\perp}$ is a conservative extension of **N4** as well as of intuitionistic logic. The explosive logic **N3** is obtained by adding to **N4** the explosion axiom $\sim p \rightarrow (p \rightarrow q)$. Notice that by putting $\perp :=\sim (p_0 \rightarrow p_0)$ one can prove in **N3** the additional axioms of $\mathbf{N4}^{\perp}$.

In the second section, the logic **N4** is characterized via Fidel structures [29]. This is direct generalization of M. Fudel's result for **N3** obtained in [29]. Fidel structures are implicative lattices augmented with a family of unary predicates.

In the third section, we describe a semantics for **N4** with the help of twist-structures over implicative lattices (see [28, 110]). The completeness result will follow from the equivalence of Fidel structures and twist-structures, also established in this section. A twist-structure is an algebraic structure defined over the Cartesian square of an implicative lattice, the operations of this structure agrees with the operations of the underlying implicative lattice on the first component and are "twisted" on the second component.

Further, in Section 4 of this chapter, we prove that the class of algebras isomorphic to twist-structures admits a lattice theoretical definition. We distinguish the class of **N4**-lattices, prove that any twist-structure is an **N4**-lattice and that any **N4**-lattice $\mathcal{A}$ is isomorphic to a twist-structure over $\mathcal{A}_{\bowtie}$, the implicative lattice defined as quotient of $\mathcal{A}$ wrt to a congruence of a special form. These results imply that **N4** is characterized by **N4**-lattices.

In the next section, it is proved that **N4**-lattices form a variety $\mathcal{V}_{\mathbf{N4}}$ such that the lattice $\mathcal{E}\mathbf{N4}$ of **N4**-extensions is dually isomorphic to the lattice of subvarieties of $\mathcal{V}_{\mathbf{N4}}$.

In the last section of Chapter 8, we transfer all these results to the logic $\mathbf{N4}^{\perp}$ and the lattice of its extensions $\mathcal{E}\mathbf{N4}^{\perp}$. In this case, the twist-structures are defined over Heyting algebras and for any $\mathbf{N4}^{\perp}$-lattice $\mathcal{A}$, the quotient

$\mathcal{A}_{\bowtie}$ is also a Heyting algebra. We call $\mathcal{A}_{\bowtie}$ the basic Heyting algebra of an $\mathbf{N4}^{\perp}$-lattice $\mathcal{A}$.

In Chapter 9, we develop the origins of the algebraic theory of $\mathbf{N4}^{\perp}$-lattices necessary to study the lattice of extensions of the logic $\mathbf{N4}^{\perp}$. In particular, $\mathbf{N4}^{\perp}$-lattices are represented in the form of Heyting algebras with distinguished filter and ideal. We define a pair of adjoint functors between categories of $\mathbf{N4}^{\perp}$-lattices and of Heyting algebras. We prove that if a homomorphism of basic algebras can be lifted to $\mathbf{N4}^{\perp}$-lattices, it can be done in a unique way. It is shown that congruences on an $\mathbf{N4}^{\perp}$-lattice are in one-to-one correspondence with implicative filters and that the lattices of congruences of an $\mathbf{N4}^{\perp}$-lattices and of its basic algebra are isomorphic. As a consequence, we describe subdirectly irreducible $\mathbf{N4}^{\perp}$-lattices as lattices with subdirectly irreducible basic algebra. Finally, in terms of the above-mentioned representation, we formulate an embeddability criterion and describe the quotients of $\mathbf{N4}^{\perp}$-lattices.

In the last chapter, we study the structure of the lattice of $\mathbf{N4}^{\perp}$-extensions and show that it is similar to the structure of the class of **Lj**-extensions. Although the distinctions of the structures of these two classes of logics are also essential. The first of these distinctions is that $\mathbf{N4}^{\perp}$ has no contradictory extensions, whereas minimal logic has the subclass of inconsistent extensions isomorphic to the class of extensions of positive logic.

We investigate the interrelations between a logic L extending $\mathbf{N4}^{\perp}$ and its intuitionistic fragment. In the lattice $\mathcal{E}\mathbf{N4}^{\perp}$, we distinguish the subclasses Exp of explosive logics, Nor of normal logics, and Gen of logics of general form, which play the roles similar to that of classes Int, Neg, and Par in the lattice of extensions of minimal logic. The interrelations between classes Exp, Nor and Gen are investigated with the help of notions of explosive and normal counterparts for logics in Gen.

Finally, we give some first applications of the developed theory of the lattice of $\mathbf{N4}^{\perp}$-extensions. First, we completely describe the lattice of extensions of the logic $\mathbf{N4}^{\perp}\mathbf{C}$ obtained by adding the Dummett linearity axiom to $\mathbf{N4}^{\perp}$. We prove that all extensions of $\mathbf{N4}^{\perp}\mathbf{C}$ are finitely axiomatized and decidable and that given a formula, one can effectively determine which of the $\mathbf{N4}^{\perp}\mathbf{C}$-extensions is axiomatized by this formula. Second, we describe tabular, pretabular logics and logics with Graig's interpolation property in the lattice of $\mathbf{N4}^{\perp}$-extensions.

Regarding the authorship of the presented results, this book contains mainly the investigations of the author, previously published in a series of articles [66–81]. Chapter 2 and Section 8.1 have a preliminary character and here we do not carefully trace the authorship of the presented results. Except for Chapter 2 and Section 8.1, we give explicit references to all results quoted from other authors.

Acknowledgments. I am deeply indebted to Professors L.L. Maksimova and K.F. Samokhvalov for our fruitful discussions, which inspired, in fact, the beginning of this investigation. The investigations presented in the first part of the book were carried out during my stay in Toruń, at the Logic Department of Nicholas Copernicus University. I am very grateful to Prof. Jerzy Perzanowski, the head of this department, for the invitation, hospitality and helpful criticism. I want to acknowledge my deep indebtedness to the Alexander von Humboldt Foundation for granting the research fellowship at Dresden University of Technology and the return fellowship. The investigations presented in the second part of the book were carried out during this period. Finally, I am especially grateful to Prof. Heinrich Wansing, my academic host in Dresden, for the very fruitful collaboration.

Part I

Reductio ad Absurdum

Chapter 2

Minimal Logic. Preliminary Remarks

2.1 Definition of Basic Logics

A *propositional language* $\mathcal{L}$ is a finite set of logical connectives of different arities, $\mathcal{L} = \{f_1^{n_1}, \ldots, f_k^{n_k}\}$. A *propositional constant* is a connective of arity 0. Given a set of propositional variables, we define formulas of the language $\mathcal{L}$ via the standard inductive definition.

In the first part of the book we will consider logics and deductive systems formulated in the following propositional languages: the language of positive logic $\mathcal{L}^+ := \{\wedge^2, \vee^2, \rightarrow^2\}$, the language $\mathcal{L}^\perp := \mathcal{L}^+ \cup \{\perp^0\}$ extending $\mathcal{L}^+$ with the constant $\perp$ for "absurdity", and the language $\mathcal{L}^\neg := \mathcal{L}^+ \cup \{\neg^1\}$ with the symbol $\neg$ for negation. Extensions of minimal logic admit equivalent formulations in the languages $\mathcal{L}^\perp$ and $\mathcal{L}^\neg$.

If φ is a formula in some propositional language and $p_1, \ldots, p_n$ are propositional variables, the denotation $\varphi(p_1, \ldots, p_n)$ means that all propositional variables of φ are from the list $p_1, \ldots, p_n$.

By a *logic* we mean a set of formulas closed under the rules of substitution and *modus ponens*:

$$\frac{\varphi(p_1, \ldots, p_n)}{\varphi(\psi_1, \ldots, \psi_n)} \quad \text{and} \quad \frac{\varphi \quad \varphi \rightarrow \psi}{\psi}.$$

If $\varphi(\psi_1, \ldots, \psi_n)$ is obtained from $\varphi(p_1, \ldots, p_n)$ by the substitution rule, we say that it is a *particular case* or a *substitution instance* of φ. A *deductive system* is a collection of axioms and inference rules. A *theorem* of a deductive system is a formula provable in this system. We will usually define logics as sets of theorems of Hilbert style deductive systems with only the inference rules of substitution and *modus ponens.* Therefore, to define a logic it is enough to list its axioms.

For a logic L and a set of formulas X, $L + X$ denotes the least logic containing L and all formulas of X. The symbol $+$ also denotes the operation of taking the least upper bound in the lattice of logics.

With any logic L, we associate in a standard way an inference relation $\vdash_L$. For a set of formulas X and a formula φ, the relation $X \vdash_L \varphi$ means that φ can be obtained from elements of X and tautologies of L in a finite number of steps by using the rule of *modus ponens.* A set X is said to be *non-trivial wrt* L if $X \nvdash_L \varphi$ for some φ.

Let L_i be a logic in a propositional language $\mathcal{L}_i$, $i = 1, 2$, and $\mathcal{L}_1 \subseteq \mathcal{L}_2$. We say that L_2 is a *conservative extension of* L_1 if $L_1 \subseteq L_2$ and for any formula φ in the language $\mathcal{L}_1$,

$$\varphi \in L_1 \iff \varphi \in L_2.$$

In this case we say also that L_1 is an $\mathcal{L}_1$*-fragment of* L_2. In what follows by a *positive* fragment we mean an $\mathcal{L}^+$-fragment.

Denote by $\mathcal{F}^*$ the *trivial logic*, i.e., the set of all formulas in the language $\mathcal{L}^*$, $* \in \{+, \bot, \neg\}$.

We now define several important logics. In the choice of denotation we follow the book [93] by W. Rautenberg.

Positive logic **Lp** is the least logic in the language $\mathcal{L}^+$ containing the following axioms:

1. $p \to (q \to p)$
2. $(p \to (q \to r)) \to ((p \to q) \to (p \to r))$
3. $(p \wedge q) \to p$
4. $(p \wedge q) \to q$
5. $(p \to q) \to ((p \to r) \to (p \to (q \wedge r)))$
6. $p \to (p \vee q)$
7. $q \to (p \vee q)$
8. $(p \to r) \to ((q \to r) \to ((p \vee q) \to r))$

Positive logic satisfies Deduction Theorem:

$$X \cup \{\varphi\} \vdash_{\mathbf{Lp}} \psi \iff X \vdash_{\mathbf{Lp}} \varphi \to \psi.$$

To prove this theorem we need axioms 1 and 2 of positive logic and the fact that *modus ponens* is the only inference rule. All logics considered in the book satisfy these conditions, therefore, Deduction Theorem remains true for all logics considered below.

Classical positive logic $\mathbf{Lk}^+$ also is a logic in the language $\mathcal{L}^+$ and can be axiomatized modulo $\mathbf{Lp}$ by either of the following two axioms:

P. $((p \to q) \to p) \to p$ (Peirce law)

E. $p \vee (p \to q)$ (extended law of excluded middle)

The version $\mathbf{Lj}^\perp$ of *minimal logic* (or *Johansson's logic*) in the language $\mathcal{L}^\perp$ can be defined as a logic axiomatized by the axioms 1–8 above. The equivalent version of minimal logic $\mathbf{Lj}^\neg$ in the language $\mathcal{L}^\neg$ with the negation symbol can be axiomatized by the axioms 1–8 and the following axiom:

A. $(p \to q) \to ((p \to \neg q) \to \neg p)$ (*reductio ad absurdum*)

To make precise the statement on the equivalence of two versions of minimal logic, we define the translations θ from the language $\mathcal{L}^\neg$ to $\mathcal{L}^\perp$ and ρ from $\mathcal{L}^\perp$ to $\mathcal{L}^\neg$ as follows.

For any $\varphi \in \mathcal{F}^\neg$, let $\theta(\varphi)$ be a formula in the language $\mathcal{L}^\perp$ obtained from φ by replacing each subformula of the form $\neg\psi$ by the subformula $\psi \to \perp$. For any $\varphi \in \mathcal{F}^\perp$, we denote by $\rho(\varphi)$ a formula in the language $\mathcal{L}^\neg$ obtained from φ by replacing every occurrence of $\perp$ by the subformula $\neg(p \to p)$, where p is some fixed propositional variable. For a set of formulas $X \subseteq \mathcal{F}^\neg$, denote by $\theta(X)$ the set $\{\theta(\varphi) \mid \varphi \in X\}$. Respectively, for $X \subseteq \mathcal{F}^\perp$, put $\rho(X) := \{\rho(\varphi) \mid \varphi \in X\}$.

Proposition 2.1.1 *The following statements hold.*

1. *For an arbitrary set of formulas $X \subseteq \mathcal{F}^\neg$ and for any formula $\varphi \in \mathcal{F}^\neg$,*

   $$X \vdash_{\mathbf{Lj}^\neg} \varphi \text{ if and only if } \theta(X) \vdash_{\mathbf{Lj}^\perp} \theta(\varphi).$$

   Moreover, $\mathbf{Lj}^\neg \vdash \varphi \leftrightarrow \rho\theta(\varphi)$ *for any formula* φ.

2. *For an arbitrary set of formulas $X \subseteq \mathcal{F}^\perp$ and for any formula $\varphi \in \mathcal{F}^\perp$,*

   $$X \vdash_{\mathbf{Lj}^\perp} \varphi \text{ if and only if } \rho(X) \vdash_{\mathbf{Lj}^\neg} \rho(\varphi).$$

   Moreover, $\mathbf{Lj}^\perp \vdash \varphi \leftrightarrow \theta\rho(\varphi)$ *for any formula* φ.

Thus, the translations defined above preserve all deductive properties and the subsequent application of two translations results in a formula equivalent to the original one. Due to these facts we pass freely in the following from the language $\mathcal{L}^{\perp}$ to the language $\mathcal{L}^{\neg}$ and vice versa. We will omit the superscripts in the denotation of minimal logic and will not explicitly indicate with which version of minimal logic or of its extension we are dealing at the time. And we write $\mathcal{F}$ for either $\mathcal{F}^{\perp}$ or $\mathcal{F}^{\neg}$.

Intuitionistic logic **Li** and *minimal negative logic* **Ln** can be axiomatized modulo minimal logic in the language $\mathcal{L}^{\perp}$ as follows:

$$\mathbf{Li} = \mathbf{Lj} + \{\perp \to p\}, \quad \mathbf{Ln} = \mathbf{Lj} + \{\perp\};$$

and in the language $\mathcal{L}^{\neg}$ as follows:

$$\mathbf{Li} = \mathbf{Lj} + \{\neg p \to (p \to q)\}, \quad \mathbf{Ln} = \mathbf{Lj} + \{\neg p\}.$$

Classical logic **Lk**, *logic of classical refutability* **Le**, and *maximal negative logic* **Lmn** can be axiomatized modulo intuitionistic logic **Li**, minimal logic **Lj**, and negative logic **Ln** respectively, via either the Peirce law **P** or the extended law of excluded middle **E**.

$$\mathbf{Lk} = \mathbf{Li} + \{p \vee (p \to q)\}, \ \mathbf{Le} = \mathbf{Lj} + \{p \vee (p \to q)\}, \ \mathbf{Lmn} = \mathbf{Ln} + \{p \vee (p \to q)\}.$$

The positive fragments of **Lk**, **Le**, and **Lmn** coincide with classical positive logic, whereas the positive fragments of **Li**, **Lj**, and **Ln** coincide with positive logic.

$$\mathbf{Lk}^{+} = \mathbf{Lk} \cap \mathcal{F}^{+} = \mathbf{Le} \cap \mathcal{F}^{+} = \mathbf{Lmn} \cap \mathcal{F}^{+},$$

$$\mathbf{Lp} = \mathbf{Li} \cap \mathcal{F}^{+} = \mathbf{Lj} \cap \mathcal{F}^{+} = \mathbf{Ln} \cap \mathcal{F}^{+}.$$

All logics introduced above except for positive and classical positive logics are extensions of minimal logic. The class of all non-trivial extensions of the logic **Lj** we denote by Jhn, the class of all extensions by Jhn^{+}. Clearly, the class of logics Jhn^{+} forms a lattice, where $L_1 + L_2$ is the least upper bound of logics L_1 and L_2, and the intersection $L_1 \cap L_2$ is the greatest lower bound. For an arbitrary logic L, the lattice of its extensions with lattice operations $+$ and $\cap$ we denote as $\mathcal{E}L$. Notice that $\mathcal{E}L$ is a complete lattice. For any family $\{L_i \mid i \in I\}$ of logics in $\mathcal{E}L$, the intersection $\bigcap_{i \in I} L_i$ is a logic and it extends L. Obviously, $\bigcap_{i \in I} L_i$ is the greatest logic contained in all logics L_i. For this reason, in $\mathcal{E}L$, there also exists the *sum of logics* $\Sigma_{i \in I} L_i$, i.e., the least logic containing all logics L_i, $i \in I$.

Recall several important formulas provable in **Lp** and **Lj**.

Proposition 2.1.2 *The following formulas are provable in* **Lp***:*

1. $p \to p$ *(the identity law).*
2. $(p \vee q) \leftrightarrow (q \vee p)$, $(p \wedge q) \leftrightarrow (q \wedge p)$ *(the commutativity of* $\vee$ *and* $\wedge$*).*
3. $(p \vee (q \vee r)) \leftrightarrow ((p \vee q) \vee r)$, $(p \wedge (q \wedge r)) \leftrightarrow ((p \wedge q) \wedge r)$ *(the associativity of* $\vee$ *and* $\wedge$*).*
4. $(p \vee (q \wedge r)) \leftrightarrow ((p \vee q) \wedge (p \vee r))$, $(p \wedge (q \vee r)) \leftrightarrow ((p \wedge q) \vee (p \wedge r))$ *(the distributivity laws).*
5. $(p \to (q \to r)) \leftrightarrow (q \to (p \to r))$ *(the permutation law).*
6. $(p \to (p \to q)) \leftrightarrow (p \to q)$ *(the contraction law).*
7. $(p \to (q \to r)) \leftrightarrow ((p \wedge q) \to r)$ *(import and export of the premiss).*
8. $((p \to q) \vee r) \to ((p \vee r) \to (q \vee r))$.
9. $((p \to q) \wedge r) \leftrightarrow ((p \wedge r) \to (q \wedge r)) \wedge r$.

□

Proposition 2.1.3 *The following formulas are provable in* **Lj***:*

1. $\neg\neg(p \vee \neg p)$,
2. $(p \to \neg q) \to (q \to \neg p)$,
3. $(p \to q) \to (\neg q \to \neg p)$,
4. $(\neg p \vee \neg q) \to \neg(p \wedge q)$,
5. $\neg(p \vee q) \leftrightarrow (\neg p \wedge \neg q)$,
6. $p \to \neg\neg p$,
7. $\neg(p \wedge \neg p)$,
8. $(p \vee q) \to \neg(\neg p \wedge \neg q)$,
9. $(p \wedge q) \to \neg(\neg p \vee \neg q)$,
10. $(p \to q) \to \neg(p \wedge \neg q)$

For the *proof* of this and the previous proposition the reader may consult one or another traditional textbook in classical logic and observe that the standard proofs of formulas listed in these propositions use only axioms of **Lj** or **Lp** respectively. It is also not hard to prove all these formulas directly or with the help of Deduction Theorem.

□

The next proposition gives some information on the results of adjoining different new axioms to **Lj**.

Proposition 2.1.4 [98]

1. *The equality* $\mathbf{Lk} = \mathbf{Lj} + \{\varphi\}$ *holds, where* φ *is one of the following formulas:*

   (a) $\neg\neg p \to p$,

   (b) $(\neg p \to q) \to (\neg q \to p)$,

   (c) $(\neg p \to \neg q) \to (q \to p)$,

   (d) $\neg(\neg p \wedge \neg q) \to (p \vee q)$,

   (e) $\neg(\neg p \vee \neg q) \to (p \wedge q)$,

   (f) $\neg(p \wedge \neg q) \to (p \to q)$.

2. $\mathbf{Lk} \neq \mathbf{Lj} + \{p \vee \neg p\} = \mathbf{Lj} + \{(p \to q) \to (\neg p \vee q)\}$.

3. $\mathbf{Lj} + \{p \vee \neg p\} \neq \mathbf{Lj} + \{\neg p \vee \neg\neg p\} = \mathbf{Lj} + \{\neg(p \wedge q) \to (\neg p \vee \neg q)\}$.

4. $\mathbf{Li} = \mathbf{Lj} + \{(\neg p \vee q) \to (p \to q)\}$.

The next proposition shows how to construct axioms of an intersection of logics. In fact, we repeat the proof of Miura's result [63] for intersections of superintuitionistic logics (see also [16, p. 111]). We call the formula $\varphi(p_1, \dots, p_m) \vee \psi(p_{n+1}, \dots, p_{n+m})$ the *repeatedless disjunction* of the formulas $\varphi(p_1, \dots, p_n)$ and $\psi(p_1, \dots, p_m)$ and denote it by $\varphi \underline{\vee} \psi$.

Proposition 2.1.5 *Let* $\mathbf{L} \in \{\mathbf{Lp}, \mathbf{Lj}\}$, $L_1 = \mathbf{L} + \{\varphi_i \mid i \in I\}$, *and* $L_2 := \mathbf{L} + \{\psi_j \mid j \in J\}$. *Then* $L_1 \cap L_2 = \mathbf{L} + \{\varphi_i \underline{\vee} \psi_j \mid i \in I, j \in J\}$.

Proof. Suppose $\chi \in L_1 \cap L_2$. By Deduction Theorem and the properties of $\wedge$ (see Proposition 2.1.2) we have $\bigwedge_{i \in I'} \varphi_i' \to \chi \in \mathbf{L}$ and $\bigwedge_{j \in J'} \psi_j' \to \chi \in \mathbf{L}$, where $I' \subseteq I$, $J' \subseteq J$, I' and J' are finite, and every φ_i' and ψ_j' are substitution instances of φ_i and ψ_j respectively. Using axiom 8 of positive logic and the distributivity laws, we then obtain

$$\bigwedge_{i\in I',j\in J'} (\varphi_i' \vee \psi_j') \to \chi \in \mathbf{L},$$

from which $\chi \in \mathbf{L} + \{\varphi_i \underline{\vee} \psi_j \mid i \in I, j \in J\}$.

Conversely, assume that $\chi \in \mathbf{L} + \{\varphi_i \underline{\vee} \psi_j \mid i \in I, j \in J\}$. Then χ is derivable in $\mathbf{L}$ from some finite set of substitution instances $\varphi_i' \vee \psi_j'$ of axioms of this logic. Using axioms 6 and 7 of positive logic we can also derive χ from the set of φ_i' as well as from the set of ψ_j'. Consequently, $\chi \in L_1 \cap L_2$.

□

Proposition 2.1.6 *The lattices $\mathcal{E}\mathbf{Lp}$ and $\mathcal{E}\mathbf{Lj}$ are distributive. Moreover, the intersection distributes with the infinite sum in these lattices.*

Proof. Let $\mathbf{L} \in \{\mathbf{Lp}, \mathbf{Lj}\}$. We prove only that

$$L \cap \Sigma_{i\in I} L_i = \Sigma_{i\in I}(L \cap L_i),$$

$L, L_i \in \mathcal{E}\mathbf{L}$. Assume $L = \mathbf{L} + \Gamma$ and $L_i = \mathbf{L} + \Delta_i$ for $i \in I$.

$$\begin{aligned} L \cap \Sigma_{i\in I} L_i &= (\mathbf{L}+\Gamma) \cap (\mathbf{L} + \textstyle\bigcup_{i\in I} \Delta_i) \\ &= \mathbf{L} + \{\varphi \underline{\vee} \psi \mid \varphi \in \Gamma, \psi \in \textstyle\bigcup_{i\in I} \Delta_i\} \\ &= \mathbf{L} + \textstyle\bigcup_{i\in I} \{\varphi \underline{\vee} \psi \mid \varphi \in \Gamma, \psi \in \Delta_i\} \\ &= \Sigma_{i\in I}(\mathbf{L} + \{\varphi \underline{\vee} \psi \mid \varphi \in \Gamma, \psi \in \Delta_i\}) \\ &= \Sigma_{i\in I}(\mathbf{L}+\Gamma) \cap (\mathbf{L} + \Delta_i) \end{aligned}$$

□

2.2 Algebraic Semantics

In this section, we give necessary definitions and facts concerning the algebraic semantics for propositional logics. Detailed information can be found in [92, 93].

Let $\mathcal{L} = \{f_1^{n_1}, \ldots, f_k^{n_k}\}$ be a propositional language. An *algebra* $\mathcal{A} = \langle A, f_1^{\mathcal{A}}, \ldots, f_k^{\mathcal{A}}\rangle$ *of* the language $\mathcal{L}$ is a set, where the connectives of $\mathcal{L}$ are interpreted as operations of respective arities, $f_i^{\mathcal{A}} : A^{n_i} \longrightarrow A$. The set A is the *universe* of $\mathcal{A}$ and is denoted $|\mathcal{A}|$. We write $a \in \mathcal{A}$ instead of $a \in |\mathcal{A}|$.

If $\mathcal{A}_1, \ldots, \mathcal{A}_n$ are algebras of the same language, then the direct product $\mathcal{A}_1 \times \ldots \times \mathcal{A}_n$ is defined as an algebra whose universe is the direct product of universes $|\mathcal{A}_1| \times \ldots \times |\mathcal{A}_n|$ and the operations are componentwise. Note that $\pi_i : |\mathcal{A}_1| \times \ldots \times |\mathcal{A}_n| \longrightarrow |\mathcal{A}_i|$, the projection onto the ith coordinate, determines an epimorphism of $\mathcal{A}_1 \times \ldots \times \mathcal{A}_n$ onto $\mathcal{A}_i$.

By $\mathcal{A} \hookrightarrow \mathcal{B}$ we denote that the algebra $\mathcal{A}$ is embeddable into $\mathcal{B}$, i.e., that there exists a monomorphism $h : \mathcal{A} \to \mathcal{B}$.

If $\mathcal{K}$ is a class of algebras, we denote by $\mathsf{I}(\mathcal{K})$ the class of algebras isomorphic to algebras from $\mathcal{K}$, by $\mathsf{H}(\mathcal{K})$ the class of homomorphic images of algebras from $\mathcal{K}$, by $\mathsf{S}(\mathcal{K})$ the class of algebras embeddable into algebras from $\mathcal{K}$, and, finally, $\mathsf{Up}(\mathcal{K})$ denotes the class of algebras isomorphic to ultraproducts [14] of algebras from $\mathcal{K}$.

A *matrix* is, as usual, a pair $\mathcal{M} = \langle \mathcal{A}, D^{\mathcal{A}} \rangle$, where $\mathcal{A}$ is an algebra and $D^{\mathcal{A}} \subseteq |\mathcal{A}|$ is the set of distinguished elements of this matrix. In cases where $D^{\mathcal{A}} = \{1\}$ is one-element, we write $\langle \mathcal{A}, 1 \rangle$ instead of $\langle \mathcal{A}, \{1\} \rangle$ and identify, in fact, a matrix with an algebra in a language with an additional constant 1. A *valuation* in an algebra $\mathcal{A}$ is defined as a mapping from the set of propositional variables into $|\mathcal{A}|$. A valuation extends to the set of all formulas in a homomorphic way. A formula φ is said to be *true* on a matrix $\mathcal{M} = \langle \mathcal{A}, D^{\mathcal{A}} \rangle$, $\mathcal{M} \models \varphi$, if for any $\mathcal{A}$-valuation v, $v(\varphi) \in D^{\mathcal{A}}$. An identity $\varphi = \psi$ is *true* on an algebra $\mathcal{A}$, $\mathcal{A} \models \varphi = \psi$, if $v(\varphi) = v(\psi)$ for any $\mathcal{A}$-valuation v. The set $L(\mathcal{M}) := \{\varphi \mid \mathcal{M} \models \varphi\}$ is the *logic* of a matrix $\mathcal{M}$ and the set $Eq(\mathcal{A}) := \{\varphi = \psi \mid \mathcal{A} \models \varphi = \psi\}$ is the *equational theory* of an algebra $\mathcal{A}$. For a class $\mathcal{K}$ of matrices (algebras), we define $L(\mathcal{K}) := \bigcap\{L(\mathcal{M}) \mid \mathcal{M} \in \mathcal{K}\}$ $(Eq(\mathcal{K}) := \bigcap\{Eq(\mathcal{A}) \mid \mathcal{A} \in \mathcal{K}\})$.

The direct product of matrices $\mathcal{M}_1 = \langle \mathcal{A}_1, D^{\mathcal{A}_1} \rangle, \ldots, \mathcal{M}_n = \langle \mathcal{A}_n, D^{\mathcal{A}_n} \rangle$ is defined as

$$\mathcal{M}_1 \times \ldots \times \mathcal{M}_n = \langle \mathcal{A}_1 \times \ldots \times \mathcal{A}_n, D^{\mathcal{A}_1} \times \ldots \times D^{\mathcal{A}_n} \rangle.$$

Since the operations on the direct product are componentwise, we have

$$L(\mathcal{M}_1 \times \ldots \times \mathcal{M}_n) = L\mathcal{M}_1 \cap \ldots \cap L\mathcal{M}_n.$$

In this part of the book we deal mainly with matrices having one distinguished element. Let $\mathcal{A}$ be an algebra of the language $\mathcal{L}^+ \cup \{1\}$ $(\mathcal{L}^\perp \cup \{1\}$, $\mathcal{L}^\neg \cup \{1\})$. Note that $\mathcal{A} \models \varphi$ is equivalent to $\varphi = 1 \in Eq(\mathcal{A})$. Elements of $L\mathcal{A}$ are called *identities of* $\mathcal{A}$ in this case.

An algebra $\mathcal{A}$ is a *model for* a logic L if $L \subseteq L\mathcal{A}$. If $L = L\mathcal{A}$, we say that $\mathcal{A}$ is a *characteristic model for* L. It is clear that the class of models for some logic L forms a variety. Write $\mathcal{A} \models L$ if $\mathcal{A}$ is a model of L. Denote $Mod(L) := \{\mathcal{A} \mid \mathcal{A} \models L\}$.

Proposition 2.2.1 [93] *Every* **Lj**-*extension has a characteristic model.*

The reader is expected to be familiar with lattices and with distributive lattices.

If $\mathcal{A} = \langle A, \wedge, \vee \rangle$ is a lattice, then the lattice ordering $\leq_{\mathcal{A}}$ is defined by the condition $a \leq_{\mathcal{A}} b \iff a \wedge b = a$. If $a, b \in \mathcal{A}$ and $a \leq_{\mathcal{A}} b$, we denote by $[a, b]$ an interval wrt the lattice ordering with end points a and b, i.e., $[a, b] := \{c \in \mathcal{A} \mid a \leq_{\mathcal{A}} c \leq_{\mathcal{A}} b\}$. In what follows we omit the lower index in the denotation of the ordering if it does not lead to a confusion.

For $a \leq b$ and $c \in [a, b]$, an element d is said to be a *complement of* c *in the interval* $[a, b]$ if $c \vee d = b$ and $c \wedge d = a$. Recall that if the lattice $\mathcal{A}$ is distributive and $c \in \mathcal{A}$ has a complement in $[a, b] \subseteq \mathcal{A}$, then this complement is unique.

An algebra $\mathcal{A} = \langle A, \wedge, \vee, \rightarrow, 1 \rangle$ is called an *implicative lattice* if $\langle A, \wedge, \vee, 1 \rangle$ is a lattice with the greatest element 1 and such that for any $a, b \in A$ there exists a supremum $\bigvee\{x \mid a \wedge x \leq b\}$ equal to the value of the *implication* (or *relative pseudo-complement*) operation $a \rightarrow b$. Here $\leq$ denotes the lattice ordering of $\mathcal{A}$. All implicative lattices form a variety and the logic of this variety is **Lp** [92].

Proposition 2.2.2 [92] *Let $\mathcal{A}$ be an implicative lattice and $a, b \in \mathcal{A}$. Then the following holds.*

1) $a \rightarrow b = 1$ *iff* $a \leq b$;

2) $a = b$ *iff* $a \rightarrow b = 1$ *and* $b \rightarrow a = 1$;

3) $b \leq a \rightarrow b$;

4) $a \wedge (a \rightarrow b) = a \wedge b$.

□

By a *j-algebra* we mean an algebra $\mathcal{A} = \langle A, \wedge, \vee, \rightarrow, \bot, 1 \rangle$ of the language $\mathcal{L}^{\bot} \cup \{1\}$ such that $\langle A, \wedge, \vee, \rightarrow, 1 \rangle$ is an implicative lattice and the constant $\bot$ is interpreted as an arbitrary element of the universe A. Minimal logic **Lj** corresponds to the variety of j-algebras [92], which we denote as $\mathcal{V}_j$. Equivalently, we can define j-algebras in the language $\mathcal{L}^{\neg} \cup \{1\}$ as implicative lattices with the negation operation satisfying the property $a \rightarrow \neg b = b \rightarrow \neg a$. These equivalent definitions are related as follows: $\neg a = a \rightarrow \bot$, $\bot = \neg 1$.

A *Heyting algebra* is a j-algebra with the least element $\bot$. Intuitionistic logic **Li** is the logic of the variety $\mathcal{V}_i$ of Heyting algebras [92].

A *negative algebra* is a j-algebra with $\bot = 1$. Obviously, negative algebras are distinguished in the variety of j-algebras via the identity $\bot$. Therefore, minimal negative logic **Ln** is the logic of the variety $\mathcal{V}_n$ of negative j-algebras.

An arbitrary variety $\mathcal{V}$ of implicative lattices, j-algebras, Heyting algebras, or negative algebras determines a logic $L\mathcal{V}$ extending **Lp**, **Lj**, **Li**, or respectively **Ln**. More exactly, let $Sub(V)$ denote the lattice of subvarieties of an arbitrary variety V. For a logic $L \in \mathsf{Jhn}^+$, define a variety of j-algebras

$$V(L) := \{\mathcal{A} \mid \mathcal{A} \in \mathcal{V}_j,\ \mathcal{A} \models L\},$$

and for a variety $V \in Sub(\mathcal{V}_j)$, define a logic

$$L(V) := \{\varphi \mid \mathcal{A} \models \varphi \text{ for all } \mathcal{A} \in V\}.$$

It is clear that for any $L \in \mathsf{Jhn}^+$ and $V \in Sub(\mathcal{V}_j)$ we have

$$V(L) \in Sub(\mathcal{V}_j) \text{ and } L(V) \in \mathsf{Jhn}^+.$$

Moreover, the following statement holds.

Theorem 2.2.3 *The mappings*

$$V : \mathsf{Jhn}^+ \to Sub(\mathcal{V}_j) \text{ and } L : Sub(\mathcal{V}_j) \to \mathsf{Jhn}^+$$

are mutually inverse dual lattice isomorphisms. The restrictions $V\restriction_{Sub(\mathcal{V}_i)}$ *and* $L\restriction_{\mathcal{E}\mathbf{Li}}$ *are mutually inverse dual isomorphisms of the lattices* $\mathcal{E}\mathbf{Li}$ *and* $Sub(\mathcal{V}_i)$*, whereas* $V\restriction_{Sub(\mathcal{V}_n)}$ *and* $L\restriction_{\mathcal{E}\mathbf{Ln}}$ *are mutually inverse dual isomorphisms of the lattices* $\mathcal{E}\mathbf{Ln}$ *and* $Sub(\mathcal{V}_n)$*.*

□

By a dual isomorphism of lattices $\mathcal{A}_1$ and $\mathcal{A}_2$ we mean an isomorphism from the lattice $\mathcal{A}_1$ onto the lattice $(\mathcal{A}_2)^{op}$ with the inverse ordering.

For a j-algebra $\mathcal{A}$ and a Heyting algebra $\mathcal{B}$ we denote by $\mathcal{A}\oplus\mathcal{B}$ the direct sum of these algebras. It is a j-algebra in which the unit element of $\mathcal{A}$ is identified with the zero of $\mathcal{B}$, and for any $a \in \mathcal{A}$ and $b \in \mathcal{B}$, we have $a \leq b$.

Recall that a non-empty subset F of an implicative lattice (a j-algebra) $\mathcal{A}$ is a *filter* on $\mathcal{A}$ if it satisfies the following conditions: 1) for any $x, y \in F$, $x \wedge y \in F$; 2) for any $x \in F$ and $y \in \mathcal{A}$, if $x \leq y$, then $y \in F$. Denote by $\mathcal{F}(\mathcal{A})$ the set of all filters on $\mathcal{A}$. If $X \subseteq \mathcal{A}$, then $\langle X\rangle$ denotes a *filter generated by* the set X, i.e., the least filter on $\mathcal{A}$ containing X. It is clear that

$$\langle X\rangle = \{a \in \mathcal{A} \mid b_1 \wedge \ldots \wedge b_n \leq a \text{ for some } b_1, \ldots, b_n \in X\}.$$

Instead of $\langle\{a\}\rangle$ we write $\langle a\rangle$.

For a Heyting algebra $\mathcal{A}$, denote by $F_d(\mathcal{A})$ its *filter of dense elements* and by $\mathcal{R}(\mathcal{A})$ *the Boolean algebra of its regular elements.* Recall that

$$F_d(\mathcal{A}) = \{a \in \mathcal{A} \mid \neg\neg a = 1\} = \{a \vee \neg a \mid a \in \mathcal{A}\},$$

$$\mathcal{R}(\mathcal{A}) = \{a \in \mathcal{A} \mid a \vee \neg a = 1\} = \{a \in \mathcal{A} \mid \neg\neg a = a\},$$

and $\mathcal{R}(\mathcal{A}) \cong \mathcal{A}/F_d(\mathcal{A})$.

Let $\mathcal{A}$ be an implicative lattice (a j-algebra). For any congruence θ on $\mathcal{A}$, we define a filter $F_\theta := \{a \in \mathcal{A} \mid a\theta 1_\mathcal{A}\}$. For any filter F on $\mathcal{A}$, define a congruence $\theta_F := \{(a,b) \mid a \to b, b \to a \in F\}$. It is clear that $\theta = \theta_{F_\theta}$ and $F = F_{\theta_F}$. We have thus defined a one-to-one correspondence between the set of congruences and the set of filters on the implicative lattice (j-algebra) $\mathcal{A}$. Notice that for an identity congruence $Id_\mathcal{A}$, $F_{Id_\mathcal{A}} = \{1_\mathcal{A}\}$.

We define a *subdirectly irreducible* algebra $\mathcal{A}$ as an algebra, which has minimal non-identity congruence (comp. [14]). Taking into account the above correspondence between filters and congruences on implicative lattices and j-algebras, we arrive at the following statement.

Proposition 2.2.4 *An implicative lattice (a j-algebra) is subdirectly irreducible if and only if*

$$\{1_\mathcal{A}\} \neq \bigcap\{F \mid F \textit{ is a filter on } \mathcal{A},\ F \neq \{1_\mathcal{A}\}\}.$$

□

An element $\star_\mathcal{A}$ of an implicative lattice (a j-algebra) $\mathcal{A}$ is called an *opremum*, if $\star_\mathcal{A} \neq 1$ and for any $a \in \mathcal{A}$, the inequality $a \neq 1$ implies $a \leq \star_\mathcal{A}$.

Proposition 2.2.5 *An implicative lattice (a j-algebra) $\mathcal{A}$ is subdirectly irreducible iff it has an opremum.*

□

For Heyting algebras, a similar result was stated by C.M. McKay [62]. It can be transferred to implicative lattices and j-algebras in a trivial way.

Due to the well-known Birkhoff theorem [14], any algebra $\mathcal{A}$ is isomorphic to a subdirect product of subdirectly irreducible algebras (being homomorphic images of $\mathcal{A}$). This immediately implies that every variety is completely determined by its subdirectly irreducible algebras. Let $Mod_{fsi}(L)$ denote the set of finitely generated subdirectly irreducible models of a logic L. In view of the correspondence between logics and varieties, we have the following

Proposition 2.2.6 *Let* L_1 *and* L_2 *be logics extending* **Lp** (**Lj**). *We have* $L_1 = L_2$ *if and only if* $Mod_{fsi}(L_1) = Mod_{fsi}(L_2)$.

□

We call a *Peirce algebra* an implicative lattice satisfying the identity **P** (or, equivalently, **E**).

Let $\mathbf{2}^P = \langle \{0, 1\}, \wedge, \vee, \rightarrow, 1\rangle$ be a two-element Peirce algebra.

Proposition 2.2.7 $L\mathbf{2}^P = \mathbf{Lk}^+$.

Proof. It is clear that $\mathbf{Lk}^+ \subseteq L\mathbf{2}^P$. To prove the inverse inclusion we show that there is only one subdirectly irreducible Peirce algebra, $\mathbf{2}^P$.

Let $\mathcal{A}$ be a Peirce algebra with more than two elements. We show that for any $1 \neq a \in \mathcal{A}$ there exists a filter $F_a \neq \{1\}$ on $\mathcal{A}$ with $a \notin F_a$.

Take $1 \neq a \in \mathcal{A}$. There is also a $b \in \mathcal{A}$ with $1 \neq b \neq a$. If $b \not\leq a$, then $a \notin \langle b\rangle$. Assuming $b \leq a$, we consider an element $a \rightarrow b$. Since $a \neq b$, we have $a \wedge a \not\leq b$ and $a \wedge 1 \not\leq b$. By definition of relative pseudo-complement we conclude $a \not\leq a \rightarrow b$ and $a \rightarrow b \neq 1$, i.e., $a \notin \langle a \rightarrow b\rangle$ and $\langle a \rightarrow b\rangle \neq \{1\}$.

We have thus constructed a collection $\{F_a \mid a \in \mathcal{A}\}$ of filters on $\mathcal{A}$ such that

$$\bigcap_{a\in\mathcal{A}} F_a = \{1\},$$

and $F_a \neq \{1\}$ for all $a \in \mathcal{A}$. By Proposition 2.2.4 this means that $\mathcal{A}$ is not subdirectly irreducible.

□

Let $\mathbf{2} = \langle \{0, 1\}, \vee, \wedge, \rightarrow, 0, 1\rangle$ be a two-element Heyting algebra, which is a characteristic model for classical logic, $L\mathbf{2} = \mathbf{Lk}$.

By $\mathbf{2}' = \langle \{0, 1\}, \vee, \wedge, \rightarrow, 1, 1\rangle$ we denote a two-element negative algebra.

Proposition 2.2.8 $L\mathbf{2}' = \mathbf{Lmn}$.

Proof. Obviously, $\bot$ and $((p \rightarrow q) \rightarrow p) \rightarrow p$ are identities of $\mathbf{2}'$, and so $L\mathbf{2}' \supseteq \mathbf{Lmn}$. The inverse inclusion can be stated similar to Proposition 2.2.7.

□

Proposition 2.2.9 [93] *The logic* **Lj** *has exactly two maximal non-trivial extensions,* **Lk** *and* **Lmn**. *Every non-trivial* **Lj***-extension is contained in one of them.*

Proof. Consider an arbitrary non-trivial extension L of **Lj** and its characteristic model $\mathcal{A}$, $L = L\mathcal{A}$. Obviously, $\mathcal{A}$ is not one-element. If $\bot_{\mathcal{A}} = 1_{\mathcal{A}}$, then for every $a \in \mathcal{A}$, $a \neq 1$, the set $\{a, \bot_\mathcal{A}\}$ is the universe of a subalgebra isomorphic to $\mathbf{2}'$. Consequently, $L\mathcal{A} \subseteq L\mathbf{2}'$. If $\bot_\mathcal{A} \neq 1_\mathcal{A}$, then the subalgebra with the universe $\{\bot_\mathcal{A}, 1_\mathcal{A}\}$ is isomorphic to $\mathbf{2}$, whence $L\mathcal{A} \subseteq \mathbf{Lk}$.

□

Recall several facts from the universal algebra. A variety $\mathcal{V}$ is called *congruence distributive* if for any algebra $\mathcal{A} \in \mathcal{V}$, the lattice $Con(\mathcal{A})$ of congruences of algebra $\mathcal{A}$ is distributive. A variety $\mathcal{V}$ is called *congruence permutable* if for any algebra $\mathcal{A} \in \mathcal{V}$, the congruences are permutable wrt composition. In this case the join of two congruences coincide with their composition $\theta_1 \vee \theta_1 = \theta_1 \circ \theta_2$ for any $\theta_1, \theta_2 \in Con(\mathcal{A})$. An *arithmetic variety* is a variety, which is congruence permutable and congruence distributive. According to Pixley's theorem (see [14]) a variety $\mathcal{V}$ is arithmetic if and only if there exists a term $m(x, y, z)$ such that the identities

$$m(x, y, x) = m(x, y, y) = m(y, y, x) = x$$

hold in $\mathcal{V}$.

Proposition 2.2.10 *The variety of j-algebras is arithmetic.*

Proof. In case of j-algebras, as well as in case of Heyting algebras (see [14]), we can use the term

$$m(x, y, z) := ((x \to y) \to z) \wedge ((z \to y) \to x) \wedge (x \vee z)$$

to establish that the varieties of j-algebras and Heyting algebras are arithmetic. The verification is straightforward.

□

Let us consider an ω-generated free j-algebra $\mathcal{A}_\omega$ and its congruence lattice $Con(\mathcal{A}_\omega)$, which is distributive by the last proposition. Moreover, congruences of $Con(\mathcal{A}_\omega)$ are permutable wrt the composition. Elements of $\mathcal{A}_\omega$ can be identified with classes of equivalence of formulas wrt $\mathbf{Lj}$,

$$|\mathcal{A}_\omega| = \{[\varphi] \mid \varphi \in \mathcal{F}\},$$

where $[\varphi] := \{\psi \mid \varphi \leftrightarrow \psi \in \mathbf{Lj}\}$. With any $L \in \mathsf{Jhn}^+$ we associate the congruence

$$\theta_L := \{([\varphi_0], [\varphi_1]) \mid \varphi_0 \leftrightarrow \varphi_1 \in L\}.$$

Clearly, the mapping $L \mapsto \theta_L$ is one-to-one and preserves the ordering. Consequently, to prove that it is a lattice embedding it is enough to check that for any $L_0, L_1 \in \mathsf{Jhn}^+$, the congruences $\theta_{L_0} \wedge \theta_{L_1}$ and $\theta_{L_0} \vee \theta_{L_1}$ also have the form θ_L for a suitable L. Observe that θ_L is closed under substitution, i.e., if $[\varphi_0]\theta_L[\varphi_1]$, then $[\varphi_0(\psi_1,\dots,\psi_n)]\theta_L[\varphi_1(\psi_1,\dots,\psi_n)]$ for any $\psi_1,\dots,\psi_n$. It can easily be seen that if $\theta \in Con(\mathcal{A}_\omega)$ is closed under substitution, then $L_\theta = \{\varphi \mid [\varphi]\theta 1\}$, where 1 is the class of **Lj**-tautologies, is a logic from Jhn^+ and $\theta = \theta_{L_\theta}$.

In this way, it is enough to check that $\theta_{L_0} \wedge \theta_{L_1}$ and $\theta_{L_0} \vee \theta_{L_1}$ are closed under substitution. We consider only the non-trivial case of $\theta_{L_0} \vee \theta_{L_1}$. Since $\mathcal{A}_\omega$ is congruence permutable, $\theta_{L_0} \vee \theta_{L_1} = \theta_{L_0} \circ \theta_{L_1}$. So, $[\varphi_0]\theta_{L_0} \vee \theta_{L_1}[\varphi_1]$ if and only if there is a formula ψ such that $[\varphi_0]\theta_{L_0}[\psi]$ and $[\psi]\theta_{L_1}[\varphi_1]$. This immediately implies that $\theta_{L_0} \vee \theta_{L_1}$ is closed under substitution.

So, the set of congruences of the form θ_L forms a lattice. It is easy to see that $L \mapsto \theta_L$ is an order isomorphism of the lattice Jhn^+ and the lattice of congruences of the form θ_L. If two lattices are isomorphic as orders, they are isomorphic as lattices too. We have thus proved in an algebraic way the distributivity of Jhn^+.

Corollary 2.2.11 *The lattice* Jhn^+ *is distributive.*

2.3 Kripke Semantics

In conclusion of this chapter we say a few words on the Kripke-style semantics for minimal logic and its extensions.

A *j-frame* is a triple $\mathcal{W} = \langle W, \sqsubseteq, Q\rangle$, where W is a set of possible worlds, $\sqsubseteq$ is an accessibility relation such that $\langle W, \sqsubseteq\rangle$ is an ordinary Kripke frame for intuitionistic logic, i.e., a partially ordered set, and $Q \subseteq W$ is a cone (upward closed set) with respect to $\sqsubseteq$, which we call the cone of *abnormal worlds*. Worlds lying out of Q are called *normal*. A *valuation* v of a j-frame $\mathcal{W}$ is a mapping from the set of propositional variables to the set of cones of the ordering $\langle W, \sqsubseteq\rangle$. A *model* $\mu = \langle \mathcal{W}, v\rangle$ is a pair consisting of a j-frame and its valuation. Say in this case that μ is a model on $\mathcal{W}$.

The forcing relation between models and formulas is defined in exactly the same way as for ordinary Kripke frames. The only exception is the case of constant $\bot$. More precisely, we define the relation $\mu \models_x \varphi$, where $\mu = \langle \mathcal{W}, v\rangle$ is a model, $\mathcal{W} = \langle W, \sqsubseteq, Q\rangle$, $x \in W$, and φ is a formula, by induction on the structure of formulas as follows. For a propositional variable p, put

$$\mu \models_x p \iff x \in v(p).$$

And further,

$$\begin{array}{lcl} \mu \models_x \varphi \wedge \psi & \Longleftrightarrow & \mu \models_x \varphi \text{ and } \mu \models_x \psi; \\ \mu \models_x \varphi \vee \psi & \Longleftrightarrow & \mu \models_x \varphi \text{ or } \mu \models_x \psi; \\ \mu \models_x \varphi \to \psi & \Longleftrightarrow & \forall y \in W(x \sqsubseteq y \;\Rightarrow\; (\mu \models_y \varphi \;\Rightarrow\; \mu \models_y \psi)). \end{array}$$

We did not consider yet the case of constant $\bot$, and we put

$$\mu \models_x \bot \quad \Longleftrightarrow \quad x \in Q.$$

In particular, for a negated formula $\neg\varphi$ considered as an abbreviation for $\varphi \to \bot$, we have

$$\mu \models_x \neg\varphi \quad \Longleftrightarrow \quad \forall y \in W(x \sqsubseteq y \;\Rightarrow\; (\mu \models_y \varphi \;\Rightarrow\; y \in Q)).$$

Read $\mu \models_x \varphi$ as "a formula φ is true at a world (or at a point) x in a model μ". A formula φ is *true on a model* $\mu = \langle \mathcal{W}, v \rangle$, $\mu \models \varphi$, if $\mu \models_x \varphi$ holds for all $x \in W$. A formula φ is *true on a j-frame* $\mathcal{W}$, $\mathcal{W} \models \varphi$, if it is true on a model $\langle \mathcal{W}, v \rangle$ for an arbitrary valuation v of the j-frame $\mathcal{W}$. A formula φ is *valid on the class* $\mathcal{K}$ of Kripke j-frames if $\mathcal{W} \models \varphi$ for any j-frame $\mathcal{W} \in \mathcal{K}$.

Let $\mathcal{W} = \langle W, \sqsubseteq, Q \rangle$ be a j-frame and let $K \subseteq W$ be a cone wrt $\sqsubseteq$. We define a j-frame $\mathcal{W}^K$ in the following way: $\mathcal{W}^K := \langle K, \sqsubseteq^K, Q^K \rangle$, where $\sqsubseteq^K := \sqsubseteq \cap K^2$, $Q^K := Q \cap K$. If $\mu = \langle \mathcal{W}, v \rangle$ is a model on $\mathcal{W}$, then $\mu^K := \langle \mathcal{W}^K, v^K \rangle$, where $v^K(p) := v(p) \cap K$ for all propositional variables p.

If $[x]\uparrow := \{y \in W \mid x \sqsubseteq y\}$ is a cone generated by x, we write $\mathcal{W}^x$ and μ^x instead of $\mathcal{W}^{[x]\uparrow}$ and $\mu^{[x]\uparrow}$ respectively.

Lemma 2.3.1 *Let $\mathcal{W} = \langle W, \sqsubseteq, Q \rangle$ be an arbitrary j-frame, μ a model on $\mathcal{W}$, and $K \subseteq W$ a cone. For any $x \in K$ and an arbitrary formula φ, we have*

$$\mu \models_x \varphi \quad \Longleftrightarrow \quad \mu^K \models_x \varphi.$$

In particular,

$$\mathcal{W} \models \varphi \quad \Longrightarrow \quad \mathcal{W}^K \models \varphi.$$

We say that a j-frame $\mathcal{W}$ is a *model for* a logic $L \in \mathsf{Jhn}$, $\mathcal{W} \models L$, if $\mathcal{W} \models \varphi$ for all $\varphi \in L$. For a class of j-frames $\mathcal{K}$ we put

$$L\mathcal{K} := \{\varphi \mid \forall \mathcal{W} \in \mathcal{K}\ (\mathcal{W} \models \varphi)\}.$$

A logic $L \in \mathsf{Jhn}$ is *characterized* by a class of j-frames $\mathcal{K}$ if $L = L\mathcal{K}$.

We call a j-frame $\mathcal{W} = \langle W, \sqsubseteq, Q\rangle$ *normal* if $Q = \varnothing$, i.e., if all worlds of this frame are normal. It is clear that normal j-frames can be identified with ordinary Kripke frames for intuitionistic logic. We call a j-frame $\mathcal{W} = \langle W, \sqsubseteq, Q\rangle$ *abnormal* if $Q = W$, i.e., if all worlds are abnormal. Finally, a j-frame $\mathcal{W} = \langle W, \sqsubseteq, Q\rangle$ will be called *identical* if the accessibility relation $\sqsubseteq$ coincides with the identity relation on W, $\sqsubseteq = Id_W$.

Proposition 2.3.2 [98]

1. *Minimal logic* **Lj** *is characterized by the class of all j-frames.*
2. *Intuitionistic logic* **Li** *is characterized by the class of all normal j-frames.*
3. *Minimal negative logic* **Ln** *is characterized by the class of all abnormal j-frames.*
4. *Logic of classical refutability* **Le** *is characterized by the class of all identical j-frames.*
5. *Classical logic* **Lk** *is characterized by the class of all identical normal j-frames.*
6. *Maximal negative logic* **Lmn** *is characterized by the class of all identical abnormal j-frames.*

Further, we define several special classes of j-frames. Let $\mathcal{W} = \langle W, \sqsubseteq, Q\rangle$ be a j-frame. We say that $\mathcal{W}$ is *separated* if

$$\forall x, y \in W((x \notin Q \wedge y \in Q) \Rightarrow x \sqsubseteq y).$$

And we say that $\mathcal{W}$ is *closed* if

$$\forall x, y \in W((x \notin Q \wedge y \in Q) \Rightarrow \neg(x \sqsubseteq y)).$$

Denote by Sep the class of all separated j-frames and by Cl the class of all closed j-frames.

Proposition 2.3.3 [98] *The logic* $\mathbf{Lj}+\{(p \to \bot)\vee(\bot \to p)\}$ *is characterized by the class Sep, and the logic* $\mathbf{Lj} + \{\bot \vee (\bot \to p)\}$ *is characterized by Cl.*

A j-frame $\mathcal{W} = \langle W, \sqsubseteq, Q\rangle$ is called *dense* if

$$\forall x \in W(x \notin Q \Rightarrow \exists y \sqsupseteq x \forall z \sqsupseteq y(z \notin Q)).$$

The class of all dense j-frames is denoted by Den.

Proposition 2.3.4 [124] *The logic* $\mathbf{Lj} + \{\neg\neg(\bot \to p)\}$ *is characterized by the class Den.*

Chapter 3

Logic of Classical Refutability[1]

We start the investigation of the class of **Lj**-extensions with the logic of classical refutability $\mathbf{Le} = \mathbf{Lj} + \{((p \to q) \to p) \to p\}$. This important logic arose in the work of different authors and with different motivations. It was introduced for the first time in the P. Bernays review [10] of H.B. Curry's articles [20, 21]. P. Bernays observed that one can obtain a new logical system, namely **Le**, by extending axiom schemes of classical positive logic $\mathbf{Lk}^+$ to the language with additional constant $\bot$ in the same way as one can obtain minimal logic **Lj** by extending axiom schemes of positive logic to the language $\mathcal{L}^\bot$. Two years later, the system equivalent to **Le** was introduced in S. Kanger's work [42]. Kanger's reason for defining **Le** is that "...it constitutes a weakened classical calculus in the same sense as the minimal calculus is a weakened intuitionistic calculus". Further, this logic was studied by S. Kripke [48], who stated, in particular, the equation $\mathbf{Le} = \mathbf{Lk} \cap \mathbf{Lmn}$. The name "logic of classical refutability" was suggested in the H.B. Curry monograph [22]. In [22, Ch.6, Sec.A], one can find the discussion of this name. In [98], K. Segerberg studied the Kripke-style semantics for numerous extensions of minimal logic, and among them for **Le** [98, p.46]. J. Porte [84, 85] investigated interrelations between logic of classical refutability and Łukasiewicz's modal logic [52, 53]. His results will play an essential part in Chapter 7, where we will study the generalized version of negation as reduction to absurdity. In [85], it was stated, in particular, that **Le** is a four-valued logic. The same four-element matrix for **Le** will be introduced in Section 3.1 in a different way, as a simplest characteristic model for **Le**.

[1]Parts of this chapter were originally published in [67, 68].

Another time, the logic **Le** arose under the name of Carnot's logic CAR in the work [19] by N.C.A. da Costa and J.-Y. Béziau to explicate some ideas of Lasare Carnot. The equality $CAR = \mathbf{Le}$ was stated by I. Urbas [108].

The author studies in [66] also led to the system **Le**, and it arises here in a rather unexpected way, from the investigation of the constructivity concept suggested by K.F. Samokhvalov [97]. In [66], it was proved that **Le** coincides with the logic of all exactly constructive systems in the sense of K.F. Samokhvalov. We also established in [66] the maximality property for **Le**. Adjoining to **Le** a new classical tautology gives classical logic, and adjoining a new maximal negative tautology results in maximal negative logic. In this respect, **Le** is similar to the logic P^1 suggested by A. Sette [102], the first example of maximal paraconsistent logic. The maximality property of **Le** will be presented in Section 3.1. In the conclusion to this section we show that the class Jhn of all nontrivial extensions of minimal logic is divided into three intervals: the interval of well-known intermediate logics lying between the intuitionistic and the classical logics; the interval of negative logics lying between minimal and maximal negative logics and the interval of properly paraconsistent logics, which all lie between minimal logic and **Le**.

The results of Section 3.2 were inspired by A. Karpenko article [43], where isomorphs of classical logic into three-valued Bochvar's logic $\mathbf{B}_3$ were considered. It turns out that in **Le** one can naturally define one isomorph of classical logic and two different isomorphs of maximal negative logic. Starting from these isomorphs, we define translations of **Lmn** and **Lk** into **Le**. In the next chapter, these translations will be generalized to translations of arbitrary negative and intermediate logics into properly paraconsistent extensions of minimal logic, which allow one to define the notions of intuitionistic and negative counterparts of a paraconsistent **Lj**-extension.

3.1 Maximality Property of Le

According to Proposition 2.1.5 the intersection of logics $\mathbf{Lmn} = \mathbf{Lj} + \{\bot,\ p \vee (p \to q)\}$ and $\mathbf{Lk} = \mathbf{Lj} + \{\bot \to p,\ p \vee (p \to q)\}$ is axiomatized as follows:

$$\mathbf{Lk} \cap \mathbf{Lmn} = \mathbf{Lj} + \{p \vee (p \to q),\ \bot \vee (\bot \to p)\}.$$

The second formula is a substitution instance of the first one and we have thus proved the following statement, announced for the first time by S. Kripke in 1959 [48].

Proposition 3.1.1 $\mathbf{Le} = \mathbf{Lk} \cap \mathbf{Lmn}$.

This representation of **Le** allows one to make the following observations.

Lemma 3.1.2 *1. The following formulas are in* **Le***:*

$$p \vee \neg p,\ \ \neg(p \wedge \neg p),\ \ \neg(p \wedge q) \leftrightarrow (\neg p \vee \neg q),\ \ \neg(p \vee q) \leftrightarrow (\neg p \wedge \neg q).$$

2. **Le** *does not contain* $\neg\neg p \to p$ *and* $\neg p \to (p \to q)$*.*

Proof. 1. These formulas are classical tautologies, which can be easily deduced in maximal negative logic using the scheme $\neg\varphi$.

2. Assuming $\mathbf{Le} \vdash \neg\neg p \to p$ we have $\mathbf{Lmn} \vdash \neg\neg p \to p$. But $\mathbf{Lmn} \vdash \neg\neg p$, hence $\mathbf{Lmn} \vdash p$. The substitution rule implies that any formula is provable in **Lmn**, a contradiction.

If $\mathbf{Lmn} \vdash \neg p \to (p \to q)$, then $\mathbf{Lmn} \vdash p \to q$. Substituting $\neg p$ for p in the latter formula we again have $\mathbf{Lmn} \vdash q$.

□

Now we consider models for **Le**.

We call $\mathcal{A} = \langle A, \vee, \wedge, \to, \bot, 1\rangle$ a *Peirce-Johansson algebra* (*pj-algebra*) if $\langle A, \vee, \wedge, \to, 1\rangle$ is a Peirce algebra and the constant $\bot$ is interpreted as an arbitrary element of the universe A. These algebras provide a semantics for the logic **Le**. Recall that Peirce algebras provide algebraic semantics for $\mathbf{Lk}^+$ and that **Le** can be considered as an expansion of $\mathbf{Lk}^+$ to the language $\mathcal{L}^\bot$. These facts immediately imply the following

Proposition 3.1.3 *A j-algebra* $\mathcal{A} = \langle A, \vee, \wedge, \to, \bot, 1\rangle$ *is a model for* **Le** *if and only if* $\mathcal{A}$ *is a pj-algebra.*

□

List some simple properties of *pj*-algebras.

Proposition 3.1.4 *Let* $\mathcal{A} = \langle A, \wedge, \vee, \to, \bot, 1\rangle$ *be a pj-algebra.*

1. *The interval* $[\bot, 1]_{\mathcal{A}}$ *forms a subalgebra of* $\mathcal{A}$*, which is a Boolean algebra.*

2. *For any* $a \in A$*, if* $a \leq \bot$*, then* $\neg a = 1$*.*

3. *If* $\bot \neq 1$ *and* $[\bot, 1]_{\mathcal{A}} \neq A$*, then* $\mathcal{A}$ *contains an element incomparable with* $\bot$*.*

Proof. All statements of the proposition can easily be deduced from the definition of the relative pseudo-complement. Consider, for example, the last statement.

3. There exists an element a under $\bot$ by assumption. Then $\bot \to a$ is incomparable with $\bot$ in view of the equality $\bot \vee (\bot \to a) = 1$.

□

Proposition 3.1.5 *Let $\mathcal{A}$ be a pj-algebra. Then either $\mathcal{A}$ is a model for* **Lmn**, *or $\mathcal{A}$ is a model for* **Lk**, *or $L\mathcal{A} \subseteq$* **Le**, *i.e., $\mathcal{A}$ is a characteristic model for* **Le**.

Proof. Let $\mathcal{A}$ be a *pj*-algebra. If $\bot = \neg 1 = 1$, then $\neg a = 1$ for any $a \in \mathcal{A}$ by Item 2 of the previous proposition. This means that $\mathcal{A}$ is a model for **Lmn**.

Assume $\bot \neq 1$, then $[\bot, 1]_{\mathcal{A}}$ is a non-trivial Boolean algebra. If $[\bot, 1]_{\mathcal{A}} = \mathcal{A}$, then $\mathcal{A}$ is a model for classical logic.

Finally, assume that $\bot \neq 1$ and $[\bot, 1]_{\mathcal{A}} \neq \mathcal{A}$. In this case, $\mathcal{A}$ contains a subalgebra isomorphic to **2**, whence, $L\mathcal{A} \subseteq L\mathbf{2} = \mathbf{Lk}$. Consider the filter $\langle \bot \rangle$ and the corresponding quotient algebra. Since $\neg a \geq 0$ for any $a \in \mathcal{A}$, the algebra $\mathcal{A}/\langle \bot \rangle$ satisfies the identity $\neg p = 1$. Therefore, $L(\mathcal{A}/\langle \bot \rangle) \supseteq \mathbf{Ln}$. Moreover, $((p \to q) \to p) \to p$ is an identity of $\mathcal{A}/\langle \bot \rangle$ as a quotient of $\mathcal{A}$. Consequently, $L(\mathcal{A}/\langle \bot \rangle) = \mathbf{Lmn}$, and we have $L\mathcal{A} \subseteq \mathbf{Lk} \cap \mathbf{Lmn} = \mathbf{Le}$.

□

Consider the lattice $\mathbf{4}' = \langle \{0, 1, -1, a\}, \leq \rangle$, where $-1 \leq a \leq 1, -1 \leq 0 \leq 1$, and the elements a and 0 are incomparable.

It is a Peirce algebra. Interpreting $\bot$ as 0 turns it into a *pj*-algebra. By the previous proposition we obtain.

Corollary 3.1.6 $\mathbf{4}'$ *is a characteristic model for* **Le**.

Proof. Indeed, $\mathbf{4}'$ is neither a Boolean algebra, nor a negative algebra.

□

Remark. The *j*-algebra $\mathbf{4}'$ is the simplest among characteristic models for **Le**. Propositions 3.1.4 and 3.1.5 easily imply that any characteristic model for **Le** must contain at least four elements. Indeed, the unity element differs from $\bot$, there is a third element under $\bot$, and there is a fourth element incomparable with $\bot$.

Now we are ready to prove the maximality property for **Le**.

Theorem 3.1.7 *Let $\varphi \notin$ **Le**. There are three possible cases:*

1. **Le** $+ \{\varphi\}$ *is trivial;*
2. **Le** $+ \{\varphi\} =$ **Lmn***;*
3. **Le** $+ \{\varphi\} =$ **Lk**.

Proof. Assume that **Le**$+ \{\varphi\}$ is non-trivial and $\mathcal{A}$ is its characteristic model. The inclusion $L\mathcal{A} \subseteq$ **Le** fails, since φ is not in **Le**. By Proposition 3.1.5 we have either $L\mathcal{A} =$ **Lk** or $L\mathcal{A} =$ **Lmn**.

□

We now make an important observation on the structure of the class Jhn of all nontrivial extensions of **Lj**. Let $\mathsf{Int} := \{L \mid L \in \mathsf{Jhn},\ \bot \to p \in L\}$ be the class of all *intermediate logics*; let $\mathsf{Neg} := \{L \mid L \in \mathsf{Jhn},\ \neg p \in L\}$ be the class of all *negative logics*, i.e., Neg consists of logics with a degenerate negation. Finally, let $\mathsf{Par} := \mathsf{Jhn} \setminus (\mathsf{Int} \cup \mathsf{Neg})$ be the class of all *properly paraconsistent logics*. Obviously, the class Jhn is a disjoint union of the classes Int, Neg, and Par. It is well known that $L \in \mathsf{Int}$ if and only if **Li** $\subseteq L \subseteq$ **Lk**. It turns out that the other two classes also form intervals in the lattice Jhn^+.

Proposition 3.1.8 *Let $L \in \mathsf{Jhn}^+$. Then the following equivalences hold:*

1. $L \in \mathsf{Neg}$ *if and only if* **Ln** $\subseteq L \subseteq$ **Lmn***;*
2. $L \in \mathsf{Par}$ *if and only if* **Lj** $\subseteq L \subseteq$ **Le**.

Proof. 1. If $L \in \mathsf{Neg}$, then **Ln** is contained in L by definition. At the same time, adding the axiom $\neg\neg p \to p$ to L leads to a trivialization, therefore, L cannot be extended to **Lk**, consequently, $L \subseteq$ **Lmn** by Proposition 2.2.9.

2. Let $L \in \mathsf{Par}$, and let $\mathcal{A}$ be a characteristic model for L. Since $L \notin \mathsf{Neg}$, the inequality $\bot \neq 1$ holds in $\mathcal{A}$, hence $\{\bot, 1\}$ is a nontrivial Boolean algebra and a subalgebra of $\mathcal{A}$. Consequently, $L \subseteq$ **Lk**. Further, $L \notin \mathsf{Int}$, therefore, the quotient $\mathcal{A}/\langle\bot\rangle$ is nontrivial and has the greatest element $\bot$, hence it contains a two-element subalgebra isomorphic to $\mathbf{2}'$. The latter means that $L \subseteq$ **Lmn**. Thus, $L \subseteq$ **Le** $=$ **Lk** $\cap$ **Lmn**.

□

3.2 Isomorphs of Le

The term “isomorph” was used in the first monograph on multi-valued logics [94], and now it looks a bit archaic. Let L_1 and L_2 be logics and let L_2 be given via its logical matrix. Due to N. Rescher [94], an *isomorph* of the logic

L_1 in the logic L_2 is a definition of a matrix for L_1 in the matrix for L_2 with the help of term operations.

We can define a translation of L_1 into L_2 whenever some isomorph of L_1 in L_2 is given. The relations between interdefinability of logical matrices and mutual translations of logics was studied in detail in the works by P. Wojtylak [122, 123]. However, it will be quite enough for our goal to use the old notion of the isomorph.

The interest of the author in isomorphs, which can be defined inside logic of classical refutability, is connected with the talk of A. Karpenko at the First World Congress on Paraconsistency (see [43]), in which he considered isomorphs of classical logic in three-valued Bochvar's logic $\mathbf{B}_3$ given via so-called "internal" and "external" connectives. He also tried to argue that different isomorphs of classical logic in the given many-valued logic determine the paraconsistent structure of this logic. The logic **Le** as well as all other extensions of minimal logic takes a borderline position among paraconsistent logics, and this fact naturally gives rise to the question of isomorphs, which can be defined inside the logic **Le**. It turns out that there is only one natural isomorph of classical logic in **Le**. At the same time, there are two isomorphs of maximal negative logic. As we will see in the next chapter, the translations of **Lk** and **Lmn** defined by these isomorphs can be used to define translations of intermediate and negative logics into arbitrary paraconsistent extensions of **Lj**. In this sense, the studying of isomorphs of the logic **Le** plays the key role in the investigation of the class of **Lj**-extensions.

Consider the 4-element matrix $\mathbf{4}' = \langle\{1, 0, a, -1\}, \vee, \wedge, \rightarrow, \neg, \{1\}\rangle$ for **Le** and define the mapping $\varepsilon(x) := \neg\neg x$.

x	$\varepsilon(x)$
1	1
a	1
0	0
-1	0

The operations $\vee_\varepsilon$, $\wedge_\varepsilon$, $\rightarrow_\varepsilon$, and $\neg_\varepsilon$ are defined as follows:

$$x *_\varepsilon y := \varepsilon(x * y) = \varepsilon(x) * \varepsilon(y), \quad * \in \{\vee, \wedge, \rightarrow\},$$

$$\neg_\varepsilon x := \neg\varepsilon(x),$$

These operations determine an isomorph of classical logic in **Le**, which we denote $\mathbf{Lk}_\varepsilon$. The fact that $\mathbf{Lk}_\varepsilon$ really is an isomorph of **Lk** can easily be verified by considering the truth tables for the above operations.

$\wedge_\varepsilon$	1	a	0	-1
1	1	1	0	0
a	1	1	0	0
0	0	0	0	0
-1	0	0	0	0

$\vee_\varepsilon$	1	a	0	-1
1	1	1	1	1
a	1	1	1	1
0	1	1	0	0
-1	1	1	0	0

$\rightarrow_\varepsilon$	1	a	0	-1
1	1	1	0	0
a	1	1	0	0
0	1	1	1	1
-1	1	1	1	1

$\neg_\varepsilon$	
1	0
a	0
0	1
-1	1

As we can see, rows and columns corresponding to elements 1 and a are identical. The same holds for elements 0 and -1. Thus, identifying these pairs of elements we obtain two-valued truth tables for operations of classical logic. It is not hard to check with the help of the above truth tables that the mapping ε defines an epimorphism from $\mathbf{4}'$ onto the two-element Boolean algebra $\mathbf{2}_\varepsilon$ with the universe $\{1, 0\}$. It is also clear that $\mathbf{2}_\varepsilon$ is a subalgebra of $\mathbf{4}'$.

Lemma 3.2.1 *The mapping* $\varepsilon : \mathbf{4}' \rightarrow \mathbf{2}_\varepsilon$*, where* $\mathbf{2}_\varepsilon$ *is a subalgebra of* $\mathbf{4}'$ *with the universe* $\{1, 0\}$*, is an epimorphism.*

□

We now consider the mapping $\delta(x) := \neg\neg x \rightarrow x$, which acts on the set of truth-values as follows.

x	$\delta(x)$
1	1
a	a
0	1
-1	a

As above, define the operations $\vee_\delta$, $\wedge_\delta$, $\rightarrow_\delta$, $\neg_\delta$:

$$x *_\delta y := \delta(x * y) = \delta(x) * \delta(y), \quad * \in \{\vee, \wedge, \rightarrow\},$$

$$\neg_\delta x := \neg\delta(x).$$

The truth tables of these operations look as follows.

$\wedge_\delta$	1	a	0	-1
1	1	a	1	a
a	a	a	a	a
0	1	a	1	a
-1	a	a	a	a

$\vee_\delta$	1	a	0	-1
1	1	1	1	1
a	1	a	1	a
0	1	1	1	1
-1	1	a	1	a

$\rightarrow_\delta$	1	a	0	-1
1	1	a	1	a
a	1	1	1	1
0	1	a	1	a
-1	1	1	1	1

$\neg_\delta$	
1	1
a	1
0	1
-1	1

As we can see, the pairs of elements 1 and 0, a and -1 are indiscernible with respect to the introduced operations. Identifying these elements, we obtain truth tables of the two-element negative algebra $\mathbf{2}_\delta$ with the universe $\{1, a\}$. The algebra $\mathbf{2}_\delta$ is a characteristic model for maximal negative logic **Lmn**. This fact allows one to conclude that the introduced operations define an isomorph of **Lmn** into **Le**, which we denote $\mathbf{Lmn}_\delta$. Moreover, one can check that the mapping δ preserves the operations of $\mathbf{4}'$.

Lemma 3.2.2 *The mapping* $\delta : \mathbf{4}' \rightarrow \mathbf{2}_\delta$*, where* $\mathbf{2}_\delta$ *is a two-element negative algebra with the universe* $\{1, a\}$*, is an epimorphism.*

□

Note that $\mathbf{2}_\delta$ is not a subalgebra of $\mathbf{4}'$, though it is an implicative sublattice of $\mathbf{4}'$.

Finally, we define the mapping $\tau(x) := x \wedge \bot$ (its action on the truth values of **Le** is in the table below)

x	$\tau(x)$
1	0
a	-1
0	0
-1	-1

and the operations

$$x \vee_\tau y := \tau(x \vee y) = \tau(x) \vee \tau(y),$$

$$x \wedge_\tau y := \tau(x \wedge y) = \tau(x) \wedge \tau(y),$$

$$x \to_\tau y := \tau(x \to y) = \tau(\tau(x) \to \tau(y)),$$

$$\neg_\tau x := \tau(\neg x).$$

Consider the truth tables of these operations.

$\wedge_\tau$	1	a	0	-1
1	0	-1	0	-1
a	-1	-1	-1	-1
0	0	-1	0	-1
-1	-1	-1	-1	-1

$\vee_\tau$	1	a	0	-1
1	0	0	0	0
a	0	-1	0	-1
0	0	0	0	0
-1	0	-1	0	-1

$\to_\tau$	1	a	0	-1
1	0	-1	0	-1
a	0	0	0	0
0	0	-1	0	-1
-1	0	0	0	0

$\neg_\tau$	
1	0
a	0
0	0
-1	0

Again, we see that the pairs of elements 1 and 0, a and -1 are indiscernible with respect to the introduced operations and that their identification yields the truth tables of the two-element negative algebra $\mathbf{2}_\tau := \langle\{0,-1\}, \vee_\tau, \wedge_\tau, \to_\tau, \neg_\tau\rangle$, where 0 plays the part of a unit element and the negation $\neg_\perp$ is identically equal to 0, the conjunction and disjunction operations are induced by the respective operations of the algebra $\mathbf{4}'$, whereas the implication $\to_\tau$ is defined as $x \to_\tau y := (x \to y) \wedge \perp$. Since $L\mathbf{2}_\tau = \mathbf{Lmn}$, we conclude that the operations $\vee_\tau$, $\wedge_\tau$, $\to_\tau$, and $\neg_\tau$ define an isomorph of **Lmn** into **Lk** with a new distinguished value 0, i.e., the set of tautologies of the matrix

$$\mathbf{4}_\tau = \langle\{1, 0, a, -1\}, \vee_\tau, \wedge_\tau, \to_\tau, \neg_\tau, \{0\}\rangle$$

with the only distinguished value 0 coincides with **Lmn**. This isomorph is denoted as $\mathbf{Lmn}_\tau$.

Again we note the following fact

Lemma 3.2.3 *The mapping $\tau : \mathbf{4}' \to \mathbf{2}_\tau$, where $\mathbf{2}_\tau$ is a two-element negative algebra with the universe $\{0, -1\}$ and unit element 0, is an epimorphism.*

□

The isomorphs defined above lead to the following translations of classical and maximal negative logics into **Le**.

Proposition 3.2.4 *For any formula φ, the following equivalences hold:*

1. $\mathbf{Lk} \vdash \varphi \iff \mathbf{Le} \vdash \neg\neg\varphi$;
2. $\mathbf{Lmn} \vdash \varphi \iff \mathbf{Le} \vdash \neg\neg\varphi \to \varphi$;
3. $\mathbf{Lmn} \vdash \varphi \iff \mathbf{Le} \vdash \bot \to (\varphi \wedge \bot) \iff \mathbf{Le} \vdash \bot \to \varphi$.

Proof. 1. Assume $\mathbf{Le} \vdash \neg\neg\varphi$. In this case $\mathbf{Lk} \vdash \neg\neg\varphi$ since $\mathbf{Lk}$ extends $\mathbf{Le}$. In $\mathbf{Lk}$, any formula is equivalent to its double negation, whence $\mathbf{Lk} \vdash \varphi$.

Let us prove the inverse implication. Suppose that a formula $\varphi = \varphi(p_1, \ldots, p_n)$ is such that $\mathbf{Lk} \vdash \varphi$, but $\mathbf{Le} \nvdash \neg\neg\varphi$. In this case, there is a $\mathbf{4}'$-valuation v such that $v(\neg\neg\varphi) \neq 1$. Due to Lemma 3.2.1 the double negation preserves the operations of $\mathbf{4}'$, and so we have

$$v(\neg\neg\varphi(p_1, \ldots, p_n)) = v(\varphi(\neg\neg p_1, \ldots, \neg\neg p_n)).$$

Let v_1 be a $\mathbf{2}_\varepsilon$-valuation with the property $v_1(p_1) := v(\neg\neg p_1), \ldots, v_1(p_n) := v(\neg\neg p_n)$. In view of the last equality, we have

$$v_1(\varphi(p_1, \ldots, p_n)) = v(\neg\neg\varphi(p_1, \ldots, p_n)) \neq 1.$$

The latter inequality means that φ is not provable in $\mathbf{Lk}$.

2. We can prove this item in the same way as was done above, using Lemma 3.2.2 instead of Lemma 3.2.1. We can also reduce this item to the next one. Indeed, $\mathbf{Le} \vdash \neg\neg\varphi \leftrightarrow \varphi \vee \bot$, whence

$$\mathbf{Le} \vdash \neg\neg\varphi \to \varphi \iff \mathbf{Le} \vdash (\varphi \vee \bot) \to \varphi \iff \mathbf{Le} \vdash \bot \to \varphi.$$

3. Let $\mathbf{Le} \vdash \bot \to \varphi$. Then $\mathbf{Lmn} \vdash \bot \to \varphi$ since $\mathbf{Lmn}$ extends $\mathbf{Le}$. In view of $\bot \in \mathbf{Lmn}$ we immediately obtain $\mathbf{Lmn} \vdash \varphi$.

To prove the inverse implication we consider a formula φ such that $\mathbf{Le} \nvdash \bot \to \varphi$ and a $\mathbf{4}'$-valuation v such that

$$v(\bot \to \varphi) = v(\bot \to (\varphi \wedge \bot)) \neq 1.$$

The latter means that $v(\varphi \wedge \bot) = -1$. Let $\varphi = \varphi(p_1, \ldots, p_n)$. Due to epimorphism properties of $\tau(x) = x \wedge \bot$ (see Lemma 3.2.3) we obtain

$$v(\varphi(p_1, \ldots, p_n) \wedge \bot) = v(\varphi(p_1 \wedge \bot, \ldots, p_n \wedge \bot)).$$

Consider a $\mathbf{2}_\tau$-valuation v_1 such that $v_1(p_1) := v(p_1 \wedge \bot), \ldots, v_1(p_n) := v(p_n \wedge \bot)$. Then $v_1(\varphi) = v(\varphi \wedge \bot) = -1$, which refutes the provability $\mathbf{Lmn} \vdash \varphi$.

□

Chapter 4

The Class of Extensions of Minimal Logic[1]

In this chapter, we assign to every properly paraconsistent extension L of minimal logic an intermediate logic L_{int} and negative logic L_{neg} called intuitionistic and negative counterparts of L, respectively. It will be proved that the negative counterpart L_{neg} explicates the structure of contradictions of paraconsistent logic L. We show that both counterparts L_{int} and L_{neg} are faithfully embedded into the original logic L. Finally, we investigate a question: to what extent is a logic $L \in \mathsf{Par}$ determined by its counterparts? As a first step, we study paraconsistent extensions of the logic

$$\mathbf{Le}' := \mathbf{Li} \cap \mathbf{Ln} = \mathbf{Lj} + \{\bot \vee (\bot \rightarrow p)\}.$$

The class of extensions of this logic has a nice property that every logic $L \in \mathcal{E}\mathbf{Le}' \cap \mathsf{Par}$ is uniquely determined by its intuitionistic and negative counterparts.

4.1 Extensions of Le′

In this section, we state that properly paraconsistent extensions of $\mathbf{Le}'$ are exactly intersections of two logics, one of which is intermediate and the other is negative. Prior to this, we study the algebraic semantics for logics extending $\mathbf{Le}'$.

[1]Parts of this chapter were originally published in [70] (Nicholas Copernicus University Press, Poland) and in [76] (Elsevier, UK). Reprinted here by permission of the publishers.

For an implicative lattice $\mathcal{A} = \langle A, \vee, \wedge, \to, 1\rangle$ and $a \in A$, we put

$$A^a := \{b \in A \mid b \geq a\} \text{ and } A_a := \{b \in A \mid b \leq a\}.$$

The set A^a is obviously closed under the operations of $\mathcal{A}$ and we can define an implicative sublattice $\mathcal{A}^a$ of $\mathcal{A}$, with the universe A^a. Except for the case $a = 1$, the set A_a forms a sublattice but not an implicative sublattice of $\mathcal{A}$, because A_a is not closed under the implication ($a \to a = 1$). However, the operation $x \to_a y := (x \to y) \wedge a$ turns A_a into an implicative lattice with unit element a. Denote this implicative lattice by $\mathcal{A}_a$.

If $\mathcal{A}$ is a j-algebra and $a = \bot$, $\mathcal{A}^a$ can be treated as a Heyting algebra and $\mathcal{A}_a$ as a negative one. In the following we call Heyting algebra $\mathcal{A}^\bot$ an *upper algebra of* $\mathcal{A}$. Negative algebra $\mathcal{A}_\bot$ is a *lower algebra of* $\mathcal{A}$.

Recall one well-known fact from the theory of distributive lattices. Let $\mathcal{A}$ be a distributive lattice, a an arbitrary element of $\mathcal{A}$, and let sublattices $\mathcal{A}^a$ and $\mathcal{A}_a$ be defined as above. The mappings $\varepsilon(x) := x \vee a$ and $\tau(x) := x \wedge a$ are epimorphisms of $\mathcal{A}$ onto $\mathcal{A}^a$ and $\mathcal{A}_a$ respectively. The mapping $\lambda(x) := (x \vee a, x \wedge a)$ gives an embedding of $\mathcal{A}$ into the direct product of lattices $\mathcal{A}^a$ and $\mathcal{A}_a$.

These facts do not generally hold for implicative lattices. As before, the mapping τ is an epimorphism of implicative lattices. But $\varepsilon : \mathcal{A} \to \mathcal{A}^a$ and $\lambda : \mathcal{A} \to \mathcal{A}^a \times \mathcal{A}_a$ are an epimorphism and an embedding of implicative lattices only if some additional condition is imposed on $\mathcal{A}$. More precisely, the following assertions take place.

Proposition 4.1.1 *For an implicative lattice* $\mathcal{A}$ *and* $a \in \mathcal{A}$, *the mapping* $\tau : \mathcal{A} \to \mathcal{A}_a$, $\tau(x) = x \wedge a$, *is an epimorphism of implicative lattices.*

Proof. It follows from the definition of implication in $\mathcal{A}_a$ and the identity $(x \to y) \wedge z = ((x \wedge z) \to (y \wedge z)) \wedge z$ satisfied in all implicative lattices. The latter fact follows from Item 9 of Proposition 2.1.2.

□

Proposition 4.1.2 *Let* $\mathcal{A}$ *be an implicative lattice and* $a \in \mathcal{A}$. *The following three conditions are equivalent.*

1. *For all* $x, y \in \mathcal{A}$, *we have*

$$(x \vee a) \to (y \vee a) \leq (x \to y) \vee a.$$

2. *The mapping* $\varepsilon : \mathcal{A} \to \mathcal{A}^a$ *given by the rule* $\varepsilon(x) = x \vee a$ *is an epimorphism of implicative lattices.*

3. *The mapping* $\lambda : \mathcal{A} \to \mathcal{A}^a \times \mathcal{A}_a$ *given by the rule* $\lambda(x) = (x \vee a, x \wedge a)$ *is an isomorphism of* $\mathcal{A}$ *onto a direct product of implicative lattices* $\mathcal{A}^a \times \mathcal{A}_a$.

Proof. 1 ⇒ 2. Check that ε preserves the implication, i.e., that the equality $(x \to y) \vee a = (x \vee a) \to (y \vee a)$ holds. We have

$$\mathbf{Lp} \vdash ((p \to q) \vee r) \to ((p \vee r) \to (q \vee r))$$

by Item 8 of Proposition 2.1.2, whence the inequality $(x \to y) \vee a \leq (x \vee a) \to (y \vee a)$ is valid in any implicative lattice. The inverse inequality holds by assumption.

2 ⇒ 3. It follows easily by assumption that λ is a homomorphism of $\mathcal{A}$ into $\mathcal{A}^a \times \mathcal{A}_a$. We prove the injectivity of λ. Take an element $b \in \mathcal{A}$, it is a complement of a in the interval $[b \wedge a, b \vee a]$. Assuming $\lambda(c) = \lambda(b)$ for some $c \in \mathcal{A}$ yields that c is a complement of a in the same interval $[b \wedge a, b \vee a]$. We have $b = c$, since complements are unique in distributive lattices. Thus, it remains to prove that λ maps $\mathcal{A}$ onto $\mathcal{A}^a \times \mathcal{A}_a$.

For $x \in \mathcal{A}^a$ and $y \in \mathcal{A}_a$, we set $z := (a \to y) \wedge x$. The direct computation shows that $z \wedge a = y$ and $z \vee a = ((a \to y) \vee a) \wedge x$. Further, $(a \to y) \vee a = (a \vee a) \to (y \vee a) = a \to a = 1$ in view of the assumption that ε is a homomorphism, whence $z \vee a = x$. We have thereby proved $\lambda(z) = (x, y)$.

3 ⇒ 1. Obviously, 3 implies 2. Therefore, the desired inequality follows from the fact that ε preserves the implication.

□

Let us consider the following formulas

P. $((p \to q) \to p) \to p$

E. $p \vee (p \to q)$

D. $((p \vee r) \to (q \vee r)) \to ((p \to q) \vee r)$

We have

$$\mathbf{Lk}^+ = \mathbf{Lp} + \{\mathbf{P}\} = \mathbf{Lp} + \{\mathbf{E}\} = \mathbf{Lp} + \{\mathbf{D}\},$$

where $\mathbf{Lk}^+$ is the positive fragment of classical logic. It is well-known that $\mathbf{Lk}^+$ is axiomatized relative to positive logic by the Peirce law **P** or by the extended law of excluded middle **E**. It can be verified directly that **D** is true on the 2-element Peirce algebra $\mathbf{2}^P$. On the other hand, substituting p for r in **D**, we immediately obtain $\mathbf{Lp} + \mathbf{D} \vdash \mathbf{E}$. We have thus obtained that **D** axiomatizes $\mathbf{Lk}^+$ modulo **Lp**. Combining this fact and Proposition 4.1.2 yields a characterization of Peirce algebras in terms of mappings ε and λ.

Proposition 4.1.3 *Let $\mathcal{A}$ be an implicative lattice. The following conditions are equivalent.*

1. *$\mathcal{A}$ is a Peirce algebra.*
2. *For any $a \in \mathcal{A}$, the mapping $\varepsilon_a(x) = x \vee a$ defines an epimorphism of $\mathcal{A}$ onto $\mathcal{A}^a$.*
3. *For any $a \in \mathcal{A}$, the mapping $\lambda_a(x) = (x \vee a, x \wedge a)$ defines an isomorphism of $\mathcal{A}$ and $\mathcal{A}^a \times \mathcal{A}_a$.*

□

We now turn to the subsystem **Le**′ of **Le**, which can be axiomatized relative to **Lj** by each of the following substitution instances of **E** and **D**:

E′. $\bot \vee (\bot \to p)$.

D′. $((p \vee \bot) \to (q \vee \bot)) \to ((p \to q) \vee \bot)$.

The equality $\mathbf{Lj} + \{\mathbf{E}'\} = \mathbf{Lj} + \{\mathbf{D}'\}$ can be checked as follows. On the one hand, **E**′ follows from the instance of **D**′ obtained by replacing p for $\bot$. On the other hand, **D**′ is equivalent in **Lj** to $(p \to (q \vee \bot)) \to ((p \to q) \vee \bot)$, and the latter formula follows in **Lj** from $\bot \vee (\bot \to q)$. Indeed, $\bot$ implies $\bot$, and formulas $\bot \to q$ and $p \to (q \vee \bot)$ imply $p \to q$.

Note a curious fact that the instance of the Peirce law

$$\mathbf{P}^* = ((p \to \bot) \to p) \to p = (\neg p \to p) \to p,$$

which is known as the Clavius law, is not equivalent to the above formulas relative to **Lj**. Indeed, $\mathbf{Lj} \vdash \mathbf{P}^* \leftrightarrow (p \vee \neg p)$ as a particular case of the equivalence $\mathbf{P} \leftrightarrow \mathbf{E}$, and the logics $\mathbf{Lj} + (p \vee \neg p)$ and **Le**′ are incomparable in the lattice of **Lj**-extensions. To prove the latter assertion, consider the 3-element linearly ordered Heyting algebra **3** and 3-element j-algebra **3**′ with the universe $\{-1, \bot, 1\}$, $-1 \leq \bot \leq 1$. It can be checked directly that

$$\mathbf{3} \times \mathbf{2}' \not\models p \vee \neg p, \; \mathbf{3} \times \mathbf{2}' \models \mathbf{E}', \; \mathbf{3}' \not\models \mathbf{E}', \; \mathbf{3}' \models p \vee \neg p.$$

Consider the algebraic semantics for **Le**′.

Proposition 4.1.4 *Let $\mathcal{A}$ be a j-algebra. $\mathcal{A}$ is a model for* **Le**′ *if and only if one of the following equivalent conditions holds.*

1. *The mapping $\varepsilon(x) = x \vee \bot$ defines an epimorphism of the j-algebra $\mathcal{A}$ onto the Heyting algebra $\mathcal{A}^{\bot}$.*

2. *The mapping* $\lambda(a) = (a \vee \bot, a \wedge \bot)$ *determines an isomorphism of* j*-algebras* $\mathcal{A}$ *and* $\mathcal{A}^{\perp} \times \mathcal{A}_{\perp}$.

3. *For any* $a, b \in \mathcal{A}$ *with* $a \leq \bot \leq b$, $\bot$ *has a complement in the interval* $[a, b]$.

Proof. The inclusion $\mathbf{Le}' \subseteq L\mathcal{A}$ is equivalent to the fact that $\mathbf{D}'$ is an identity of $\mathcal{A}$, which is equivalent in its own right to Item 1 of Proposition 4.1.2 for $a = \bot$. In this way, Proposition 4.1.2 implies that each of the conditions 1, 2 characterizes models for $\mathbf{Le}'$. Proving Proposition 4.1.2 we established, in fact, that 2 implies 3. Now we check the inverse implication, which completes the proof.

Condition 3 means exactly that an embedding of distributive lattices $\lambda : \mathcal{A} \to \mathcal{A}^{\perp} \times \mathcal{A}_{\perp}$ is "onto", i.e., that λ is an isomorphism of distributive lattices $\mathcal{A}$ and $\mathcal{A}^{\perp} \times \mathcal{A}_{\perp}$. The implication is defined in terms of the ordering preserved by λ, consequently, λ also preserves the implication.

□

Corollary 4.1.5 *Let* $L \in \mathsf{Jhn}$. *Then* $\mathbf{Le}' \subseteq L \subseteq \mathbf{Le}$ *if and only if* $L = L_1 \cap L_2$, *where* $L_1 \in \mathsf{Int}$ *and* $L_2 \in \mathsf{Neg}$.

Proof. Let L be an intersection of intermediate and negative logics L_1 and L_2. Then $\mathbf{Li} \subseteq L_1$ and $\mathbf{Ln} \subseteq L_2$, whence $\mathbf{Le}' = \mathbf{Li} \cap \mathbf{Ln} \subseteq L$. It is clear that the L is neither intermediate nor negative, therefore, $L \in \mathsf{Par}$ and $L \subseteq \mathbf{Le}$.

Conversely, let $\mathbf{Le}' \subseteq L \subseteq \mathbf{Le}$ and let $\mathcal{A}$ be a characteristic model for L. By the above proposition $\mathcal{A}$ is presented as a direct product of Heyting algebra $\mathcal{A}^{\perp}$ and negative algebra $\mathcal{A}_{\perp}$, hence, $L = L\mathcal{A} = L\mathcal{A}^{\perp} \cap L\mathcal{A}_{\perp}$. It remains to note that $L\mathcal{A}^{\perp}$ is an intermediate logic and $L\mathcal{A}_{\perp}$ is a negative one.

□

4.1.1 Intuitionistic and Negative Counterparts for Extensions of $\mathbf{Le}'$

First we state one more property of models for $\mathbf{Le}'$.

Letting $\mathcal{A}$ be a j-algebra, $\mathbf{Le}' \subseteq L\mathcal{A}$, put $C_{\mathcal{A}}(\bot) := \{a \in \mathcal{A} \mid a \vee \bot = 1\}$ and decompose $\mathcal{A}$ into a direct product $\mathcal{A}^{\perp} \times \mathcal{A}_{\perp}$. Then $C_{\mathcal{A}}(\bot) = \{(1, b) \mid b \in \mathcal{A}_{\perp}\}$. Indeed, for $a = (x, y) \in \mathcal{A}^{\perp} \times \mathcal{A}_{\perp}$, we have

$$1 = a \vee \bot \iff (1, 1) = (x, y) \vee (0, 1) = (x, 1) \iff x = 1.$$

It follows immediately that the set $C_{\mathcal{A}}(\bot)$ is closed under $\vee$, $\wedge$, and $\to$. We will consider $C_{\mathcal{A}}(\bot)$ as a negative algebra with operations induced from $\mathcal{A}$ and $1 = \bot$.

Proposition 4.1.6 *Let a j-algebra $\mathcal{A}$ be a model for* **Le**$'$. *Then $C_{\mathcal{A}}(\bot) \cong \mathcal{A}_{\bot}$ and the mapping $\delta(x) = \bot \to x$ defines an epimorphism of the j-algebra $\mathcal{A}$ onto the negative algebra $C_{\mathcal{A}}(\bot)$.*

Proof. Again, we need a presentation of $\mathcal{A}$ as a direct product of Heyting and negative algebras. The isomorphism $C_{\mathcal{A}}(\bot) \cong \mathcal{A}_{\bot}$ follows from the above equality $C_{\mathcal{A}}(\bot) = \{(1, b) \mid b \in \mathcal{A}_{\bot}\}$.

Check that δ is an epimorphism. For $a = (b, c) \in \mathcal{A}^{\bot} \times \mathcal{A}_{\bot}$, we have $\delta(a) = \delta(b, c) = (0, 1) \to (b, c) = (0 \to b, 1 \to c) = (1, c)$. Consequently, for $* \in \{\vee, \wedge, \to\}$ and for any $(a, b), (c, d) \in \mathcal{A}^{\bot} \times \mathcal{A}_{\bot}$,

$$\delta((a, b) * (c, d)) = \delta((a * c, b * d)) = (1, b * d) = (1, b) * (1, d) = \delta(a, b) * \delta(c, d).$$

It remains to note that $\delta(\bot) = 1$ and $\delta(a) = a$ for any $a \in C_{\mathcal{A}}(\bot)$.

□

We are now ready to define translations of intermediate and negative logics into properly paraconsistent extensions of **Le**$'$, which are similar to translations of classical logic and maximal negative logic into **Le** defined at the end of the previous chapter.

Theorem 4.1.7 *Let L extend* **Le**$'$, *$L \in \mathsf{Par}$, and let $\mathcal{A}$ be a characteristic model for L. Set $L_1 = L\mathcal{A}^{\bot}$ and $L_2 = L\mathcal{A}_{\bot}$. Then for an arbitrary formula φ, the following equivalences hold.*

1. $L_1 \vdash \varphi \iff L \vdash \varphi \vee \bot.$
2. $L_2 \vdash \varphi \iff L \vdash \bot \to \varphi.$

Proof. 1) Assume $L_1 \vdash \varphi$ and for an $\mathcal{A}$-valuation v, compute the value $v(\varphi \vee \bot)$. By Proposition 4.1.2, $\varepsilon : \mathcal{A} \to \mathcal{A}^{\bot}$ is an epimorphism, from which we have $v(\varphi \vee \bot) = \varepsilon v(\varphi)$. Here εv denotes an $\mathcal{A}^{\bot}$-valuation obtained as a composition of v and ε. By definition $L_1 = L\mathcal{A}^{\bot}$, whence $\varepsilon v(\varphi) = 1$. We have thus proved that $v(\varphi \vee \bot) = 1$ for any $\mathcal{A}$-valuation v, i.e., $L \vdash \varphi \vee \bot$.

Conversely, let $L \vdash \varphi \vee \bot$. Every $\mathcal{A}^{\bot}$-valuation v can be treated as an $\mathcal{A}$-valuation with the property $\varepsilon v = v$. As above, we have $v(\varphi) = \varepsilon v(\varphi) = v(\varphi \vee \bot) = 1$, which immediately implies $L_1 \vdash \varphi$.

2) This proof is similar to the previous one with ε replaced by δ.

□

As follows from the theorem, the logics $L_1 := L\mathcal{A}^{\bot}$ and $L_2 := L\mathcal{A}_{\bot}$ do not depend on the choice of a characteristic model $\mathcal{A}$ for the logic L extending **Le**$'$. Indeed,

$$L_1 = \{\varphi \mid L \vdash \varphi \vee \bot\}, \; L_2 = \{\varphi \mid L \vdash \bot \to \varphi\}.$$

It is clear that $L_1 \in \mathsf{Int}$ and $L_2 \in \mathsf{Neg}$. We call the logics L_1 and L_2 defined as above *intuitionistic* and *negative counterparts* of $L \supseteq \mathbf{Le}'$ and denote them L_{int} and L_{neg} respectively. Since $\mathcal{A} \cong \mathcal{A}^{\perp} \times \mathcal{A}_{\perp}$, we have $L = L_{int} \cap L_{neg}$.

Let, on the contrary, $L = L_1 \cap L_2$, where $L_1 \in \mathsf{Int}$ and $L_2 \in \mathsf{Neg}$. For a suitable Heyting algebra $\mathcal{B}$ and for some negative algebra $\mathcal{C}$, we have $L_1 = L\mathcal{B}$ and $L_2 = L\mathcal{C}$. The direct product $\mathcal{A} = \mathcal{B} \times \mathcal{C}$ is a characteristic model for L since $L(\mathcal{B} \times \mathcal{C}) = L\mathcal{B} \cap L\mathcal{C} = L_1 \cap L_2 = L$. Moreover, $\mathcal{B} \cong \mathcal{A}^{\perp}$ and $\mathcal{C} \cong \mathcal{A}_{\perp}$, consequently, $L_1 = L_{int}$ and $L_2 = L_{neg}$. In this way, we arrive at the following statement.

Proposition 4.1.8 *The mapping $L \mapsto (L_{int}, L_{neg})$ defines a lattice isomorphism between $[\mathbf{Le}', \mathbf{Le}]$ and the direct product* $\mathsf{Int} \times \mathsf{Neg}$. *The inverse mapping is given by the rule $(L_1, L_2) \mapsto L_1 \cap L_2$.*

Proof. In fact, it was stated above that the mapping under consideration is a bijection. According to Theorem 4.1.7 it preserves an ordering. It remains to note that an order isomorphism of two lattices is a lattice isomorphism too.

□

We can now describe the class of models for $L \supseteq \mathbf{Le}'$ as follows.

Proposition 4.1.9 *Let $L \supseteq \mathbf{Le}'$. A j-algebra $\mathcal{A}$ is a model for L if and only if $\mathcal{A} \cong \mathcal{A}^{\perp} \times \mathcal{A}_{\perp}$, $\mathcal{A}^{\perp} \models L_{int}$, and $\mathcal{A}_{\perp} \models L_{neg}$.*

Proof. Let $\mathcal{A} \models L$. According to Proposition 4.1.4 the condition $L \supseteq \mathbf{Le}'$ implies $\mathcal{A} \cong \mathcal{A}^{\perp} \times \mathcal{A}_{\perp}$. Denote $L' := L\mathcal{A}$. Then $\mathcal{A}^{\perp} \models L'_{int}$ and $\mathcal{A}_{\perp} \models L'_{neg}$ by Theorem 4.1.7. In view of the previous proposition $L_{int} \subseteq L'_{int}$ and $L_{neg} \subseteq L'_{neg}$, whence $\mathcal{A}^{\perp} \models L_{int}$ and $\mathcal{A}_{\perp} \models L_{neg}$.

Conversely, let $\mathcal{A} \cong \mathcal{A}^{\perp} \times \mathcal{A}_{\perp}$, $\mathcal{A}^{\perp} \models L_{int}$, and $\mathcal{A}_{\perp} \models L_{neg}$. Then the direct product $\mathcal{A}$ is a model for the intersection $L_{int} \cap L_{neg}$. But $L \supseteq \mathbf{Le}'$, hence, $L = L_{int} \cap L_{neg}$ by Corollary 4.1.5.

□

Thus, the class of properly paraconsistent extensions of $\mathbf{Le}'$ is completely described in terms of intermediate and negative logics. It should be emphasized that the mapping defined in Proposition 4.1.8 has an essentially effective character. Theorem 4.1.7 allows one to effectively reconstruct intuitionistic and negative counterparts from the given paraconsistent L, whereas the L itself is simply an intersection of its counterparts, i.e., a formula is proved in L if and only if it is proved in both L_{int} and L_{neg}.

However, the interval $[\mathbf{Le}', \mathbf{Le}]$ constitutes a relatively small part of the class Par of all properly paraconsistent extensions of $\mathbf{Lj}$. There are many interesting logics, which do not belong to this interval. One of them is the Glivenko logic treated in the beginning of the next chapter.

4.2 Intuitionistic and Negative Counterparts for Extensions of Minimal Logic

As the first stage in studying the whole class Par we define intuitionistic and negative counterparts for an arbitrary extension of minimal logic.

For extensions of $\mathbf{Le}'$, our definitions will be equivalent to those of the previous section.

We define the following translation

$$In(\bot) = \bot,\ In(p) = p \vee \bot,\ In(\varphi * \psi) = In(\varphi) * In(\psi),$$

where p is a propositional variable, φ and ψ arbitrary formulas, and $* \in \{\vee, \wedge, \rightarrow\}$. In other words, if $\varphi = \varphi(p_0, \ldots, p_n)$, then

$$In(\varphi) = \varphi(p_0 \vee \bot, \ldots, p_n \vee \bot).$$

For $L \in \mathsf{Jhn}^+$, define

$$L_{int} := \{\varphi \mid L \vdash In(\varphi)\},\ L_{neg} := \{\varphi \mid L \vdash \bot \rightarrow \varphi\}.$$

It can easily be seen that L_{int} and L_{neg} are logics. Moreover, $\mathbf{Li} \subseteq L_{int}$ since $\bot \rightarrow (p \vee \bot) \in \mathbf{Lj}$, and $\mathbf{Ln} \subseteq L_{neg}$ in view of $\bot \rightarrow \bot \in \mathbf{Lj}$. We call L_{int} and L_{neg} *intuitionistic* and *negative counterparts of* the logic L respectively.

Notice that this definition of negative counterpart is exactly the same as the definition of negative counterparts for $\mathbf{Le}'$-extensions given in the previous section. As for L_{int}, using formula $\mathbf{D}'$ we can easily prove in $\mathbf{Le}'$ the equivalence $(\varphi \vee \bot) \leftrightarrow In(\varphi)$ for any formula φ. Therefore, if L extends $\mathbf{Le}'$, L_{int} coincides with the intuitionistic counterpart defined in the previous section.

List some simple properties of the notions introduced above.

Proposition 4.2.1 *1. For any $L \supseteq \mathbf{Lj}$, we have $L_{int} \in \mathsf{Int} \cup \{\mathcal{F}\}$, $L_{neg} \in \mathsf{Neg} \cup \{\mathcal{F}\}$, and $L \subseteq L_{int} \cap L_{neg}$. The last inclusion is not proper if and only if L extends $\mathbf{Le}'$.*

2. $L \in \mathsf{Int}$ if and only if $L \neq \mathcal{F}$, $L = L_{int}$, and $L_{neg} = \mathcal{F}$.

3. $L \in \mathsf{Neg}$ if and only if $L \neq \mathcal{F}$, $L = L_{neg}$, and $L_{int} = \mathcal{F}$.

4. *If* $\mathbf{Lj} \subseteq L^1 \subseteq L^2$, *then* $L^1_{int} \subseteq L^2_{int}$ *and* $L^1_{neg} \subseteq L^2_{neg}$.
5. *If* $L \subseteq L_1 \in \mathsf{Int}$, *then* $L_{int} \subseteq L_1$.
6. *If* $L \subseteq L_1 \in \mathsf{Neg}$, *then* $L_{neg} \subseteq L_1$.

Proof. We only prove the last two assertions.

5. If $L \vdash In(\varphi)$, then also $L_1 \vdash In(\varphi)$. Since L_1 is intermediate, we have $L_1 \vdash In(\varphi) \to \varphi$, and so $L_1 \vdash \varphi$, which implies the desired inclusion.

6. Again, from $L_1 \vdash \bot \to \varphi$ we conclude $L_1 \vdash \varphi$ since $\bot$ belongs to any negative logic.

□

We have thus proved, in particular, that L_{int} is the least intermediate logic containing L, and L_{neg} is the least negative logic with the same property. It can easily be seen that the mappings $(-)_{int} : \mathsf{Jhn}^+ \to \mathsf{Int}$ and $(-)_{neg} : \mathsf{Jhn}^+ \to \mathsf{Neg}$ can be defined as follows. For any $L \in \mathsf{Jhn}^+$, put

$$L_{int} := L + \{\bot \to p\} = L + \mathbf{Li} \text{ and } L_{neg} := L + \{\bot\} = L + \mathbf{Ln}.$$

Proposition 4.2.2 *The mappings* $(-)_{int}$ *and* $(-)_{neg}$ *are lattice epimorphisms.*

Proof. This fact easily follows from the distributivity of Jhn^+ (Proposition 2.1.6).

□

Further, we prove that upper and lower algebras associated with a given j-algebra are semantic analogs of intuitionistic and negative counterparts.

Proposition 4.2.3 *For any* j*-algebra* $\mathcal{A}$ *and formula* φ, *the following equivalences hold.*

1. $\mathcal{A} \models In(\varphi) \iff \mathcal{A}^\bot \models \varphi$.
2. $\mathcal{A} \models \bot \to \varphi \iff \mathcal{A}_\bot \models \varphi$.

Proof. 1. Assume $\mathcal{A}^\bot \models \varphi$ and prove $\mathcal{A} \models In(\varphi)$. For an $\mathcal{A}$-valuation v, define an $\mathcal{A}^\bot$-valuation v' by the rule $v'(p) := v(p) \vee \bot$. Then it follows easily that $v(In(\varphi)) = v'(\varphi)$, which immediately implies the desired conclusion.

Conversely, let $\mathcal{A} \models In(\varphi)$. For any $\mathcal{A}^\bot$-valuation v, we have $v = v'$, in particular, $v(In(\varphi)) = v(\varphi)$, which completes the proof.

2. We use the mapping $\tau(x) = x \wedge \bot$, which is an epimorphism of $\mathcal{A}$ onto $\mathcal{A}_\bot$ by Proposition 4.1.1. Note also that $\bot \to \varphi$ is equivalent to $\bot \to (\varphi \wedge \bot)$ in $\mathbf{Lj}$.

Assuming $\mathcal{A}_\bot \models \varphi$ we take an $\mathcal{A}$-valuation v and consider the composition τv, which is an $\mathcal{A}_\bot$-valuation. By epimorphism properties of τ we have $v(\varphi \wedge \bot) = \tau v(\varphi)$. But $\tau v(\varphi) = \bot$ by assumption, which yields $v(\bot \to (\varphi \wedge \bot)) = 1$. Thus, $\mathcal{A} \models \bot \to (\varphi \wedge \bot)$ by an arbitrary choice of v.

Now, we let $\mathcal{A} \models \bot \to (\varphi \wedge \bot)$. Clearly, $v = \tau v$ for any $\mathcal{A}_\bot$-valuation v. By assumption $v(\bot) \leq v(\varphi \wedge \bot) = \tau v(\varphi) = v(\varphi)$. The greatest element of $\mathcal{A}_\bot$ is $\bot$, whence $v(\varphi) = \bot$. In this way, $\mathcal{A}_\bot \models \varphi$.

□

Corollary 4.2.4 *Let $L \in \mathsf{Jhn}^+$.*

1. *If $\mathcal{A}$ is a model for L, then*

$$\mathcal{A}^\bot \models L_{int} \text{ and } \mathcal{A}_\bot \models L_{neg}.$$

2. *If $\mathcal{A}$ is a characteristic model for L, then*

$$L\mathcal{A}^\bot = L_{int} \text{ and } L\mathcal{A}_\bot = L_{neg}.$$

□

Consider classes of logics with given intuitionistic and negative counterparts. For $L_1 \in \mathsf{Int}$ and $L_2 \in \mathsf{Neg}$, we define

$$Spec(L_1, L_2) := \{L \supseteq \mathbf{Lj} \mid L_{int} = L_1,\ L_{neg} = L_2\}.$$

It is clear that for any pair of intermediate and negative logics, (L_1, L_2), the set $Spec(L_1, L_2)$ is non-empty. It contains at least the intersection $L_1 \cap L_2$. Moreover, in view of Item 1 of Proposition 4.2.1 $L_1 \cap L_2$ is the greatest element of $Spec(L_1, L_2)$. It turns out this set also contains the least element and forms an interval in the lattice of $\mathbf{Lj}$-extensions. Let

$$L_1 * L_2 := \mathbf{Lj} + \{In(\varphi),\ \bot \to \psi \mid \varphi \in L_1,\ \psi \in L_2\},$$

where $L_1 \in \mathsf{Int}$ and $L_2 \in \mathsf{Neg}$.

Proposition 4.2.5 *Let $L_1 \in \mathsf{Int}$ and $L_2 \in \mathsf{Neg}$. Then*

$$Spec(L_1, L_2) = [L_1 * L_2, L_1 \cap L_2].$$

Proof. Let $L^* := L_1 * L_2$. It follows from definition that $L_1 \subseteq L^*_{int}$ and $L_2 \subseteq L^*_{neg}$. On the other hand, for any $L \in Spec(L_1, L_2)$, we have $L^* \subseteq L$. Indeed, L contains all axioms of L^*. As noted above, $L_1 \cap L_2$ is the greatest element of $Spec(L_1, L_2)$, whence, by Item 4 of Proposition 4.2.1 the logic L^* and all logics intermediate between L^* and $L_1 \cap L_2$ also belongs to $Spec(L_1, L_2)$.

□

The logic $L_1 * L_2$, the least element of the interval $Spec(L_1, L_2)$, we call a *free combination of* logics L_1 and L_2. This name is justified by the next proposition saying that models for $L_1 * L_2$ are all j-algebras such that their upper and lower algebras are models for L_1 and L_2, respectively.

Proposition 4.2.6 *Let* $L_1 \in \mathsf{Int}$ *and* $L_2 \in \mathsf{Neg}$. *For any* j*-algebra* $\mathcal{A}$, *we have* $\mathcal{A} \models L_1 * L_2$ *if and only if* $\mathcal{A}^{\perp} \models L_1$ *and* $\mathcal{A}_{\perp} \models L_2$.

Proof. This statement easily follows from the definition of free combination and Corollary 4.2.4.

□

The next proposition allows one to write axioms for $L_1 * L_2$ relative to **Lj** given an axiomatics of L_1 relative to **Li** and of L_2 relative to **Ln**.

Proposition 4.2.7 *Let* $L_1 \in \mathsf{Int}$, $L_1 = \mathbf{Li} + \{\varphi_i \mid i \in I\}$ *and* $L_2 \in \mathsf{Neg}$, $L_2 = \mathbf{Ln} + \{\psi_j \mid j \in J\}$. *Then*

$$L_1 * L_2 = \mathbf{Lj} + \{In(\varphi_i),\ \perp \to \psi_j \mid i \in I,\ j \in J\}.$$

Proof. Denote the right-hand side of the last equality by D. The inclusion $D \subseteq L_1 * L_2$ is trivial. To state the inverse inclusion we show that $L_1 \subseteq D_{int}$ and $L_2 \subseteq D_{neg}$.

We argue for $L_2 \subseteq D_{neg}$. Note that $\mathbf{Ln} = \mathbf{Lj}_{neg}$, i.e., $\mathbf{Ln} \vdash \varphi$ if and only if $\mathbf{Lj} \vdash \perp \to \varphi$. Assume $\psi \in L_2$, then $\mathbf{Ln} \vdash (\psi'_{j_1} \wedge \ldots \wedge \psi'_{j_n}) \to \psi$ for suitable particular cases $\psi'_{j_1}, \ldots, \psi'_{j_n}$ of axioms $\psi_{j_1}, \ldots, \psi_{j_n}$, $j_1, \ldots, j_n \in J$. Whence

$$\mathbf{Lj} \vdash \perp \to ((\psi'_{j_1} \wedge \ldots \wedge \psi'_{j_n}) \to \psi).$$

The last formula implies in **Lj**

$$((\perp \to \psi'_{j_1}) \wedge \ldots \wedge (\perp \to \psi'_{j_n})) \to (\perp \to \psi),$$

from which we infer $\perp \to \psi \in D$. Consequently, $L_2 \subseteq D_{neg}$.

The remaining inclusion follows in the same way from the equality $\mathbf{Li} = \mathbf{Lj}_{int}$.

□

As we can see from Proposition 4.2.5, the class of properly paraconsistent **Lj**-extensions decomposes into a union of disjoint intervals

$$\mathsf{Par} = \bigcup \{Spec(L_1, L_2) \mid L_1 \in \mathsf{Int},\ L_2 \in \mathsf{Neg}\}.$$

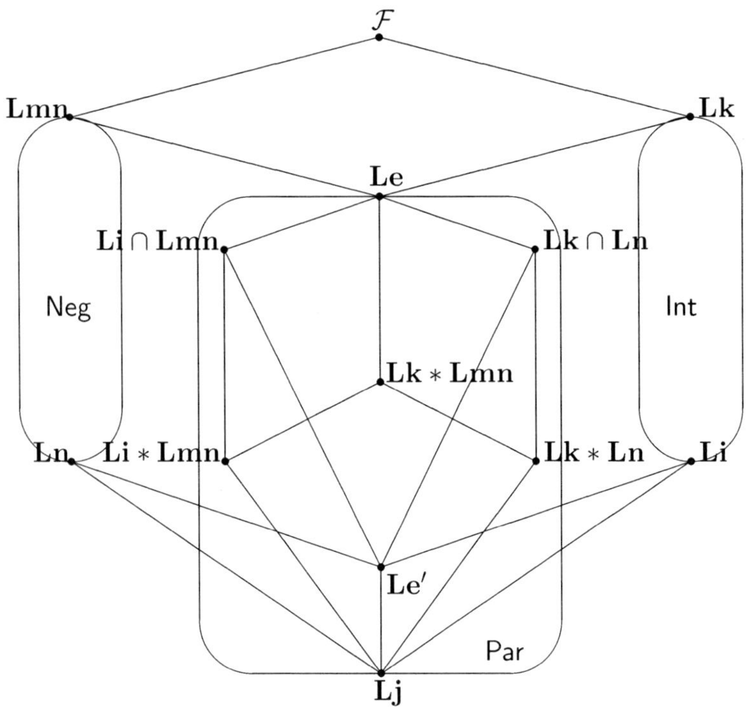

Figure 4.1

It is interesting that the upper points of these intervals also form an interval in Jhn^+, $[\mathbf{Le}', \mathbf{Le}]$. Figure 4.1 illustrates the structure of the class Jhn^+.

In this way, the investigation of the class of **Lj**-extensions is reduced to the problem: what is the structure of the interval $Spec(L_1, L_2)$ for the given intermediate logic L_1 and negative logic L_2? This problem will be treated in the subsequent sections but first we make an observation on the nature of the negative counterpart L_{neg} of a paraconsistent logic L.

4.2.1 Negative Counterparts as Logics of Contradictions

We define a contradiction operator $C(\varphi) := \varphi \wedge \neg\varphi$ and extend this operator to sets of formulas as follows. Put $C(\varnothing) := \{\bot\}$ and $C(X) := \{C(\varphi) \mid \varphi \in X\}$ for $X \neq \varnothing$. The contradiction operator is trivial in all intermediate logics. If the law *ex contadictione quodlibet* holds, we have $C(\varphi) \leftrightarrow \bot$ for any formula φ. Rejecting *ex contadictione quodlibet* we obtain the possibility to distinguish contradictions constructed with the help of different formulas. In particular, for $L \in \mathsf{Par}$ we have $L \vdash C(\varphi) \leftrightarrow \bot$ if and only if $\varphi \in L_{neg}$. Moreover, it turns out that relative to deducibility properties, the behavior of formulas in the negative counterpart L_{neg} is completely similar to the behavior of contradictions constructed with the help of these formulas in the original logic L. More precisely, the following fact takes place.

Proposition 4.2.8 *Let* $L \in \mathsf{Par}$. *For an arbitrary set of formulas* X *and for any formula* φ, *the following equivalence holds:*

$$X \vdash_{L_{neg}} \varphi \iff C(X) \vdash_L C(\varphi).$$

The *proof* of this proposition is an easy exercise on the deducibility in minimal logic.

□

We have thus proved that the contradiction operator defines a strong translation of the negative counterpart L_{neg} in a paraconsistent logic $L \in \mathsf{Par}$. This fact allows one to consider the negative counterpart L_{neg} as a logic of contradictions associated with a given paraconsistent logic L.

4.3 Three Dimensions of Par

We can see now that the class Par has a three-dimensional structure. The position of a logic L in this class is determined by its intuitionistic counterpart L_{int}, which represents reasoning in L under the additional assumption of inconsistency, or of impossibility of contradictions, and by its structure of contradictions explicated in the negative counterpart L_{neg}. When an explosive pattern of reasoning and a structure of contradictions are fixed, we have a variety of possibilities for combining them presented by the interval of logics $Spec(L_{int}, L_{neg})$. The place of L in this interval can be considered as its third coordinate in Par, the sense of which is not quite clear yet. It becomes clearer in the next chapter. Now we turn to the question of a scale for this third coordinate. Unlike first and second coordinates having absolute scales, Int and Neg respectively, the intervals $Spec(I, N)$ are mutually disjoint for different pairs of logic $I \in \mathsf{Int}$ and $N \in \mathsf{Neg}$. However, one can find natural interrelations between these scales, i.e., between the intervals of the form $Spec(I, N)$ for various $I \in \mathsf{Int}$ and $N \in \mathsf{Neg}$.

Consider two pairs of logics $\mathsf{P}_1 = (I_1, N_1)$ and $\mathsf{P}_2 = (I_2, N_2)$, where $I_1, I_2 \in \mathsf{Int}$, $N_1, N_2 \in \mathsf{Neg}$. $\mathsf{P}_1 \leq \mathsf{P}_2$ means that $I_1 \subseteq I_2$ and $N_1 \subseteq N_2$. We write also $Spec(\mathsf{P}_1)$ for $Spec(I_1, N_1)$.

Let $\mathsf{P}_1 = (I_1, N_1)$ and $\mathsf{P}_2 = (I_2, N_2)$ be such that $\mathsf{P}_1 \leq \mathsf{P}_2$. Mappings $r_{\mathsf{P}_2,\mathsf{P}_1} : Spec(\mathsf{P}_2) \to \mathsf{Par}$ and $e_{\mathsf{P}_1,\mathsf{P}_2} : Spec(\mathsf{P}_1) \to \mathsf{Par}$ are defined as follows

$$r_{\mathsf{P}_2,\mathsf{P}_1}(L) := L \cap (I_1 \cap N_1), \; e_{\mathsf{P}_1,\mathsf{P}_2}(L) := L + (I_2 * N_2).$$

Proposition 4.3.1 *Let pairs of logics* P_1 *and* P_2 *be such that* $\mathsf{P}_1 \leq \mathsf{P}_2$. *The following facts hold.*

1. *For any* $L \in Spec(\mathsf{P}_2)$, *we have* $e_{\mathsf{P}_1,\mathsf{P}_2} r_{\mathsf{P}_2,\mathsf{P}_1}(L) = L$.
2. *For any* $L \in Spec(\mathsf{P}_1)$, *we have*
$$r_{\mathsf{P}_2,\mathsf{P}_1} e_{\mathsf{P}_1,\mathsf{P}_2}(L) = L + r_{\mathsf{P}_2,\mathsf{P}_1}(I_2 * N_2)$$
3. $e_{\mathsf{P}_1,\mathsf{P}_2}$ *is a lattice epimorphism from* $Spec(\mathsf{P}_1)$ *onto* $Spec(\mathsf{P}_2)$.
4. $r_{\mathsf{P}_2,\mathsf{P}_1}$ *is a lattice monomorphism from* $Spec(\mathsf{P}_2)$ *into* $Spec(\mathsf{P}_1)$ *and*
$$r_{\mathsf{P}_2,\mathsf{P}_1}(\mathsf{P}_2) = [r_{\mathsf{P}_2,\mathsf{P}_1}(I_2 * N_2), I_1 \cap N_1].$$
5. *For any* P_3 *such that* $\mathsf{P}_2 \leq \mathsf{P}_3$, *we have*
$$e_{\mathsf{P}_1,\mathsf{P}_2} e_{\mathsf{P}_2,\mathsf{P}_3} = e_{\mathsf{P}_1,\mathsf{P}_3}, \ r_{\mathsf{P}_3,\mathsf{P}_2} r_{\mathsf{P}_2,\mathsf{P}_1} = r_{\mathsf{P}_3,\mathsf{P}_1}.$$

Proof. 1. We calculate

$$e_{\mathsf{P}_1,\mathsf{P}_2} r_{\mathsf{P}_2,\mathsf{P}_1}(L) = (L \cap (I_1 \cap N_1)) + (I_2 * N_2)$$
$$= (L + (I_2 * N_2)) \cap ((I_1 \cap N_1) + (I_2 * N_2)).$$

By Proposition 4.2.5 $I_2 * N_2$ is the least point of $Spec(\mathsf{P}_2)$, therefore, we have $L + (I_2 * N_2) = L$. Further, we need one lemma.

Lemma 4.3.2 *For any* $L \in Spec(I, N)$, $I \cap N = L + \mathbf{Le}'$.

Proof. $(\mathbf{Le}')_{int}$ equals to $\mathbf{Li}$, the least logic in Int, and $(\mathbf{Le}')_{neg} = \mathbf{Ln}$, which is the least logic in Neg. Now, it follows from Proposition 4.2.2 that $L + \mathbf{Le}'$ has the same counterparts as L. By Corollary 4.1.5 $L + \mathbf{Le}'$ coincides with the greatest point of $Spec(I, N)$.

□

Using this lemma and the obvious relation $I_1 * N_1 \subseteq I_2 * N_2$ we obtain

$$(I_1 \cap N_1) + (I_2 * N_2) = ((I_1 * N_1) + \mathbf{Le}') + (I_2 * N_2) = I_2 * N_2 + \mathbf{Le}' = I_2 \cap N_2.$$

And finally, $e_{\mathsf{P}_1,\mathsf{P}_2} r_{\mathsf{P}_2,\mathsf{P}_1}(L) = L \cap (I_2 \cap N_2) = L$.

2. The direct computation shows

$$r_{\mathsf{P}_2,\mathsf{P}_1} e_{\mathsf{P}_1,\mathsf{P}_2}(L) = (L + (I_2 * N_2)) \cap (I_1 \cap N_1)$$
$$= (L \cap (I_1 \cap N_1)) + ((I_2 * N_2) \cap (I_1 \cap N_1)) = L + r_{\mathsf{P}_2,\mathsf{P}_1}(I_2 * N_2).$$

3. It follows from the distributivity of Jhn^+ that $e_{\mathsf{P}_1,\mathsf{P}_2}$ is a lattice homomorphism. Let $L \in Spec(\mathsf{P}_1)$ and $L' := e_{\mathsf{P}_1,\mathsf{P}_2}(L) = L + (I_2 * N_2)$. By Proposition 4.2.2 $(L')_{int} = L_{int} + (I_2 * N_2)_{int} = I_1 + I_2 = I_2$. In the same way, $(L')_{neg} = N_2$. Consequently, $L' \in Spec(\mathsf{P}_2)$. The fact that $e_{\mathsf{P}_1,\mathsf{P}_2}$ is an epimorphism follows from Item 1.

4. As above, we use Proposition 4.2.2 to check that $r_{\mathsf{P}_2,\mathsf{P}_1}$ maps $Spec(\mathsf{P}_2)$ into $Spec(\mathsf{P}_1)$. This is a homomorphism due to the distributivity of Jhn^+. If $r_{\mathsf{P}_2,\mathsf{P}_1}(L_1) = r_{\mathsf{P}_2,\mathsf{P}_1}(L_2)$, then applying the formula of Item 1 we obtain $L_1 = L_2$. Thus, $r_{\mathsf{P}_2,\mathsf{P}_1}$ is a monomorphism. The equality $r_{\mathsf{P}_2,\mathsf{P}_1}(\mathsf{P}_2) = [r_{\mathsf{P}_2,\mathsf{P}_1}(I_2 * N_2), I_1 \cap N_1]$ follows from Item 2.

5. This item follows from the obvious relations $I_2 * N_2 \subseteq I_3 * N_3$ and $I_1 \cap N_1 \subseteq I_2 \cap N_2$.

□

The above proposition shows, in particular, that any interval $Spec(I, N)$ is isomorphic to an upper subinterval of $Spec(\mathbf{Li}, \mathbf{Ln})$. In this way, the latter interval can be considered as a scale for the third dimension of the class Par. Extending intuitionistic and negative counterparts, we restrict simultaneously the part of the scale that can be used to construct a logic with given counterparts. It is also worth noticing the following consequence of the last proposition.

Corollary 4.3.3 *Let* $\mathsf{P}_1 = (I_1, N_1)$ *and* $\mathsf{P}_2 = (I_2, N_2)$ *be pairs of logics such that* $\mathsf{P}_1 \leq \mathsf{P}_2$*. For any logics* $L_1, L_2 \in Spec(\mathsf{P}_2)$*,* $L_1 \neq L_2$*, there is a formula* $\varphi \in I_1 \cap N_1$ *such that* $\varphi \in (L_1 \setminus L_2) \cup (L_2 \setminus L_1)$*.*

Proof. Let $L_1, L_2 \in Spec(\mathsf{P}_2)$. If $L_1 \neq L_2$, but these logics are not distinguished by a formula $\varphi \in I_1 \cap N_1$, then $r_{\mathsf{P}_2,\mathsf{P}_1}(L_1) = r_{\mathsf{P}_2,\mathsf{P}_1}(L_2)$. By Item 4 of the previous proposition $r_{\mathsf{P}_2,\mathsf{P}_1}$ is a monomorphism, whence $L_1 = L_2$, a contradiction.

□

In particular, any two logics from $Spec(I, N)$ can be distinguished via a formula from $\mathbf{Le}' = \mathbf{Li} \cap \mathbf{Ln}$. Moreover, any logic from the interval $Spec(I, N)$ can be axiomatized by formulas from $\mathbf{Le}'$ modulo the least logic of the interval $I * N$. Indeed, for any $L \in Spec(I, N)$ we have by Item 1 of Proposition 4.3.1 $L = (L \cap \mathbf{Le}') + (I * N)$. In this way, any possible combination of intuitionistic and negative logics can be determined by corollaries of the formula $\bot \vee (\bot \to p)$.

For further investigations of the class Par, we need semantic considerations.

Chapter 5

Adequate Algebraic Semantics for Extensions of Minimal Logic[1]

The goal of this chapter is to find a representation of j-algebras, convenient for working with logics lying inside the intervals $Spec(L_1, L_2)$. We have to understand the structure of an arbitrary j-algebra $\mathcal{A}$ with given upper algebra $\mathcal{A}^{\perp}$ and lower algebra $\mathcal{A}_{\perp}$. The semantic characterization of Glivenko's logic considered in Section 5.1 prompts the solution to this problem. The desired representation is described in Section 5.2. In Section 5.3 with the help of the obtained representation we characterize the Segerberg logics and demonstrate its effectiveness in this way. Finally, in Section 5.4 we consider the Kripke semantics and define for j-frames analogs of upper and lower algebras associated with a j-algebra.

5.1 Glivenko's Logic

Consider the following substitution instance of the Peirce law:

P′. $((\perp \to p) \to \perp) \to \perp = \neg\neg(\perp \to p)$.

We call the logic $\mathbf{Lg} := \mathbf{Lj} + \{\mathbf{P'}\}$ *Glivenko's logic*. It was mentioned in [98, p. 46] that Glivenko's logic is the weakest one in which $\neg\neg\varphi$ is derivable whenever φ is derivable in classical logic. Unfortunately, this work contains neither the proof of this assertion, nor any further reference. In this section,

[1]Parts of this chapter were originally published in [69].

we present a natural algebraic proof of this statement. We also show that **Lg** is a proper subsystem of **Le**′.

Proposition 5.1.1 *1. Let $\mathcal{A}$ be a j-algebra. Then $\mathcal{A}$ is a model for* **Lg** *if and only if $\bot \vee (\bot \to a) \in F_d(\mathcal{A}^{\bot})$ for any $a \in \mathcal{A}$.*

2. Let $\mathcal{A}$ be a model for **Lg** *and $\nabla := F_d(\mathcal{A}^{\bot})$. Then the mapping $\pi(a) = (a \vee \bot)/\nabla$ defines an epimorphism of $\mathcal{A}$ onto $\mathcal{A}^{\bot}/\nabla$.*

Proof. 1. This item immediately follows from the definition of Glivenko's logic and the fact that $\neg(a \vee \bot) = a \vee \bot \to \bot = a \to \bot = \neg a$ for any j-algebra $\mathcal{A}$ and $a \in \mathcal{A}$. The last equality implies, in particular,

$$F_d(\mathcal{A}^{\bot}) = \{a \in \mathcal{A} \mid \neg\neg a = 1\}.$$

2. In fact, we need only check that π preserves the implication, i.e.,

$$(a \to b) \vee \bot/\nabla = (a \vee \bot) \to (b \vee \bot)/\nabla.$$

We have $((a \to b) \vee \bot) \to ((a \vee \bot) \to (b \vee \bot)) = 1 \in \nabla$, since the corresponding formula is provable in **Lj** (see Item 8 of Proposition 2.1.2). Further, it can be verified directly that

$$\mathbf{Lj} \vdash (\bot \to q) \to (((p \vee \bot) \to (q \vee \bot)) \to ((p \to q) \vee \bot)).$$

In view of $\mathbf{Lj} \vdash (p \to q) \to (\neg\neg p \to \neg\neg q)$ we obtain

$$\mathbf{Lj} \vdash \neg\neg(\bot \to q) \to \neg\neg(((p \vee \bot) \to (q \vee \bot)) \to ((p \to q) \vee \bot)),$$

i.e.,

$$\mathbf{Lg} \vdash \neg\neg(((p \vee \bot) \to (q \vee \bot)) \to ((p \to q) \vee \bot)).$$

By assumption $\mathcal{A}$ is a model for **Lg**, consequently,

$$((a \vee \bot) \to (b \vee \bot)) \to ((a \to b) \vee \bot) \in \nabla,$$

which completes the proof.

□

Theorem 5.1.2 *(Generalized Glivenko's Theorem.) For every logic $L \in$* Jhn, *the following conditions are equivalent.*

1. For any φ,

$$\mathbf{Lk} \vdash \varphi \iff L \vdash \neg\neg\varphi.$$

2. $L \supseteq$ **Lg** *and $L \notin$* Neg.

Proof. $1 \Rightarrow 2$. This implication is trivial.

$2 \Rightarrow 1$. Since $L \notin \mathsf{Neg}$, the logic L is contained in $\mathbf{Lk}$. If $L \vdash \neg\neg\varphi$, then $\mathbf{Lk} \vdash \neg\neg\varphi$, and so $\mathbf{Lk} \vdash \varphi$.

Assume $\mathbf{Lk} \vdash \varphi$. Let $\mathcal{A} \models L$ and v be an $\mathcal{A}$-valuation. Consider an $\mathcal{A}^{\perp}/F_d(\mathcal{A}^{\perp})$-valuation v' such that $v'(p) = v(p \vee \perp)/F_d(\mathcal{A}^{\perp})$. Using the assumption $\mathbf{Lg} \subseteq L$ and Proposition 5.1.1 we obtain $v'(\varphi) = v(\varphi \vee \perp)/F_d(\mathcal{A}^{\perp})$. On the other hand, $\mathcal{A}^{\perp}/F_d(\mathcal{A}^{\perp})$ is a Boolean algebra and $\mathbf{Lk} \vdash \varphi$, therefore, $v'(\varphi) = 1$. In this way, $v(\varphi \vee \perp) \in F_d(\mathcal{A}^{\perp})$, i.e., $v(\neg\neg(\varphi \vee \perp)) = v(\neg\neg\varphi) = 1$. Since $\mathcal{A}$ and v are arbitrary, we obtain $L \vdash \neg\neg\varphi$.

□

Let us prove that Glivenko's logic does not belong to the class of $\mathbf{Le}'$-extensions. To this end it will be enough to show that Glivenko's logic has models different from direct products of Heyting and negative algebras.

Proposition 5.1.3 *Let $\mathcal{A}$ be a model for $\mathbf{Le}'$, and $\mathcal{B}$ a Heyting algebra. Then $\mathcal{A} \oplus \mathcal{B}$ is a model for $\mathbf{Lg}$.*

Proof. It follows from two facts. For all $a \in \mathcal{A} \oplus \mathcal{B}$, we have $\perp \vee (\perp \rightarrow a) \in \mathcal{B}$. All elements of $\mathcal{B}$ are dense in $(\mathcal{A} \oplus \mathcal{B})^{\perp}$.

Corollary 5.1.4 *The inclusion $\mathbf{Lg} \subset \mathbf{Le}'$ is proper.*

Proof. Indeed, according to Proposition 4.1.4 the algebra $\mathcal{A} \oplus \mathcal{B}$ is not a model for $\mathbf{Le}'$ if $\mathcal{B}$ is a non-trivial Heyting algebra. But this is a model of Glivenko's logic by the previous proposition.

□

5.2 Representation of j-Algebras

In this section we give a convenient representation of j-algebras, which allows one to describe classes of models for logics lying inside intervals of the form $[L_1 * L_2, L_1 \cap L_2]$, where $L_1 \in \mathsf{Int}$ and $L_2 \in \mathsf{Neg}$. We know that an intersection $L_1 \cap L_2$ of intermediate and negative logics is characterized by the class of all direct products of the form $\mathcal{A} \times \mathcal{B}$, where $\mathcal{A}$ is a Heyting algebra being a model for the logic L_1 and $\mathcal{B}$ is a negative algebra modelling L_2. Indeed, due to Corollary 4.1.5 the intersection $L_1 \cap L_2$ extends $\mathbf{Le}'$ and each model $\mathcal{A}$ for $L_1 \cap L_2$ is isomorphic to the direct product $\mathcal{A}^{\perp} \times \mathcal{A}_{\perp}$ by Proposition 4.1.4. It remains to note that by Proposition 4.1.8 we have $L_1 = (L_1 \cap L_2)_{int}$ and $L_2 = (L_1 \cap L_2)_{neg}$. Thus, Proposition 4.1.9 implies $\mathcal{A}^{\perp} \models L_1$ and $\mathcal{A}_{\perp} \models L_2$.

At the same time, a free combination of logics, $L_1 * L_2$, is characterized by the class of all j-algebras $\mathcal{A}$ such that the upper algebra $\mathcal{A}^{\perp}$ is a model for L_1 and the lower algebra $\mathcal{A}_{\perp}$ models L_2 (see Proposition 4.2.6). At this point the following question arises. If a Heyting algebra $\mathcal{B}$ and a negative algebra $\mathcal{C}$ are given, what is the difference between an arbitrary j-algebra $\mathcal{A}$ with the condition $\mathcal{A}^{\perp} \cong \mathcal{B}$ and $\mathcal{A}_{\perp} \cong \mathcal{C}$ and the direct product of algebras $\mathcal{B} \times \mathcal{C}$? Proposition 5.1.1 allows us to assume that elements of the form $\perp \vee (\perp \to a)$, where $a \in \mathcal{A}_{\perp}$, play a special part in the structure of $\mathcal{A}$.

Proposition 5.2.1 *Let $\mathcal{A}$ be a j-algebra and a mapping $f_{\mathcal{A}} : \mathcal{A}_{\perp} \to \mathcal{A}^{\perp}$ is given by the rule $f_{\mathcal{A}}(x) = \perp \vee (\perp \to x)$. Then the following two conditions are met.*

1. *The mapping $f_{\mathcal{A}} : \mathcal{A}_{\perp} \to \mathcal{A}^{\perp}$ is a semilattice homomorphism preserving the meet $\wedge$ and the greatest element, $f_{\mathcal{A}}(\perp) = 1$;*
2. *The embedding $\lambda_{\perp} : \mathcal{A} \to \mathcal{A}^{\perp} \times \mathcal{A}_{\perp}$, where $\lambda_{\perp}(x) = (x \vee \perp, x \wedge \perp)$, has the following image*

$$\lambda_{\perp}(\mathcal{A}) = \{(x, y) \mid x \leq f_{\mathcal{A}}(y),\ x \in \mathcal{A}^{\perp},\ y \in \mathcal{A}_{\perp}\}.$$

Proof. 1. For brevity, we omit the lower index in denotation $f_{\mathcal{A}}$. We have $f(\perp) = \perp \vee (\perp \to \perp) = 1$. Further,

$$f(y_1) \wedge f(y_2) = (\perp \vee (\perp \to y_1)) \wedge (\perp \vee (\perp \to y_2)) =$$

$$= \perp \vee ((\perp \to y_1) \wedge (\perp \to y_2)) = \perp \vee (\perp \to (y_1 \wedge y_2)) = f(y_1 \wedge y_2).$$

We have thus verified that f is a semilattice homomorphism preserving the meet and the unit element.

2. If $a \in \mathcal{A}$, then $(a \vee \perp, a \wedge \perp) \in \lambda_{\perp}(\mathcal{A})$ and the inequality $a \vee \perp \leq \perp \vee (\perp \to (a \wedge \perp))$ holds. The latter can be checked, for example, by proving in **Lj** the formula $p \vee \perp \to \perp \vee (\perp \to p \wedge \perp)$. Thus, the inclusion

$$\lambda_{\perp}(\mathcal{A}) \subseteq \{(x, y) \mid x \leq f(y),\ x \in \mathcal{A}^{\perp},\ y \in \mathcal{A}_{\perp}\}$$

is proved. Now, let $x, y \in \mathcal{A}$, $x \geq \perp$, $y \leq \perp$, and $x \leq \perp \vee (\perp \to y)$. We show that there is an element $a \in \mathcal{A}$ such that $x = a \vee \perp$ and $y = a \wedge \perp$. Put $a := x \wedge (\perp \to y)$, then

$$a \vee \perp = (\perp \vee x) \wedge (\perp \vee (\perp \to y)) = x \wedge (\perp \vee (\perp \to y)) = x,$$

moreover, $a \wedge \bot = x \wedge (\bot \to y) \wedge \bot = \bot \wedge (\bot \to y) = y$. The inverse inclusion is also checked.

□

As we can see from the above proposition, every j-algebra $\mathcal{A}$ determines a triple $(\mathcal{A}^{\bot}, \mathcal{A}_{\bot}, f_{\mathcal{A}})$ consisting of a Heyting algebra, a negative algebra and a semilattice homomorphism. Now, let us take a triple $(\mathcal{B}, \mathcal{C}, f : \mathcal{C} \to \mathcal{B})$, where $\mathcal{B}$ is an arbitrary Heyting algebra, $\mathcal{C}$ a negative algebra and f a semilattice homomorphism from $\mathcal{C}$ to $\mathcal{B}$ preserving the meet and the greatest element. Starting from this triple we try to construct a j-algebra $\mathcal{A}$, the upper and lower algebras of which are isomorphic to $\mathcal{B}$ and $\mathcal{C}$ respectively, and the mapping $f_{\mathcal{A}}$ is induced in a natural way by the homomorphism f.

Define a lattice $\mathcal{B} \times_f \mathcal{C}$ as a sublattice of the direct product $\mathcal{B} \times \mathcal{C}$ with the universe

$$|\mathcal{B} \times_f \mathcal{C}| := \{(x, y) \mid x \in \mathcal{B},\ y \in \mathcal{C},\ x \le f(y)\}.$$

This is really a sublattice of $\mathcal{B} \times \mathcal{C}$, because f preserves the meet and, hence, the ordering, which easily implies the relation $f(y_1) \vee f(y_2) \le f(y_1 \vee y_2)$. From the latter immediately follows that the set $|\mathcal{B} \times_f \mathcal{C}|$ is closed under componentwise lattice operations on the direct product of lattices. As we can see from the proposition below, this lattice can be considered as a j-algebra.

Proposition 5.2.2 *Let $\mathcal{B}$, $\mathcal{C}$, f, and $\mathcal{A} := \mathcal{B} \times_f \mathcal{C}$ be as above. The lattice $\mathcal{A}$ has a natural structure of j-algebra, where the relative pseudo-complement operation is given by the rule*

$$(x_1, y_1) \to (x_2, y_2) = ((x_1 \to x_2) \wedge f(y_1 \to y_2),\ y_1 \to y_2),$$

$1_{\mathcal{A}} = (1_{\mathcal{B}}, \bot_{\mathcal{C}})$, and $\bot_{\mathcal{A}} = (\bot_{\mathcal{B}}, \bot_{\mathcal{C}})$. Moreover, $\mathcal{B} \cong \mathcal{A}^{\bot}$, $\mathcal{C} \cong \mathcal{A}_{\bot}$, and these isomorphisms are given by the rules $x \mapsto (x, \bot_{\mathcal{C}})$, $x \in \mathcal{B}$, and $y \mapsto (\bot_{\mathcal{B}}, y)$, $y \in \mathcal{C}$, respectively. Finally, for all $y \in \mathcal{C}$, we have

$$(f(y), \bot_{\mathcal{C}}) = \bot_{\mathcal{A}} \vee (\bot_{\mathcal{A}} \to (\bot_{\mathcal{B}}, y)) = f_{\mathcal{A}}((\bot_{\mathcal{B}}, y)).$$

Proof. First, we check that the relative pseudo-complement is well defined. Let $b_1, b_2 \in \mathcal{B}$, $c_1, c_2 \in \mathcal{C}$, $b_1 \le f(c_1)$, and $b_2 \le f(c_2)$. The element $(b_1, c_1) \to (b_2, c_2)$, if it is defined, is the greatest among all elements (x, y) such that $x \le f(y)$ and $(b_1, c_1) \wedge (x, y) \le (b_2, c_2)$. This is equivalent to relations $x \le (b_1 \to b_2) \wedge f(y)$ and $y \le c_1 \to c_2$. Taking into account that f preserves the ordering, we immediately obtain that the element

$$((b_1 \to b_2) \wedge f(c_1 \to c_2),\ c_1 \to c_2)$$

is the desired relative pseudo-complement.

All other relations, except the last, are trivial. Check the last relation using the obtained formula for relative pseudo-complement. We have

$$\bot_{\mathcal{A}} \vee (\bot_{\mathcal{A}} \to (\bot_{\mathcal{B}}, y)) = (\bot_{\mathcal{B}}, \bot_{\mathcal{C}}) \vee ((\bot_{\mathcal{B}}, \bot_{\mathcal{C}}) \to (\bot_{\mathcal{B}}, y)) =$$

$$= (\bot_{\mathcal{B}}, \bot_{\mathcal{C}}) \vee (1_{\mathcal{B}} \wedge f(\bot_{\mathcal{C}} \to y), \bot_{\mathcal{C}} \to y) = (\bot_{\mathcal{B}}, \bot_{\mathcal{C}}) \vee (f(y), y) = (f(y), \bot_{\mathcal{C}}).$$

□

As we can see from the above considerations, to define a class of j-algebras characterizing some extension L of minimal logic, we must choose a class of Heyting algebras and a class of negative algebras isomorphic to upper and lower algebras respectively, associated with models for L. In this way, we fix intuitionistic and negative counterparts of the logic L. Moreover, to determine the place of L inside the interval $[L_{int} * L_{neg}, L_{int} \cap L_{neg}]$, we have to distinguish in one or another way the class of admissible homomorphisms from negative algebras into Heyting ones. If no restrictions are imposed on the class of homomorphisms, we obtain a free combination of intermediate and negative logics characterized by the selected classes of Heyting and negative algebras (see Proposition 4.2.6). If we admit only homomorphisms identically equal to the unit element, we obtain the intersection $L_{int} \cap L_{neg}$. Indeed, a j-algebra $\mathcal{A} \times_f \mathcal{B}$ coincides with the direct product $\mathcal{A} \times \mathcal{B}$ if and only if $f(y) = 1$ for all $y \in \mathcal{B}$.

5.3 Segerberg's Logics and their Semantics

It is interesting to consider logics different from intersections and free combinations of intermediate and negative logics, i.e., logics lying inside intervals of the form $Spec(L_1, L_2)$. In this section, using the representation for j-algebras obtained above, we describe an algebraic semantics for logics studied previously by K. Segerberg [98], who characterized these logics in terms of Kripke semantics. Except for **Lj** K. Segerberg [98] considered logics obtained by adding to **Lj** one or several axioms from the list below.

I. $\bot \to p$

K. $\neg p \vee \neg\neg p$

X. $p \vee \neg p$

L. $(p \to q) \vee (q \to p)$

E. $p \vee (p \to q)$

$\mathbf{L'}$. $\neg p \vee (\bot \to p)(= (p \to \bot) \vee (\bot \to p))$

$\mathbf{E'}$. $\bot \vee (\bot \to p)$

$\mathbf{Q}$. $\bot$

$\mathbf{L_N}$. $(p \to q \vee \bot) \vee (q \to p \vee \bot)$

$\mathbf{L_1^Q}$. $\bot \to (p \to q) \vee (q \to p)$

$\mathbf{L_2^Q}$. $(\bot \to (p \to q)) \vee (\bot \to (q \to p))$

$\mathbf{E_1^Q}$. $\bot \to p \vee (p \to q)$

$\mathbf{E_2^Q}$. $(\bot \to p) \vee (\bot \to (p \to q))$

$\mathbf{P'}$. $\neg\neg(\bot \to p)(= ((\bot \to p) \to \bot) \to \bot)$

We may combine these axioms, which gives rise to a large number of new logics. Some of these logics have traditional denotation, for example $\mathbf{Li} = \mathbf{Lj} + \{\mathbf{I}\}$, and others have not. Due to this fact, we need some notational conventions.

If some logic is obtained from the logic already having a denotation, say L, by adding some axiom denoted by a capital letter, say $\mathbf{X}$, then the denotation of this new logic will be obtained by joining the corresponding small letter to the existing denotation, $L\mathbf{x} := L + \{\mathbf{X}\}$. Of course, in this way one logic may obtain different denotations. According to this convention we have, for example, $\mathbf{Lji} = \mathbf{Li}$, $\mathbf{Lje} = \mathbf{Le}$, $\mathbf{Ljq} = \mathbf{Ln}$, $\mathbf{Lix} = \mathbf{Ljix} = \mathbf{Lk}$, and finally, $\mathbf{Ljp'} = \mathbf{Lg}$.

We shall say a few words on how the above list of axioms arises. The Kripke semantics for extensions of $\mathbf{Lj}$ was described in Chapter 2. Recall that any j-frame is divided into two parts consisting of normal worlds and of abnormal worlds. The first axiom $\mathbf{I}$ distinguish the class of j-frames in which all worlds are normal. The next two axioms are the well-known law of excluded middle $\mathbf{X}$ and weak law of excluded middle $\mathbf{K}$. These axioms impose some restrictions on the accessibility relation only in the normal part of a j-frame [98]. It must be identical in case of $\mathbf{X}$ and directed in case of $\mathbf{K}$. The Dummett linearity axiom $\mathbf{L}$ and the extended law of excluded middle $\mathbf{E}$ define properties of accessibility relation in the whole frame. A j-frame satisfying $\mathbf{L}$ is linear, whereas in a j-frame satisfying $\mathbf{E}$ the accessibility relation is identical. The next two axioms, $\mathbf{L'}$ and $\mathbf{E'}$, are particular cases of $\mathbf{L}$ and $\mathbf{E}$, respectively. They do not impose any restrictions on either the normal

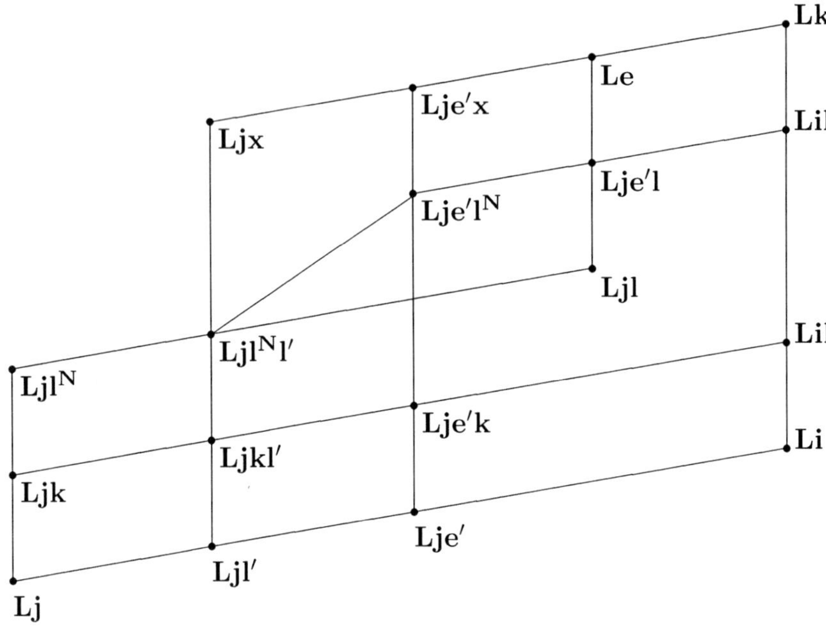

Figure 5.1

or abnormal part of a j-frame, but they define the way in which the cone of abnormal worlds is situated in the whole frame (see Proposition 2.3.3).

The interrelations between logics obtained by joining to **Lj** one or several axioms reviewed up to this point are presented in Figure 5.1.

Note that this diagram (as well as the diagram presented in Figure 5.2 below) respects only the ordering, but not the lattice structure of Jhn^+. All logics presented at the diagram are distinct, and a logic L_1 is contained in a logic L_2 if and only if there is a path leading from L_1 to L_2, which at every point is either rising or horizontal and directed to the right.

To explain the explicit irregularities of the above diagram K. Segerberg put some new axioms into consideration, which are not as natural at first glance as the axioms considered up to this point. "But as long as we cannot account for the irregularities in the above diagram, we cannot claim to understand the situation fully" [98, p. 41].

As we can see from the above the axiom **X** can be considered as a relativization of the axiom **E** to the normal part of a j-frame. Indeed, the axiom **E** imposes the condition to be identical on the accessibility relation, whereas **X** imposes essentially the same condition "to be identical" but on the accessibility relation restricted to the normal part of a j-frame. The next six axioms in the list are the axiom **Q** distinguishing the class of abnormal

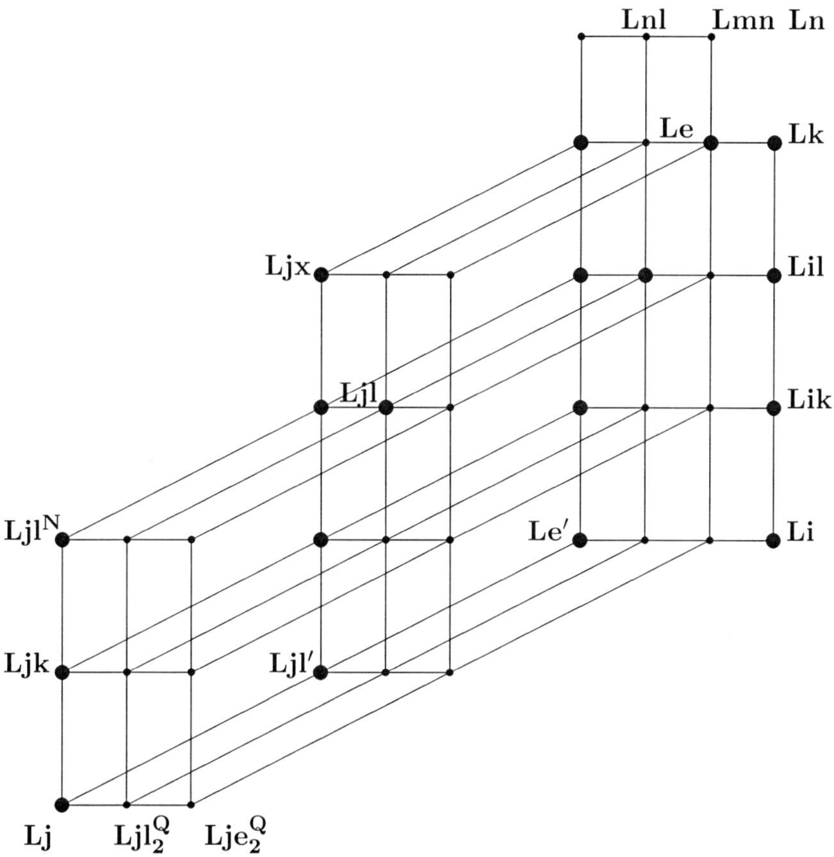

Figure 5.2

j-frames and relativizations of axioms **E** and **L** to the normal or to the abnormal part of a j-frame. The axiom $\mathbf{L_N}$ is a restriction of **L** to normal worlds. The axioms $\mathbf{L_1^Q}$ and $\mathbf{L_2^Q}$ are variants of relativization of **L** to the abnormal part of a j-frame. Relativizing **E** to abnormal words K. Segerberg also suggests two variants, $\mathbf{E_1^Q}$ and $\mathbf{E_2^Q}$.

The last axiom in the list, $\mathbf{P}'$, is similar to $\mathbf{E}'$ and $\mathbf{L}'$ because it restricts only the way of combination of normal and abnormal parts of a j-frame (see Proposition 2.3.4). This axiom, as well as axioms $\mathbf{L_1^Q}$ and $\mathbf{E_1^Q}$ lie out of the main line of considerations in [98].

If we exclude from the above list the axioms $\mathbf{P}'$, $\mathbf{L_1^Q}$, and $\mathbf{E_1^Q}$, the logics that can be constructed via adjoining to **Lj** the other axioms from the list form the beautiful diagram presented in Figure 5.2. The logics of Figure 5.1 are depicted in this diagram by bigger circles. The way, in which these logics are situated in Figure 5.2, explains the irregularities of the previous diagram. Only a few points on the diagram are endowed with the names of corresponding logics. The other logics are obtained via a combination of axioms of explicitly designated logics and one can easily reconstruct which logic corresponds to one or another point on the diagram. For example, the non-designated logics lying on the horizontal line ended with **Lik** are the following: **Ljke′**, $\mathbf{Ljke'l_2^Q}$, $\mathbf{Ljke'e_2^Q}$ (from left to right). We also note the equalities **Ljl** = $\mathbf{Ljl'l^Nl_2^Q}$ and **Le** = $\mathbf{Lje'xe_2^Q}$. As we will see the equality **Ljl** = $\mathbf{Ljl'l^Nl_1^Q}$ does not hold. So using axiom $\mathbf{L_1^Q}$ instead of $\mathbf{L_2^Q}$ results in a diagram of logics, which is not as regular as that of Figure 5.2. This explains the choice of K. Segerberg between variants of relativization of the axiom **L** to abnormal worlds.

In this diagram there are only four intermediate logics, namely, the logics lying on the vertical line from **Li** to **Lk**. The three negative logics on the diagram are those lying on the horizontal line from **Ln** to **Lmn**. All other logics on the diagram belong to the class Par. They form a three-dimensional figure, the dimensions of which, as we can see later, correspond to the three parameters, which determine the position of a paraconsistent logic in the class Par.

To better explain this correspondence we turn to the algebraic semantics of Segerberg's logics.

Recall that a *Stone* algebra is a Heyting algebra satisfying the identity **K**. Let $\mathcal{A}$ be a Heyting (negative) algebra. We call $\mathcal{A}$ a *Heyting (negative) l-algebra* if $\mathcal{A} \models (p \to q) \vee (q \to p)$.

Proposition 5.3.1 *Let $\mathcal{A}$ be an arbitrary j-algebra. The following equivalences hold.*

1. $\mathcal{A} \models$ **Ljk** *if and only if $\mathcal{A}^\perp$ is a Stone algebra.*
2. $\mathcal{A} \models$ **Ljx** *if and only if $\mathcal{A}^\perp$ is a Boolean algebra.*
3. $\mathcal{A} \models$ **Ljl′** *if and only if $f_\mathcal{A}(\mathcal{A}_\perp) \subseteq \mathcal{R}(\mathcal{A}^\perp)$.*
4. $\mathcal{A} \models$ **Ljl** *if and only if $\mathcal{A}^\perp$ and $\mathcal{A}_\perp$ are l-algebras, $f_\mathcal{A}(\mathcal{A}_\perp) \subseteq \mathcal{R}(\mathcal{A}^\perp)$, and for all $y_1, y_2 \in \mathcal{A}_\perp$, we have $f_\mathcal{A}(y_1 \to y_2) \vee f_\mathcal{A}(y_2 \to y_1) = 1$.*
5. $\mathcal{A} \models$ **Lg** *if and only if $f_\mathcal{A}(\mathcal{A}_\perp) \subseteq F_d(\mathcal{A}^\perp)$.*
6. $\mathcal{A} \models \mathbf{Ljl^N}$ *if and only if $\mathcal{A}^\perp$ is a Heyting l-algebra.*

7. $\mathcal{A} \models \mathbf{Ljl_1^Q}$ *if and only if* $\mathcal{A}_\bot$ *is a negative* l*-algebra.*

8. $\mathcal{A} \models \mathbf{Ljl_2^Q}$ *if and only if* $\mathcal{A}_\bot$ *is a negative* l*-algebra and for all* $y_1, y_2 \in \mathcal{A}_\bot$*, we have* $f_\mathcal{A}(y_1 \to y_2) \vee f_\mathcal{A}(y_2 \to y_1) = 1$.

9. $\mathcal{A} \models \mathbf{Lje_1^Q}$ *if and only if* $\mathcal{A}_\bot$ *is a negative Peirce algebra.*

10. $\mathcal{A} \models \mathbf{Lje_2^Q}$ *if and only if* $\mathcal{A}_\bot$ *is a negative Peirce algebra and for all* $y_1, y_2 \in \mathcal{A}_\bot$*, we have* $f_\mathcal{A}(y_1) \vee f_\mathcal{A}(y_1 \to y_2) = 1$.

Proof. 1. Let $\mathcal{A} \models \mathbf{Ljk}$. We represent $\mathcal{A}$ in the form $\mathcal{A}^\bot \times_{f_\mathcal{A}} \mathcal{A}_\bot$, take an arbitrary element $(x, y) \in \mathcal{A}$, and compute

$$((x, y) \to (\bot, \bot)) \vee (((x, y) \to (\bot, \bot)) \to (\bot, \bot)) = (x \to \bot, y \to \bot) \vee$$

$$\vee((x \to \bot, y \to \bot) \to (\bot, \bot)) = (\neg x, \bot) \vee ((\neg x, \bot) \to (\bot, \bot)) =$$

$$= (\neg x, \bot) \vee (\neg\neg x, \bot) = (\neg x \vee \neg\neg x, \bot) = (1, \bot).$$

The latter identity is satisfied if and only if the identity $\neg x \vee \neg\neg x = 1$ is true on $\mathcal{A}^\bot$, i.e., if and only if $\mathcal{A}^\bot$ is a Stone algebra.

2. This item can also be proved via a direct computation.

3. Let $(x, y) \in \mathcal{A}^\bot \times_{f_\mathcal{A}} \mathcal{A}_\bot$. The direct computation shows

$$((x, y) \to (\bot, \bot)) \vee ((\bot, \bot) \to (x, y)) = ((x \to \bot) \vee f(y), \bot).$$

Here after, we omit the lower index in the denotation $f_\mathcal{A}$ if it does not lead to confusion. As we can see, $\mathbf{L}'$ is an identity of $\mathcal{A}$ if and only if for all $x \in \mathcal{A}^\bot$, $y \in \mathcal{A}_\bot$, $x \leq f(y)$, the equality $(x \to \bot) \vee f(y) = 1_{\mathcal{A}^\bot}$ holds. In particular, we have $(f(y) \to \bot) \vee f(y) = \neg f(y) \vee f(y) = 1$, i.e., each element of the form $f(y)$ is regular. The inverse implication immediately follows from the above and the fact that the implication is descending with respect to the first argument. Indeed, if for some $y \in \mathcal{A}_\bot$ we have $(f(y) \to \bot) \vee f(y) = 1_{\mathcal{A}^\bot}$, then for all $x \in \mathcal{A}^\bot$, $x \leq f(y)$, we also have $(x \to \bot) \vee f(y) = 1_{\mathcal{A}^\bot}$.

4. Assume that $\mathcal{A} \models \mathbf{Ljl}$. In this case, the upper algebra $\mathcal{A}^\bot$ is a Heyting l-algebra as a subalgebra of $\mathcal{A}$. The inclusion $f_\mathcal{A}(\mathcal{A}_\bot) \subseteq \mathcal{R}(\mathcal{A}^\bot)$ holds by Item 3, because $\mathbf{L}'$ is a substitution instance of $\mathbf{L}$. Further, recall that the implication $\to_\bot$ of $\mathcal{A}_\bot$ is defined via the implication $\to$ of $\mathcal{A}$ as $x \to_\bot y = (x \to y) \wedge \bot$. Calculate

$$(x \to_\bot y) \vee (y \to_\bot x) = ((x \to y) \wedge \bot) \vee ((y \to x) \wedge \bot) =$$

$$((x \to y) \vee (y \to x)) \wedge \bot = 1 \wedge \bot = \bot.$$

Thus, $\mathcal{A}_\perp$ is a negative l-algebra. To check the last of the conditions listed in this item take arbitrary elements $y_1, y_2 \in \mathcal{A}_\perp$ and represent them in the form $(\perp, y_1)$ and $(\perp, y_2)$. We have

$$(1, \perp) = ((\perp, y_1) \to (\perp, y_2)) \vee ((\perp, y_2) \to (\perp, y_1)) =$$

$$(f(y_1 \to y_2) \vee f(y_2 \to y_1), (y_1 \to y_2) \vee (y_2 \to y_1)),$$

in particular, $f(y_1 \to y_2) \vee f(y_2 \to y_1) = 1$.

Prove the inverse implication. Let $\mathcal{A}^\perp$ and $\mathcal{A}_\perp$ be l-algebras, and let $f_\mathcal{A}(\mathcal{A}_\perp) \subseteq \mathcal{R}(\mathcal{A}^\perp)$ and for all $y_1, y_2 \in \mathbf{A}_\perp$, we have $f_\mathcal{A}(y_1 \to y_2) \vee f_\mathcal{A}(y_2 \to y_1) = 1$. Take $(x_1, y_1), (x_2, y_2) \in \mathcal{A}$ and using the formula for implication calculate

$$((x_1, y_2) \to (x_2, y_2)) \vee ((x_2, y_2) \to (x_1, y_1)) = (h, (y_1 \to y_2) \vee (y_2 \to y_1)).$$

The second component of the last pair equals $\perp = 1_{\mathcal{A}_\perp}$, because $\mathcal{A}_\perp$ is a negative l-algebra, whereas the first component has the following form:

$$h = ((x_1 \to x_2) \vee (x_2 \to x_1)) \wedge (f(y_1 \to y_2) \vee (x_2 \to x_1)) \wedge$$

$$((x_1 \to x_2) \vee f(y_2 \to y_1)) \wedge (f(y_1 \to y_2) \vee f(y_2 \to y_1)).$$

From our assumptions we immediately infer that first and last conjunctive terms of the last expression are equal to the unit element. In this way, we obtain that the satisfiability of $\mathbf{L}$ on $\mathcal{A}$ is equivalent to the condition:

$$\text{for all } (x_1, y_1), (x_2, y_2) \in \mathcal{A},\ (x_1 \to x_2) \vee f(y_2 \to y_1) = 1_{\mathcal{A}^\perp}.$$

Taking into account the facts that the implication is descending in the first argument and ascending in the second and that $x \leq f(y)$ for all $(x, y) \in \mathcal{A}$, we obtain the chain of inequalities $(x_1 \to x_2) \vee f(y_2 \to y_1) \geq (x_1 \to \perp) \vee f(\perp \to y_1) \geq (f(y_1) \to \perp) \vee f(y_1) = \neg f(y_1) \vee f(y_1) = 1$. The latter equality holds due to the condition that every element of the form $f(y)$ is regular.

Items 5–10 can be checked via direct computation.

□

Corollary 5.3.2 *1.* $\mathbf{Ljk} = \mathbf{Lik} * \mathbf{Ln}$.

2. $\mathbf{Ljx} = \mathbf{Lk} * \mathbf{Ln}$.

3. *For all* $L_1 \in \mathsf{Int}$ *and* $L_2 \in \mathsf{Neg}$*, we have* $(L_1 * L_2)\mathbf{p}', (L_1 * L_2)\mathbf{l}' \in$ *Spec*(L_1, L_2) *and the following equality holds*

$$(L_1 * L_2)\mathbf{p}' + (L_1 * L_2)\mathbf{l}' = L_1 \cap L_2.$$

   In particular, $\mathbf{Le}' = \mathbf{Lg} + \mathbf{Ljl}'$.

4. *For every* $L_1 \in \mathsf{Int}$ *and* $L_2 \in \mathsf{Neg}$ *such that* $L_1 \neq \mathbf{Lk}$*, the logics* $(L_1 * L_2)\mathbf{p}'$ *and* $(L_1 * L_2)\mathbf{l}'$ *are different from the endpoints of the interval Spec*(L_1, L_2)*. At the same time, if* $L_1 = \mathbf{Lk}$*, we have* $(\mathbf{Lk} * L_2)\mathbf{p}' = \mathbf{Lk} \cap L_2$ *and* $(\mathbf{Lk} * L_2)\mathbf{l}' = \mathbf{Lk} * L_2$.

5. $\mathbf{Ljl^N} = \mathbf{Lil} * \mathbf{Ln}$.

6. $\mathbf{Ljl_1^Q} = \mathbf{Li} * \mathbf{Lnl}$.

7. $\mathbf{Ljl_2^Q} \in$ *Spec*$(\mathbf{Li}, \mathbf{Lnl})$, $\mathbf{Ljl_2^Q} \neq \mathbf{Li} * \mathbf{Lnl}$, $\mathbf{Ljl_2^Q} \neq \mathbf{Li} \cap \mathbf{Lnl}$.

8. $\mathbf{Lje_1^Q} = \mathbf{Li} * \mathbf{Lmn}$.

9. $\mathbf{Lje_2^Q} \in$ *Spec*$(\mathbf{Li}, \mathbf{Lmn})$, $\mathbf{Lje_2^Q} \neq \mathbf{Li} * \mathbf{Lmn}$, $\mathbf{Lje_2^Q} \neq \mathbf{Li} \cap \mathbf{Lmn}$.

10. *The logic* $\mathbf{Ljl}$ *is a proper extension of* $(\mathbf{Lil} * \mathbf{Lnl})\mathbf{l}' = \mathbf{Ljl'l^Nl_1^Q}$.

Proof. Items 1,2,5,6,8 easily follow from Propositions 4.2.3 and 4.2.7 and suitable items of the last proposition.

3. By Item 3 of Proposition 5.3.1 all elements of the form $\bot \vee (\bot \to a)$ are regular in models of the logic $(L_1 * L_2)\mathbf{l}'$. On the other hand, in models of the logic $(L_1 * L_2)\mathbf{p}'$ all elements of this form are dense, as follows from Item 5 of Proposition 5.3.1. Thus, in models of the least upper bound of logics $(L_1 * L_2)\mathbf{p}'$ and $(L_1 * L_2)\mathbf{l}'$ elements of the form $\bot \vee (\bot \to a)$ are regular and dense simultaneously, i.e., they are all equal to the unit element. Consequently, models of the considered least upper bound are exactly direct products of the form $\mathcal{B} \times \mathcal{C}$, where $\mathcal{B} \models L_1$ and $\mathcal{C} \models L_2$, whence we immediately obtain the desired equality by Proposition 4.1.4 and Corollary 4.1.5.

4. The assertion of this item is true due to the fact that for any Heyting algebra $\mathcal{A}$ the following three conditions are equivalent: $\mathcal{A}$ is a Boolean algebra; the unit element is the only dense element of $\mathcal{A}$; all elements of $\mathcal{A}$ are regular.

7. By Item 8 of Proposition 5.3.1 the logic $\mathbf{Ljl_2^Q}$ belongs to *Spec*$(\mathbf{Li}, \mathbf{Lnl})$. Consider a model $\mathcal{A}$ for the free combination $\mathbf{Li} * \mathbf{Lnl}$ structured as follows. An upper algebra $\mathcal{A}^\perp$ is arbitrary; a lower algebra $\mathcal{A}_\perp$ is a 4-element negative

Peirce algebra with universe $\{\bot, a, b, 0\}$, where $0 \leq a \leq \bot$, $0 \leq b \leq \bot$, and elements a and b are incomparable; $f(\bot) = 1$, $f(x) = \bot$ for $x \neq \bot$. Calculate

$$f(a \to b) \vee f(b \to a) = f(b) \vee f(a) = \bot \vee \bot = \bot,$$

which proves that $\mathbf{Ljl_2^Q}$ differs from $\mathbf{Li} * \mathbf{Lnl}$.

We now point out a model for $\mathbf{Ljl_2^Q}$ different from the direct product of Heyting and negative algebras. This will prove that $\mathbf{Ljl_2^Q}$ is not equal to the intersection of logics **Li** and **Lnl**. Let $\mathcal{B}$ and $\mathcal{C}$ be Heyting and negative l-algebras respectively, which are isomorphic as implicative lattices, and let $f : \mathcal{C} \to \mathcal{B}$ be an arbitrary lattice isomorphism. It is not hard to check that $\mathcal{B} \times_f \mathcal{C}$ is the desired model of $\mathbf{Ljl_2^Q}$.

9. The fact that $\mathbf{Lje_2^Q}$ belongs to the interval $Spec(\mathbf{Li}, \mathbf{Lmn})$ follows from Item 10 of Proposition 5.3.1. Examples of j-algebras showing that $\mathbf{Lje_2^Q}$ differs from the endpoints of the indicated interval can be constructed in a way similar to that of Item 7.

10. This item can also be proved in a way similar to Item 7. As a counterexample showing that the indicated extension is proper we may take the j-algebra $\mathcal{A}$ from Item 7 with the additional restriction that $\mathcal{A}^{\bot}$ is a Heyting l-algebra.

□

Now we have enough information about j-algebras modelling Segerberg's axioms and we can come back to the analysis of Figure 5.2. We denote by Neg the line passing trough the logics **Ln** and **Lmn** and by Int the line passing through the logics **Li** and **Lk**. Recall that logics lying on the line Int (Neg) form the intersection of the class $\mathcal{D}$ of logics presented in Figure 5.2 with the class Int (respectively, with the class Neg), $\mathcal{D} \cap \mathsf{Int} = Int$ and $\mathcal{D} \cap \mathsf{Neg} = Neg$. These lines play a part of the coordinates for the three-dimensional part of Figure 5.2, which we denote by Par, $Par = \mathcal{D} \cap \mathsf{Par}$. For any logic $L \in Par$ we can naturally define its projections $I(L)$ and $N(L)$ to the lines Int and Neg respectively. For example,

$$I(\mathbf{Lj}) = \mathbf{Li}, N(\mathbf{Lj}) = \mathbf{Ln}, I(\mathbf{Ljl}) = \mathbf{Lil}, N(\mathbf{Ljl}) = \mathbf{Lnl},$$

$$I(\mathbf{Ljx}) = \mathbf{Lk}, N(\mathbf{Ljx}) = \mathbf{Ln}.$$

Using Proposition 5.3.1 and Corollary 5.3.2 we can easily check that for all logics $L \in Par$ the equalities

$$I(L) = L_{int} \text{ and } N(L) = L_{neg}$$

take place. Thus, for any line $\mathcal{L}$ on the diagram, which is parallel to the line $(\mathbf{Lj}, \mathbf{Le}')$, the logics of this line have fixed intuitionistic and negative counterparts, say L_1 and L_2. And so we have $\mathcal{L} = \mathcal{D} \cap Spec(L_1, L_2)$.

We stated in this way that the three dimensions of the part *Par* of Figure 5.2 exactly correspond to the three parameters determining a position of a logic in the class Par. One coordinate of a logic L is its intuitionistic counterpart $L_{int} \in \mathsf{Int}$, the second coordinate is its negative counterpart $L_{neg} \in \mathsf{Neg}$, and the third coordinate corresponds to a position of L inside the interval $Spec(L_{int}, L_{neg})$, which is determined in turn by the class of admissible semilattice homomorphisms from models of L_{neg} to models of L_{int}.

At this point we note one obvious defect of Figure 5.2. Let us consider the planes in the part *Par* of the figure parallel to the plane with points $\mathbf{Lj}$, $\mathbf{Ljk}$, and $\mathbf{Ljl}_1^{\mathbf{Q}}$. There are three such planes. We denote by $\mathcal{P}j$ the plane containing the point $\mathbf{Lj}$, by $\mathcal{P}l$ the plane containing the point $\mathbf{Ljl}$, and, finally, by $\mathcal{P}e$ the plane containing the point $\mathbf{Le}$. If we follow the geometrical analogues sketched above, we would expect that all logics belonging to one of the planes $\mathcal{P}j$, $\mathcal{P}l$, $\mathcal{P}e$ will define the same class of admissible homomorphisms. But this holds only for the plane $\mathcal{P}e$. For any logic $L \in \mathcal{P}e$ we have $\bot \vee (\bot \to p) \in L$, and so $L = L_{int} \cap L_{neg}$ is the greatest point of the interval $Spec(L_{int}, L_{neg})$, which is determined by the class of homomorphisms identically equal to the unit element.

Let us consider the plane $\mathcal{P}j$. Elements of this plane are the least points in the sets $Par \cap Spec(L_1, L_2)$, where $L_1 \in \{\mathbf{Li}, \mathbf{Lik}, \mathbf{Lil}\}$ and $L_2 \in \{\mathbf{Ln}, \mathbf{Lln}, \mathbf{Lmn}\}$. As we know from Proposition 4.2.5 the least point of $Spec(L_1, L_2)$ is the free combination $L_1 * L_2$ of logics L_1 and L_2. Moreover, for free combinations all semilattice homomorphisms from models of negative counterpart to models of intuitionistic counterpart are admissible. However, only three points of $\mathcal{P}j$, namely, the logics $\mathbf{Lj}$, $\mathbf{Ljk}$, and $\mathbf{Ljl}^{\mathbf{N}}$ are free combinations of their intuitionistic and negative counterparts (see Items 1 and 5 of Corollary 5.3.2). Logics $\mathbf{Ljl}_2^{\mathbf{Q}}$ and $\mathbf{Lje}_2^{\mathbf{Q}}$ are proper extensions of free combinations $\mathbf{Li} * \mathbf{Lnl}$ and $\mathbf{Li} * \mathbf{Lmn}$ respectively, as it follows from Items 7 and 9 of Corollary 5.3.2. Regarding the remaining four logics in $\mathcal{P}j$, we can easily modify the proofs of Items 7 and 9 of Corollary 5.3.2 to show that the restrictions, which axioms $\mathbf{L}_2^{\mathbf{Q}}$ and $\mathbf{E}_2^{\mathbf{Q}}$ impose on the class of admissible semilattice homomorphisms remain non-trivial, even if the intuitionistic counterpart of a logic satisfies axioms $\mathbf{K}$ or $\mathbf{L}^{\mathbf{N}}$ (see also Propositions 5.3.4 and 5.3.5 below).

In case of the plane $\mathcal{P}l$ we have a similar situation. Only the logics in the leftmost vertical line have the class of admissible semilattice homomorphisms with a range contained in the set of regular elements of an upper algebra (see Item 3 of Proposition 5.3.1). The other logics are characterized by narrower classes of admissible homomorphisms (see Propositions 5.3.4 and 5.3.5).

The indicated defect can easily be overcome if we replace the axioms $\mathbf{L_2^Q}$ and $\mathbf{E_2^Q}$ by $\mathbf{L_1^Q}$ and $\mathbf{E_1^Q}$ respectively. As follows from Items 7 and 9 of Proposition 5.3.1, these axioms do not impose any restrictions on the class of admissible homomorphisms and restrict only the class of lower algebras. These axioms can thus be considered as an adequate relativization of the axioms **L** and **E** to the negative counterpart of a logic. After the above-mentioned replacement and deleting axiom **L**, we obtain a diagram of logics having exactly the same configuration as that of Figure 5.2.

As we have seen above, the axioms $\mathbf{L_2^Q}$ and $\mathbf{E_2^Q}$ impose restrictions on the classes of lower algebras of their models and simultaneously on the classes of admissible homomorphisms from the lower algebras of their models to the upper ones. We can separate these restrictions. As follows from Proposition 5.3.1 axioms $\mathbf{L_1^Q}$ and $\mathbf{E_1^Q}$ restrict the classes of lower algebras in the same way as axioms $\mathbf{L_2^Q}$ and $\mathbf{E_2^Q}$ respectively, and have no influence on the classes of admissible homomorphisms. On the other hand, as follows from the next proposition, the axioms

$$\mathbf{F_1}.\ \bot \vee (\bot \to (p \to q)) \vee (\bot \to (q \to p)) (= \bot \vee \mathbf{L_2^Q})$$

$$\mathbf{F_2}.\ \bot \vee (\bot \to p) \vee (\bot \to (p \to q)) (= \bot \vee \mathbf{E_2^Q})$$

will restrict the classes of admissible homomorphisms in the same way as was done by axioms $\mathbf{L_2^Q}$ and $\mathbf{E_2^Q}$ respectively, and will not change the classes of lower algebras.

Proposition 5.3.3 *Let $\mathcal{A}$ be an arbitrary j-algebra. The following equivalences hold.*

1. $\mathcal{A} \models \mathbf{Ljf_1}$ *if and only if we have* $f_{\mathcal{A}}(y_1 \to y_2) \vee f_{\mathcal{A}}(y_2 \to y_1) = 1$ *for all* $y_1, y_2 \in \mathcal{A}_\bot$.
2. $\mathcal{A} \models \mathbf{Ljf_2}$ *if and only if we have* $f_{\mathcal{A}}(y_1) \vee f_{\mathcal{A}}(y_1 \to y_2) = 1$ *for all* $y_1, y_2 \in \mathcal{A}_\bot$.

This statement can be proved via a direct computation. It is clear that $\mathbf{Ljl_2^Q} = \mathbf{Ljl_1^Q f_1}$, $\mathbf{Lje_2^Q} = \mathbf{Lje_1^Q f_2}$, and $\mathbf{Ljl} = \mathbf{Ljl'l^N l_1^Q f_1}$.

Let us consider the class $\mathcal{D}_1$ consisting of logics which can be obtained by adjoining to **Lj** some subset of the following set of axioms

$$\{\mathbf{I}, \mathbf{Q}, \mathbf{K}, \mathbf{X}, \mathbf{L}, \mathbf{E}, \mathbf{L'}, \mathbf{E'}, \mathbf{P'}, \mathbf{L^N}, \mathbf{L_1^Q}, \mathbf{E_1^Q}, \mathbf{F_1}, \mathbf{F_2}\}.$$

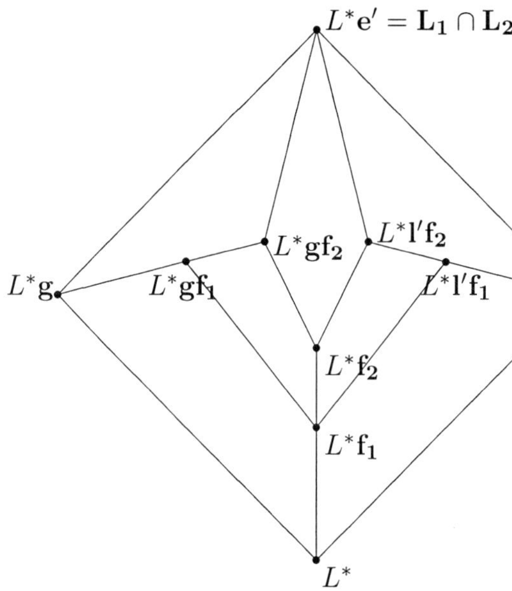

Figure 5.3

In this way, we take into account all properties involved in Segerberg's axioms. Obviously, $\mathcal{D} \subseteq \mathcal{D}_1$. At the same time, $\mathcal{D}$ satisfies the condition that for any $L_1 \in \mathsf{Int} \cap \mathcal{D}$ and $L_2 \in \mathsf{Neg} \cap \mathcal{D}$ the intersection $Spec(L_1, L_2) \cap \mathcal{D}$ is linearly ordered. In case of $\mathcal{D}_1$, this condition fails as we can see from the propositions below.

Proposition 5.3.4 *Let* $L_1 \in \{\mathbf{Li}, \mathbf{Lik}, \mathbf{Lil}\}$, $L_2 \in \{\mathbf{Ln}, \mathbf{Lnl}\}$, *and let* $L^* := L_1 * L_2$. *The set of logics* $Spec(L_1, L_2) \cap \mathcal{D}_1$ *forms an upper semilattice, shown on Figure 5.3.*

In the course of proving this and subsequent propositions, we will construct various j-algebras to check the interrelations between different logics. The following Heyting and negative algebras will play the part of breaks in our constructions: $\mathbf{2}$ and $\mathbf{2}'$ are two-element Heyting and negative algebras; $\mathbf{3^H}$ and $\mathbf{3^N}$ are three-element Heyting and negative algebras, the elements of which are linearly ordered; finally, $\mathbf{4^H}$ and $\mathbf{4^N}$ are four-element Heyting and negative algebras respectively, whose implicative lattices are Peirce algebras.

For any Heyting algebra $\mathcal{B}$, negative algebra $\mathcal{C}$, and for any j-algebra constructed from them $\mathcal{B} \times_f \mathcal{C}$, we will identify an element b of $\mathcal{B}$ (c of $\mathcal{C}$) with the corresponding element $(b, \bot)$ of the upper algebra $(\mathcal{B} \times_f \mathcal{C})^{\bot}$ ($(\bot, c)$ of the lower algebra $(\mathcal{B} \times_f \mathcal{C})_{\bot}$).

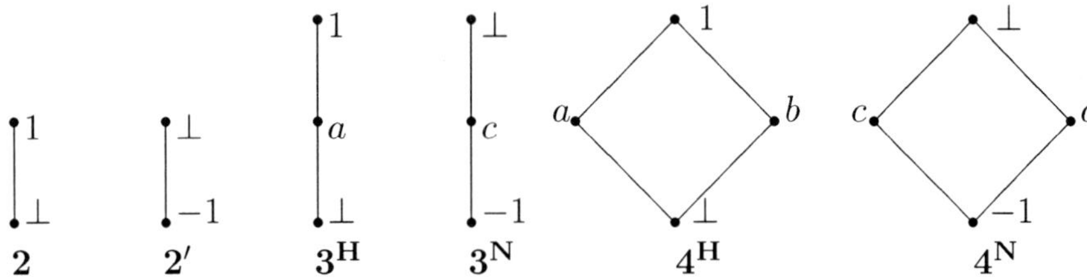

Figure 5.4

Proof (of Proposition 5.3.4.) First of all, we note that due to our assumption $L_1 \neq \mathbf{Lk}$. This fact together with Items 3 and 4 of Corollary 5.3.2, implies that logics $L^*\mathbf{g}$ and $L^*\mathbf{l}'$ are different from the endpoints of the interval $Spec(L_1, L_2)$ and the least upper bound of these logics coincides with the greatest point of the interval, $L^*\mathbf{g} + L^*\mathbf{l}' = L^*\mathbf{e}'$, which means, in particular, that $L^*\mathbf{g}$ and $L^*\mathbf{l}'$ are incomparable.

Let us consider the logics $L^*\mathbf{f_1}$ and $L^*\mathbf{f_2}$. Take an arbitrary model $\mathcal{A}$ for $L^*\mathbf{f_2}$. Due to Proposition 5.3.3 we have $f_{\mathcal{A}}(y_1) \vee f_{\mathcal{A}}(y_1 \to y_2) = 1$ for all $y_1, y_2 \in \mathcal{A}_\perp$. Since $y_1 \leq y_2 \to y_1$, we have $f_{\mathcal{A}}(y_1) \leq f_{\mathcal{A}}(y_2 \to y_1)$, and also $f_{\mathcal{A}}(y_2 \to y_1) \vee f_{\mathcal{A}}(y_1 \to y_2) = 1$ for all $y_1, y_2 \in \mathcal{A}_\perp$. In view of Proposition 5.3.3 the latter means that $\mathcal{A}$ is a model for $L^*\mathbf{f_1}$, and we have the inclusion $L^*\mathbf{f_1} \subseteq L^*\mathbf{f_2}$.

Let us consider a j-algebra $\mathcal{A}_1 := \mathbf{3^H} \times_{f_1} \mathbf{3^N}$, where $f_1 : \mathbf{3^N} \to \mathbf{3^H}$ is a uniquely defined implicative lattice isomorphism (see Figure 5.5, in which the structures of algebras constructed in this and the next proposition are presented). For any $y_1, y_2 \in (\mathcal{A}_1)_\perp$ we have

$$f_1(y_1 \to y_2) \vee f_1(y_2 \to y_1) = (f_1(y_1) \to f_1(y_2)) \vee (f_1(y_2) \to f_1(y_1)) = 1$$

since $\mathbf{3^H} \models (p \to q) \vee (q \to p)$. Thus, $\mathcal{A}_1 \models L^*\mathbf{f_1}$. Now we take the elements $-1, c \in \mathbf{3^N}$. It is clear that $f_1(-1) = \perp$ and that $f_1(c) = a$ (see Figure 5.4). We have

$$f_1(c) \vee f_1(c \to -1) = f_1(c) \vee f_1(-1) = a \vee \perp = a \neq 1.$$

This means that $\mathcal{A}_1$ is not a model for $L^*\mathbf{f_2}$, and so the inclusion $L^*\mathbf{f_1} \subset L^*\mathbf{f_2}$ is proper.

Consider j-algebras $\mathcal{A}_2 := \mathbf{2} \times_{f_2} \mathbf{4^N}$, where $f_2(\perp) = 1$ and $f_2(x) = \perp$ for $x < \perp$, and $\mathcal{A}_3 := \mathbf{4^H} \times_{f_3} \mathbf{4^N}$, where f_3 is an implicative lattice isomorphism between $\mathbf{4^N}$ and $\mathbf{4^H}$. As in Items 7 and 9 of Corollary 5.3.2 we can show that $\mathcal{A}_2$ is a model of L^*, but is not a model of $L^*\mathbf{f_1}$, respectively, that $\mathcal{A}_3$ is a model of $L^*\mathbf{f_2}$, but is not a model of $L^*\mathbf{e}'$. We have thus proved that $L^*\mathbf{f_1}$ and $L^*\mathbf{f_2}$ are different from the endpoints of the interval $Spec(L_1, L_2)$.

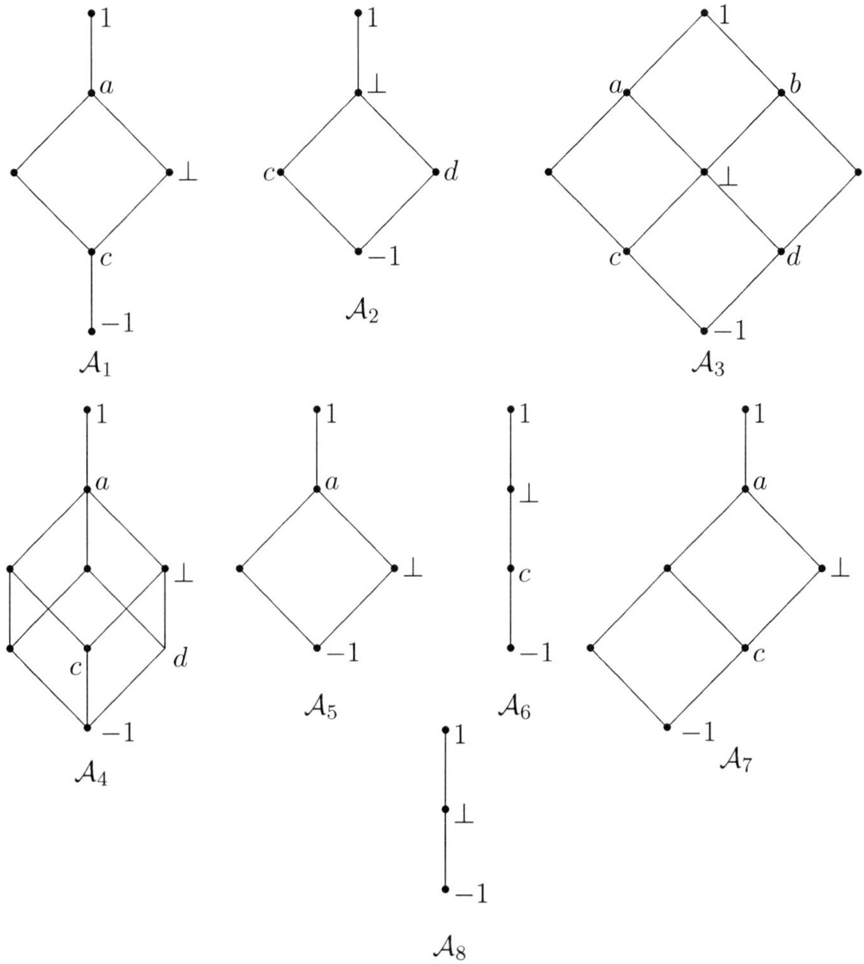

Figure 5.5

Now we check that each of the logics $L^*\mathbf{f_1}$ and $L^*\mathbf{f_2}$ is incomparable with either of the logics $L^*\mathbf{g}$ or $L^*\mathbf{l'}$. The j-algebras $\mathcal{A}_1$ and $\mathcal{A}_3$ are models of $L^*\mathbf{f_1}$ and $L^*\mathbf{f_2}$ respectively, but theirs are not models of $L^*\mathbf{g}$, which implies that $L^*\mathbf{g}$ is not contained in either of the logics $L^*\mathbf{f_1}$ or $L^*\mathbf{f_2}$. Define j-algebra $\mathcal{A}_4$ as $\mathbf{3^H} \times_{f_4} \mathbf{4^N}$, where $f_4(\bot) = 1$ and $f_4(x) = a$ for $x < \bot$. $\mathcal{A}_4$ is a model for $L^*\mathbf{g}$, since the element a is dense in $\mathbf{3^H}$, but it is not a model for $L^*\mathbf{f_1}$, in which case it is not also a model for $L^*\mathbf{f_2}$. Indeed, for $c, d \in \mathbf{4^N}$, we have $f_4(c \to d) \vee f_4(d \to c) = f_4(d) \vee f_4(c) = a \vee a = a$. We have thus proved that $L^*\mathbf{f_1}$ and $L^*\mathbf{f_2}$ are incomparable with $L^*\mathbf{g}$.

The algebra $\mathcal{A}_2$ provides a counterexample, demonstrating that either of the logics $L^*\mathbf{f_1}$ or $L^*\mathbf{f_2}$ is not contained in $L^*\mathbf{l}'$. To state that the inverse inclusions also fail we consider a j-algebra $\mathcal{A}_5 := \mathbf{3^H} \times_{f_5} \mathbf{2}'$, where $f_5(-1) = a$. This is not a model for $L^*\mathbf{l}'$ since a is not regular. At the same time, the direct calculation shows that $\mathcal{A}_5 \models L^*\mathbf{f_2}$. In this way, $L^*\mathbf{l}'$ is not contained in $L^*\mathbf{f_2}$, moreover, it is not contained in $L^*\mathbf{f_1}$.

The above facts on incomparability of logics imply, in particular, that $L^*\mathbf{gf_i}$ is a proper extension of $L^*\mathbf{f_i}$ and of $L^*\mathbf{g}$, $i = 1, 2$, and that $L^*\mathbf{l'f_i}$ is a proper extension of $L^*\mathbf{f_i}$ and of $L^*\mathbf{l}'$, $i = 1, 2$. So, it remains to verify that the inclusions $L^*\mathbf{gf_1} \subseteq L^*\mathbf{gf_2} \subseteq L^*\mathbf{e}'$ and $L^*\mathbf{l'f_1} \subseteq L^*\mathbf{l'f_2} \subseteq L^*\mathbf{e}'$ are proper.

A j-algebra $\mathcal{A}_6 := \mathbf{3^N} \oplus \mathbf{2}$ ($\cong \mathbf{2} \times_{f_6} \mathbf{3^N}$, where $f_6(x) = \bot$ for $x < \bot$) shows that $L^*\mathbf{l'f_1} \subseteq L^*\mathbf{l'f_2}$ is a proper inclusion. It is a model for $L^*\mathbf{lf_1}$ since for any $y_1, y_2 \in \mathbf{3^N}$, either $y_1 \to y_2 = \bot$ or $y_2 \to y_1 = \bot$. On the other hand, $f_6(c) \vee f_6(c \to -1) = \bot \vee \bot = \bot$. Note that $\mathcal{A}_3$ is a model for $L^*\mathbf{l'f_2}$ different from the direct product of $\mathbf{4^H}$ and $\mathbf{4^N}$. This proves that $L^*\mathbf{e}'$ is a proper extension of $L^*\mathbf{l'f_2}$.

Finally, consider the algebras $\mathcal{A}_7 := \mathbf{3^H} \times_{f_7} \mathbf{3^N}$, where $f_7(x) = a$ for $x < \bot$, and $\mathcal{A}_5$ defined above. The first of these algebras is a counterexample showing that the inclusion $L^*\mathbf{gf_1} \subseteq L^*\mathbf{gf_2}$ is proper. The second algebra can be used to check that $L^*\mathbf{e}'$ is a proper extension of $L^*\mathbf{gf_2}$.

□

Proposition 5.3.5 *Let $L_1 \in \{\mathbf{Li}, \mathbf{Lik}, \mathbf{Lil}\}$ and $L^* := L_1 * \mathbf{Lmn}$. The set of logics $Spec(L_1, \mathbf{Lmn}) \cap \mathcal{D}_1$ forms an upper semilattice shown on the semilattice diagram in Figure 5.6.*

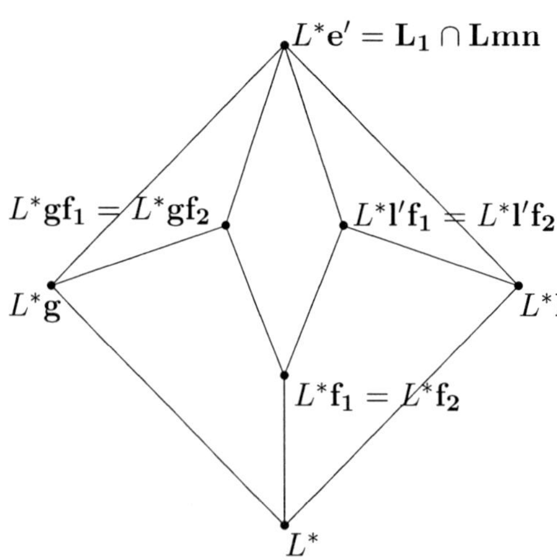

Figure 5.6

Proof. As in the previous proposition, we have the assumption $L_1 \neq \mathbf{Lk}$, which implies that logics $L^*\mathbf{g}$ and $L^*\mathbf{l}'$ are different from the endpoints of the interval $Spec(L_1, L_2)$, are incomparable and their upper bound coincides with the greatest point of the interval.

We argue to prove the equality $L^*\mathbf{f_1} = L^*\mathbf{f_2}$. The inclusion $L^*\mathbf{f_1} \subseteq L^*\mathbf{f_2}$ was stated above. Let us check the inverse inclusion. Take an arbitrary model $\mathcal{A}$ of $L^*\mathbf{f_1}$, which means that $f_{\mathcal{A}}(x \to y) \vee f_{\mathcal{A}}(y \to x) = 1$ for all $x, y \in \mathcal{A}_\perp$. By assumption $\mathcal{A}_\perp$ satisfies the Peirce law, and so for any $x, y \in \mathcal{A}_\perp$, we have $x = (x \to y) \to x$. On the other hand, in any j-algebra we have the identity $x \to y = x \to (x \to y)$. In this way, for any $x, y \in \mathcal{A}_\perp$, we have

$$f_{\mathcal{A}}(x) \vee f_{\mathcal{A}}(x \to y) = f_{\mathcal{A}}((x \to y) \to x) \vee f_{\mathcal{A}}(x \to (x \to y)) = 1,$$

which proves the desired equality.

The lower algebras of j-algebras $\mathcal{A}_2$, $\mathcal{A}_3$, $\mathcal{A}_4$, and $\mathcal{A}_5$ defined in Proposition 5.3.4 are models for **Lmn** and so these algebras can be used in the following reasoning. In particular, j-algebras $\mathcal{A}_2$ and $\mathcal{A}_3$ can be used to check that the logic $L^*\mathbf{f_1}$ lies inside the interval $Spec(L_1, \mathbf{Lmn})$.

With the help of $\mathcal{A}_4$ and $\mathcal{A}_8 := \mathbf{2}' \oplus \mathbf{2}$ we can show that the logics $L^*\mathbf{f_1}$ and $L^*\mathbf{g}$ are incomparable. $\mathcal{A}_4$ is a model for $L^*\mathbf{g}$, but not for $L^*\mathbf{f_1}$. Conversely, $\mathcal{A}_8$ is a model for $L^*\mathbf{f_1}$, but not for $L^*\mathbf{g}$.

In a similar way, one can use algebras $\mathcal{A}_2$ and $\mathcal{A}_5$ to check that logics $L^*\mathbf{f_1}$ and $L^*\mathbf{l}'$ are incomparable.

We are left to check that the following inclusions are proper: $L^*\mathbf{f_1 g} \subseteq L^*\mathbf{e}'$ and $L^*\mathbf{f_1 l}' \subseteq L^*\mathbf{e}'$. The suitable counterexamples are provided by algebras $\mathcal{A}_5$ and $\mathcal{A}_3$, respectively.

□

We have not yet considered the case when the intuitionistic counterpart coincides with the classical logic. It turns out that only in this case sets of the form $Spec(L_1, L_2) \cap \mathcal{D}_1$ are linearly ordered with respect to inclusion.

Proposition 5.3.6 *Let* $L_2 \in \{\mathbf{Ln}, \mathbf{Lnl}, \mathbf{Lmn}\}$*, and let* $L^* := \mathbf{Lk} * L_2$*. The sets of logics* $Spec(\mathbf{Lk}, L_2) \cap \mathcal{D}_1$ *have the structure presented in Figure 5.7.*

Proof. First, consider the case $L_2 \in \{\mathbf{Ln}, \mathbf{Lnl}\}$. Algebras $\mathcal{A}_6$, $\mathcal{A}_8$, and $\mathcal{A}_2$ can be used to verify that the inclusions $L^* \subset L^*\mathbf{f_1}$, $L^*\mathbf{f_1} \subset L^*\mathbf{f_2}$, and respectively $L^*\mathbf{f_2} \subset L^*\mathbf{e}'$ are proper.

In case $L_2 = \mathbf{Lmn}$ we may again use $\mathcal{A}_8$ and $\mathcal{A}_2$ to check the corresponding relations between logics.

□

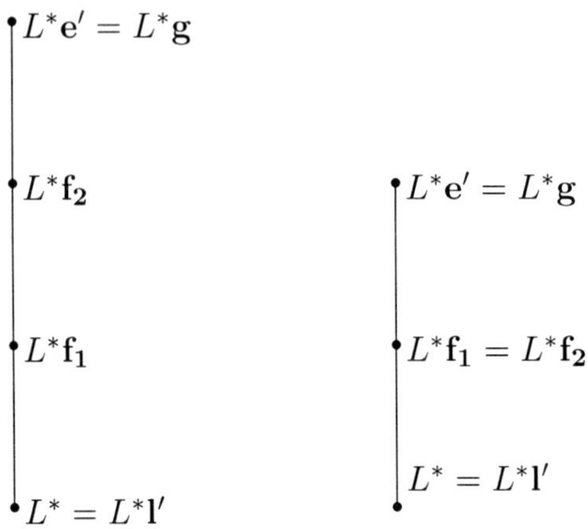

$L_2 \in \{\mathbf{Ln}, \mathbf{Lnl}\}$ $\qquad$ $L_2 = \mathbf{Lmn}$

Figure 5.7

5.4 Kripke Semantics for Paraconsistent Extensions of Lj[2]

In this section we define analogs of upper and lower algebras associated with a given j-algebra for j-frames.

For an arbitrary j-frame $\mathcal{W} = \langle W, \sqsubseteq, Q\rangle$ we define the following frames

$$\mathcal{W}^{(+)} := \langle W \setminus Q, \sqsubseteq \cap (W \setminus Q)^2, \varnothing\rangle,$$
$$\mathcal{W}^{(-)} := \langle Q, \sqsubseteq \cap Q^2, Q\rangle.$$

It is obvious that $\mathcal{W}^{(+)}$ is a model for intuitionistic logic and $\mathcal{W}^{(-)}$ is a model for minimal negative logic.

Remark. For any j-frame $\mathcal{W}$ and any formula φ, the translation $In(\varphi)$ is true on j-frame $\mathcal{W}^{(-)}$,

$$\mathcal{W}^{(-)} \models In(\varphi).$$

This fact can be checked via an easy induction on the structure of formulas.

Lemma 5.4.1 *Let $\mathcal{W}$ be an arbitrary j-frame, v a valuation of $\mathcal{W}^{(+)}$, and let v' be a valuation of $\mathcal{W}$ such that for any propositional variable p we have*

[2]The content of this section was originally published in [76] (Elsevier, UK). Reprinted here by permission of the publisher.

$v(p) = v'(p) \cap (W \setminus Q)$. Then for any formula φ and for an arbitrary element $x \in W \setminus Q$ the following equivalence holds

$$\langle \mathcal{W}, v' \rangle \models_x In(\varphi) \iff \langle \mathcal{W}^{(+)}, v \rangle \models_x \varphi.$$

Proof. Let $\mu' := \langle \mathcal{W}, v' \rangle$ and $\mu^{(+)} := \langle \mathcal{W}^{(+)}, v \rangle$.

We argue by induction on the structure of formulas. The case of constant $\bot$ is trivial. For an arbitrary propositional variable p and $x \in W \setminus Q$ we have $\mu' \models_x p \vee \bot$ if and only if either $x \in v'(p)$ or $x \in Q$. The second alternative is impossible by assumption. Thus we have $x \in v'(p)$ and $x \in W \setminus Q$, i.e., $x \in v(p)$. The latter is equivalent to $\mu^{(+)} \models_x p$.

Now, we assume that for formulas φ and ψ and for all $x \in W \setminus Q$ the equivalences

$$\mu' \models_x In(\varphi) \iff \mu^{(+)} \models_x \varphi \quad \text{and} \quad \mu' \models_x In(\psi) \iff \mu^{(+)} \models_x \psi$$

hold. Prove that the desired equivalence takes place for the implication $\varphi \to \psi$.

Let $\mu' \models_x In(\varphi \to \psi)(= In(\varphi) \to In(\psi))$ for some $x \in W \setminus Q$. This means that for all $y \in W$, the relations $x \sqsubseteq y$ and $\mu' \models_y In(\varphi)$ imply $\mu' \models_y In(\psi)$. In view of the assumed equivalences, we have

$$\forall y \in W \setminus Q(x \sqsubseteq y \Rightarrow (\mu^{(+)} \models_y \varphi \Rightarrow \mu^{(+)} \models_y \psi)),$$

and so $\mu^{(+)} \models_x \varphi \to \psi$.

Conversely, let $\mu^{(+)} \models_x \varphi \to \psi$ for some $x \in W \setminus Q$. By assumption for all $y \in W \setminus Q$ such that $x \sqsubseteq y$, if $\mu' \models_y In(\varphi)$, then $\mu' \models_y In(\psi)$. If $y \in Q$, then $\mu' \models_y In(\varphi)$ and $\mu' \models_y In(\psi)$. Thus, for all $y \in W$ such that $x \sqsubseteq y$, we have

$$\mu' \models_y \varphi \Rightarrow \mu' \models_y \psi,$$

which means that $\mu' \models_x In(\varphi \to \psi)$.

The cases of disjunction and conjunction are trivial.

□

The next proposition demonstrates that frames $\mathcal{W}^{(+)}$ and $\mathcal{W}^{(-)}$ can be considered as analog of upper and lower algebras.

Proposition 5.4.2 *For a j-frame $\mathcal{W}$ and a formula φ, the following equivalences hold*

$$\mathcal{W} \models In(\varphi) \iff \mathcal{W}^{(+)} \models \varphi,$$

$$\mathcal{W} \models \bot \to \varphi \iff \mathcal{W}^{(-)} \models \varphi.$$

Proof. The first equivalence immediately follows from the previous Lemma.

If $\mathcal{W} \models \bot \to \varphi$, then for any valuation v of $\mathcal{W}$, φ is true in all abnormal worlds of the model $\langle \mathcal{W}, v\rangle$, which means by Lemma 2.3.1 that $\langle \mathcal{W}^{(-)}, v^Q\rangle \models \varphi$. Any valuation v of $\mathcal{W}^{(-)}$ can be considered as a valuation of $\mathcal{W}$, in which case $v = v^Q$. Thus, for all valuations v of $\mathcal{W}^{(-)}$, we have $\langle \mathcal{W}^{(-)}, v\rangle \models \varphi$, i.e., $\mathcal{W}^{(-)} \models \varphi$.

Conversely, the assumption $\mathcal{W}^{(-)} \models \varphi$ implies that for any valuation v of $\mathcal{W}$, $\langle \mathcal{W}^{(-)}, v^Q\rangle \models \varphi$. In view of Lemma 2.3.1, the latter means that for all valuations v of $\mathcal{W}$, φ is true at any abnormal world of $\langle \mathcal{W}, v\rangle$, which implies, in turn, $\langle \mathcal{W}, v\rangle \models \bot \to \varphi$.

□

The following fact immediately follows from the definition of intuitionistic and negative counterparts and from the last proposition.

Corollary 5.4.3 *Let $L \in \mathsf{Jhn}$ and $\mathcal{W} \models L$. Then*

$$\mathcal{W}^{(+)} \models L_{int} \quad \text{and} \quad \mathcal{W}^{(-)} \models L_{neg}.$$

□

For a class of j-frames $\mathcal{K}$, we define

$$\mathcal{K}^{(+)} := \{\mathcal{W}^{(+)} \mid \mathcal{W} \in \mathcal{K}\}, \quad \mathcal{K}^{(-)} := \{\mathcal{W}^{(-)} \mid \mathcal{W} \in \mathcal{K}\}.$$

Proposition 5.4.4 *Let $\mathcal{K}$ be a class of j-frames and let $L = L\mathcal{K}$. Then $L_{int} = L\mathcal{K}^{(+)}$ and $L_{neg} = L\mathcal{K}^{(-)}$.*

Proof. The inclusion $L_{int} \subseteq L\mathcal{K}^{(+)}$ follows from Corollary 5.4.3. We argue for the inverse inclusion. Take a $\varphi \notin L_{int}$, in which case $In(\varphi) \notin L$. Consequently, there exist a frame $\mathcal{W} \in \mathcal{K}$, its valuation v, and an element $x \in W$ such that $\langle \mathcal{W}, v\rangle \not\models_x In(\varphi)$. As was remarked above, a formula of the form $In(\psi)$ is true in any model at any abnormal element, therefore, $x \notin Q$. Whence, by Lemma 5.4.1 we have $\mathcal{W}^{(+)} \not\models \varphi$.

Now we turn to the second equality. Again, we have to prove only the inclusion $L\mathcal{K}^{(-)} \subseteq L_{neg}$ since the inverse inclusion follows from Corollary 5.4.3. Let $\varphi \notin L_{neg}$, i.e., $L \nvdash \bot \to \varphi$. Consider a j-frame $\mathcal{W} \in \mathcal{K}$ such that $\mathcal{W} \not\models \bot \to \varphi$. From the last relation we obtain by Proposition 5.4.2 $\mathcal{W}^{(-)} \not\models \varphi$, i.e., $\varphi \notin L\mathcal{K}^{(-)}$.

Chapter 6

Negatively Equivalent Logics[1]

In the following, by negative formulas we mean formulas of the form $\neg\varphi$. The well-known Glivenko theorem implies, in particular, that in intuitionistic and in classical logic the same negative formulas are provable. This means that intuitionistic and classical logic, as well as all intermediate logic have, in a sense, the same negation. Generalizing this relation between logics we define negatively equivalent logics as logics where the same negative formulas are inferable from the same sets of hypotheses. From the constructive point of view we need negation to refute formulas on the basis of one or another set of hypotheses, therefore, negatively equivalent logics have essentially the same negation. Unlike the class of intermediate logics, the relation of negative equivalence is non-trivial on the class Jhn^+ and in this chapter we obtain several interesting results on the structure of negative equivalence classes. Simultaneously, we prove the results on cardinality of intervals of the form $Spec(L_1, L_2)$.

6.1 Definitions and Simple Properties

Let L_1 and L_2 be logics in Jhn^+. We say that L_1 is *negatively lesser* than L_2, and write $L_1 \leq_{neg} L_2$, if for any set of formulas X and formula φ, the following implication holds:

$$X \vdash_{L_1} \neg\varphi \implies X \vdash_{L_2} \neg\varphi.$$

[1]Parts of this chapter were originally published in [75] (Springer, Netherlands). Reprinted here by permission of the publisher.

In other words, one logic is negatively lesser that the other if passing from one to the other preserves the negative consequence relation, i.e., the consequence relation of the form $X \vdash \neg\varphi$, in which the conclusion is negative.

As we can see from the proposition below, the condition of preserving the negative consequence relation can be replaced by that of preserving the class of inconsistent sets of formulas. However, the equivalence proved in this proposition is typical for the class Jhn^+, because in this class the negation is defined via the constant "absurdity", whereas the absurdity $\bot$ can be defined as a negation of tautology.

Proposition 6.1.1 *For any $L_1, L_2 \in \mathsf{Jhn}^+$, the following conditions are equivalent.*

1. *$L_1 \leq_{neg} L_2$.*
2. *For an arbitrary set of formulas X, if $X \vdash_{L_1} \bot$, then $X \vdash_{L_2} \bot$.*

Proof. 1) $\Rightarrow$ 2) If $X \vdash_{L_1} \bot$, then $X \vdash_{L_1} \neg\varphi$ for any formula φ. By the assumption that $L_1 \leq_{neg} L_2$, we have $X \vdash_{L_2} \neg\varphi$ for any φ. Take an L_2-tautology ψ, then $X \vdash_{L_2} \psi, \neg\psi$, whence $X \vdash_{L_2} \bot$.

2) $\Rightarrow$ 1) Let $X \vdash_{L_1} \neg\varphi$. Then $X \cup \{\varphi\} \vdash_{L_1} \bot$. In this case, we have $X \cup \{\varphi\} \vdash_{L_2} \bot$ by assumption, consequently, $X \vdash_{L_2} \varphi \to \bot$ by Deduction Theorem, i.e., $X \vdash_{L_2} \neg\varphi$.

□

The definition of $\leq_{neg}$ can also be re-worded as follows.

Proposition 6.1.2 *For any $L_1, L_2 \in \mathsf{Jhn}^+$, the relation $L_1 \leq_{neg} L_2$ holds if and only if for any formula φ, the following implication is true:*

$$L_1 \vdash \varphi \implies L_2 \vdash \neg\neg\varphi.$$

Proof. Let $L_1 \leq_{neg} L_2$. If $L_1 \vdash \varphi$, then $\{\neg\varphi\} \vdash_{L_1} \bot$. By the last proposition we have $\{\neg\varphi\} \vdash_{L_2} \bot$, from which we immediately obtain $L_2 \vdash \neg\neg\varphi$.

Now we assume that the right-hand side of the desired equivalence holds. Let $X \vdash_{L_1} \bot$. This means that for some formulas $\varphi_1, \ldots, \varphi_n \in X$, $L_1 \vdash (\varphi_1 \wedge \ldots \wedge \varphi_n) \to \bot$. According to our assumption $L_2 \vdash \neg\neg\neg(\varphi_1 \wedge \ldots \wedge \varphi_n)$. In view of $\mathbf{Lj} \vdash \neg\neg\neg p \leftrightarrow \neg p$ we obtain $L_2 \vdash \neg(\varphi_1 \wedge \ldots \wedge \varphi_n)$, which immediately implies that $X \vdash_{L_2} \bot$. Again, Proposition 6.1.1 allows one to conclude that L_1 is negatively lesser than L_2.

□

If one of the logics is finitely axiomatizable relative to the other, the last statement can be simplified as follows.

Proposition 6.1.3 *Let $L_1, L_2 \in \mathsf{Jhn}^+$ and $L_2 = L_1 + \{\varphi_1, \ldots, \varphi_n\}$. Then $L_2 \leq_{neg} L_1$ if and only if $\neg\neg\varphi_1, \ldots, \neg\neg\varphi_n \in L_1$.*

Proof. We consider only the non-trivial implication. Let $\neg\neg\varphi_1, \ldots, \neg\neg\varphi_n \in L_1$. Take an arbitrary set of formulas X with $X \vdash_{L_2} \bot$, then $X \cup \{\psi_1, \ldots, \psi_k\}$ is inconsistent in L_1, or equivalently,

$$X \vdash_{L_1} \neg(\psi_1 \wedge \ldots \wedge \psi_k),$$

where $\psi_1, \ldots, \psi_k$ are substitution instances of formulas from the list $\varphi_1, \ldots, \varphi_n$. We have $L_1 \vdash \neg\neg\psi_1, \ldots, \neg\neg\psi_k$ by assumption. Consider an arbitrary model $\mathcal{A} \models L_1$ and an $\mathcal{A}$-valuation v. The elements $v(\psi_1), \ldots, v(\psi_k)$ are dense in $\mathcal{A}^\perp$. Consequently, the element $v(\psi_1 \wedge \ldots \wedge \psi_k)$ is also dense, in particular, $v(\neg(\psi_1 \wedge \ldots \wedge \psi_k)) = \bot_\mathcal{A}$.

Let formulas $\theta_1, \ldots, \theta_m \in X$ be such that

$$L_1 \vdash (\theta_1 \wedge \ldots \wedge \theta_m) \to \neg(\psi_1 \wedge \ldots \wedge \psi_k).$$

In view of the above considerations, for any model $\mathcal{A} \models L_1$ and any $\mathcal{A}$-valuation v, we have $v(\theta_1 \wedge \ldots \wedge \theta_m) \leq \bot_\mathcal{A}$. Consequently, we have $L_1 \vdash (\theta_1 \wedge \ldots \wedge \theta_m) \to \bot$, which means that X is inconsistent in L_1.

□

Define the relation $\equiv_{neg}$ as an intersection of $\leq_{neg}$ and its inverse relation:

$$\equiv_{neg} := \leq_{neg} \cap (\leq_{neg})^{-1}.$$

One can easily prove

Lemma 6.1.4 *1. The relation $\leq_{neg}$ is a preordering.*

2. The relation $\equiv_{neg}$ is an equivalence.

□

In view of this lemma, logics $L_1, L_2 \in \mathsf{Jhn}^+$ with $L_1 \equiv_{neg} L_2$ will be called *negatively equivalent*. From Propositions 6.1.1 and 6.1.2 we immediately obtain

Corollary 6.1.5 *For any $L_1, L_2 \in \mathsf{Jhn}^+$, the following conditions are equivalent.*

1. *$L_1 \equiv_{neg} L_2$.*
2. *An arbitrary set of formulas X is inconsistent in L_1 if and only if it is inconsistent in L_2.*

3. For any formula φ, the following implications hold:

$$(L_1 \vdash \varphi \implies L_2 \vdash \neg\neg\varphi) \text{ and } (L_2 \vdash \varphi \implies L_1 \vdash \neg\neg\varphi).$$

Remark. Notice that the inclusion $L_1 \subseteq L_2$ implies $L_1 \leq_{neg} L_2$. Therefore, the logics L_1 and L_2 from Proposition 6.1.3 will be negatively equivalent.

Remark. Any two negative logics are negatively equivalent due to the following fact. In an arbitrary negative logic any set of formulas in inconsistent since the absurdity $\bot$ belongs to the set of logical tautologies. Any two intermediate logics are also negatively equivalent, which easily follows from Proposition 6.1.3 and Glivenko's theorem. It is well known that negation in intuitionistic logic is not constructive, from the deducibility $\mathbf{Li} \vdash \neg(\varphi \wedge \psi)$ does not follow, in a general case, that either of the formulas $\neg\varphi$ or $\neg\psi$ is provable in **Li**. We have just noted that negation in an arbitrary intermediate logic is close in some sense (negatively equivalent) to classical negation. This fact can be considered as a generalization of Glivenko's theorem and also emphasizes the non-constructive character of negation in intermediate logics.

6.2 Logics Negatively Equivalent to Intermediate Ones

In this section we consider the question: which logics from the class Jhn are negatively equivalent to intermediate ones? More exactly, let some logics $L_1 \in \mathsf{Int}$ and $L_2 \in \mathsf{Neg}$ be fixed. Which logics $L \in \mathsf{Par}$ having L_1 and L_2 as intuitionistic and negative counterparts respectively, are negatively equivalent to intermediate logic L_1 and so to an arbitrary intermediate logic? In other words, to which extent can one weaken the law *ex contradictione quodlibet* while preserving the negative equivalence?

Proposition 6.2.1 *Let $L_1 \in \mathsf{Int}$, $L_2 \in \mathsf{Neg}$, and $L \in Spec(L_1, L_2)$. The equivalence $L \equiv_{neg} L_1$ holds if and only if $(L_1 * L_2)\mathbf{p}' \subseteq L$.*

Proof. Recall that $L_1 = L + \{\bot \to p\}$. According to Proposition 6.1.3 $L \equiv_{neg} L_1$ whenever $L \vdash \neg\neg(\bot \to p)$. By definition $(L_1 * L_2)\mathbf{p}' = L_1 * L_2 + \{\neg\neg(\bot \to p)\}$.

□

Call $G(L_1, L_2) := (L_1 * L_2)\mathbf{p}'$ the *relativized Glivenko's logic wrt L_1 and L_2.*

From the last fact we easily infer the following strengthening of Generalized Glivenko's Theorem (Theorem 5.1.2).

Corollary 6.2.2 *Glivenko's logic* **Lg** *is the least logic in* Jhn, *which is negatively equivalent to* **Lk**.

□

As we can see from Item 4 of Corollary 5.3.2, the interval $[\mathbf{Lk}{*}L_2, \mathbf{Lk}{\cap}L_2]$ contains a unique paraconsistent logic negatively equivalent to **Lk**, namely $\mathbf{Lk} \cap L_2$. Note that this logic is axiomatized modulo the least logic $\mathbf{Lk} * L_2$ of the interval $Spec(\mathbf{Lk}, L_2)$ via the axiom $\bot \vee (\bot \to p)$ having essentially a non-constructive character. At the same time, if $L_1 \neq \mathbf{Lk}$, there is a proper subinterval $[G(L_1, L_2), L_1 \cap L_2]$ consisting of logics negatively equivalent to intermediate logics.

It turns out that the disjunction property can be transferred from an intuitionistic counterpart to the relativized Glivenko's logic. This fact was established by M. Stukacheva [104]. Recall that a logic L has the disjunction property if $\varphi \vee \psi \in L$ implies $\varphi \in L$ or $\psi \in L$.

Let $L \in$ Jhn. By induction on the length of formula φ we define an expression $|_L\, \varphi$ ("Kleene's slash", see [13]) as follows (further on, instead of "$|_L\, \varphi$ and $\vdash_L \varphi$" we write $\Vdash_L \varphi$):

$$
\begin{array}{lcl}
|_L\, \varphi & := & \vdash_L \varphi, \text{ where } \varphi \text{ is an atomic formula;} \\
|_L\, \varphi \wedge \psi & := & |_L\, \varphi \text{ and } |_L\, \psi; \\
|_L\, \varphi \vee \psi & := & \Vdash_L \varphi \text{ or } \Vdash_L \psi; \\
|_L\, \varphi \to \psi & := & (\Vdash_L \varphi \Rightarrow |_L\, \psi).
\end{array}
$$

Proposition 6.2.3 [104] *Let* $L_1 \in$ Int, $L_2 \in$ Neg, *and* L_1 *has the disjunction property. If* $\vdash_{L_1 * L_2} \varphi$, *then* $|_{G(L_1,L_2)}\, \varphi$.

Proof. Let $\vdash_{G(L_1,L_2)} \varphi$. By induction on the length of proof, we show that $|_{G(L_1,L_2)}\, \varphi$. In the proof we omit the lower index $G(L_1, L_2)$.

Prove that this statement holds for axioms of $G(L_1, L_2)$.

a) The case of **Lj**-axioms can be easily verified;

b) For $L_1 * L_2$-axioms of the form $\bot\!\to \psi$, where $\psi \in L_2$, the conclusion is obvious since $\nvdash_{G(L_1,L_2)} \bot$;

c) By induction of the structure of $In(\varphi)$, $\varphi \in L_1$, prove that $\vdash In(\varphi)$ implies $|In(\varphi)$.

The basis is obvious. Indeed, since L_1 is non-trivial, we have $\nvdash_{L_1} p$, i.e., $\nvdash_{G(L_1,L_2)} In(p)$;

Let $In(\varphi)$ and $In(\psi)$ be such that $\vdash In(\varphi) \Rightarrow |In(\varphi)$ and $\vdash In(\psi) \Rightarrow |In(\psi)$.

If $\vdash In(\varphi \wedge \psi)$, then $\vdash In(\varphi)$ and $\vdash In(\psi)$. By induction hypothesis we have $|In(\varphi)$ and $|In(\psi)$, which means by definition $|In(\varphi) \wedge In(\psi)$. Recall that $In(\varphi) \wedge In(\psi) = In(\varphi \wedge \psi)$.

If $\vdash In(\varphi) \vee In(\psi)$, then $\vdash In(\varphi)$ or $\vdash In(\psi)$, since L_1 satisfies the disjunction property. By induction hypothesis $\Vdash In(\varphi)$ or $\Vdash In(\psi)$, i.e., $|In(\varphi) \vee In(\psi)$.

Assume $\vdash In(\varphi) \rightarrow In(\psi)$ and $\Vdash In(\varphi)$, then $\vdash In(\psi)$ and $|In(\psi)$ by induction hypothesis.

d) It remains to prove $|\neg\neg(\bot \rightarrow p)$. By definition we have

$$|\neg\neg(\bot \rightarrow p) \iff ((\vdash (\bot \rightarrow p) \rightarrow \bot \text{ and } |(\bot \rightarrow p) \rightarrow \bot) \Rightarrow |\bot).$$

Since $\vdash \neg\neg(\bot \rightarrow p)$ and $\nvdash \bot$, we have $\nvdash (\bot \rightarrow p) \rightarrow \bot$, which means that the right-hand side implication is true.

Finally, let φ is obtained by *modus ponens* from $\psi \in G(L_1, L_2)$ and $\psi \rightarrow \varphi \in G(L_1, L_2)$. We have by induction hypothesis $|\psi$ and $|\psi \rightarrow \varphi$. Consequently, $\Vdash \psi$ implies $|\varphi$, hence, $|\varphi$.

□

Proposition 6.2.4 *Let* $L_1 \in \mathsf{Int}$, $L_2 \in \mathsf{Neg}$, *and* L_1*has the disjunction property. Then* $G(L_1, L_2)$ *has the disjunction property.*

Proof. Let $G(L_1, L_2) \vdash \varphi \vee \psi$. According to Proposition 6.2.3 we have $|_{G(L_1,L_2)} \varphi \vee \psi$, consequently, $\Vdash_{G(L_1,L_2)} \varphi$ or $\Vdash_{G(L_1,L_2)} \psi$.

□

This fact shows that we can resign the law *ex contradictione quodlibet* preserving not only the class of inconsistent sets of formulas, but also constructive properties of intuitionistic logic. But the disjunction property does not hold in all relativized Glivenko's logics. In particular, if $L_1 = \mathbf{Lk}$, we have $G(\mathbf{Lk}, L_2) = \mathbf{Lk} \cap L_2 \vdash \bot \vee (\bot \rightarrow p)$ (Item 4 of Corollary 5.3.2). Moreover, if L_1 does not possess the disjunction property, then $G(L_1, L_2)$ also does not. Indeed, let $\varphi \vee \psi$ be a corresponding counterexample, i.e., $L_1 \vdash \varphi \vee \psi$, but neither φ nor ψ are provable in L_1. In this case, the formula $In(\varphi \vee \psi)(= In(\varphi) \vee In(\psi))$ will refute the disjunction property for $G(L_1, L_2)$: $G(L_1, L_2) \vdash In(\varphi) \vee In(\psi)$, but neither $In(\varphi)$ not $In(\psi)$ are provable in $G(L_1, L_2)$.

However, one can point out an interesting weak analog of the disjunction property, which holds in all relativized Glivenko's logics $G(L_1, L_2)$ with $L_1 \neq \mathbf{Lk}$. We try to find a property that holds in all relativized Glivenko's logics, independently of constructive properties of intuitionistic counterparts. Therefore, it should be a property that is trivially satisfied in all intermediate logics, but becomes non-trivial in paraconsistent extensions of **Lj**. The property of a logic to be closed under the rule

$$\frac{\varphi \vee \bot}{\varphi}$$

can serve as an example of such property. It can be considered as a weak analog of the disjunction property, because as well as in case of the disjunction property we conclude from a deducibility of disjunction to a deducibility of disjunction term.

Proposition 6.2.5 *Let* $\mathbf{Lk} \neq L_1 \in \mathsf{Int}$, $L_2 \in \mathsf{Neg}$, *and let* φ *be an arbitrary formula. If* $G(L_1, L_2) \vdash \varphi \vee \bot$, *then* $G(L_1, L_2) \vdash \varphi$.

Proof. Let $\varphi = \varphi(p_1, \dots, p_n)$. Assume that $G(L_1, L_2) \vdash \varphi \vee \bot$, but φ is not provable in $G(L_1, L_2)$. This implies, in particular, that φ is not provable in L_2. Indeed, if $L_2 \vdash \varphi$, then $G(L_1, L_2) \vdash \bot \to \varphi$ and one can easily infer $G(L_1, L_2) \vdash \varphi$. Thus, there exists a negative algebra $\mathcal{B}$ being a model for L_2, $\mathcal{B} \models L_2$, and elements $b_1, \dots, b_n \in \mathcal{B}$ such that $\varphi(b_1, \dots, b_n) \neq \bot$.

By assumption $\mathbf{Lk} \neq L_1$, hence, there exists a Heyting algebra $\mathcal{A}$ with $\mathcal{A} \models L_1$ and a non-trivial filter of dense elements, $F_d(\mathcal{A}) \neq \{1\}$. Take an element $a \in F_d(\mathcal{A})$, $a \neq 1$, and consider a j-algebra $\mathcal{A} \times_f \mathcal{B}$, where a semilattice homomorphism f is defined as follows: $f(\bot) = 1$ and $f(x) = a$ for $x \neq \bot$. In this case, for any pair $(x, y) \in \mathcal{A} \times_f \mathcal{B}$, we have $x \leq a$ when $y \neq \bot$. Moreover, $\rho f = \{a, 1\} \subseteq F_d(\mathcal{A})$, which means that $\mathcal{A} \times_f \mathcal{B}$ is a model for $G(L_1, L_2)$ (see Proposition 5.3.1).

Compute the value of φ on the elements $(0, b_1), \dots, (0, b_n) \in \mathcal{A} \times_f \mathcal{B}$. Taking into account that the mapping $(x, y) \mapsto y$ defines an epimorphism of j-algebras $\mathcal{A} \times_f \mathcal{B} \to \mathcal{B}$ we have the equality $\varphi((0, b_1), \dots, (0, b_n)) = (x, \varphi(b_1, \dots, b_n))$, where $x \leq a$ in view of $\varphi(b_1, \dots, b_n) \neq \bot$. Thus, we have

$$\varphi((0, b_1), \dots, (0, b_n)) \vee (\bot, \bot) = (x, \bot) \neq (1, \bot),$$

which contradicts our assumption that $G(L_1, L_2) \vdash \varphi \vee \bot$.

□

Remark. It is interesting that in the class of extensions of minimal logic the inference rule $\frac{\varphi \vee \bot}{\varphi}$ is equivalent to disjunctive syllogism in the following sense. Let $L \in \mathsf{Jhn}$ and let L^d be a deductive system with the set of axioms L and the only deductive rule *modus ponens.* Adding to L^d either of the rules

$$\frac{\varphi \vee \bot}{\varphi} \quad \text{or} \quad \frac{\neg\varphi, \varphi \vee \psi}{\psi}$$

results with the deductive system having exactly the same consequence relation.

6.3 Abstract Classes of Negative Equivalence

For an arbitrary logic $L \in \mathsf{Jhn}^+$, we define

$$\nabla(L) := \{\varphi \mid \neg\neg\varphi \in L\}.$$

We now observe that the set $\nabla(L)$ is itself a logic, possibly a trivial one, and point out some simple properties of the operator $\nabla : \mathsf{Jhn}^+ \to \mathsf{Jhn}^+$.

Proposition 6.3.1 *For an arbitrary* $L \in \mathsf{Jhn}^+$, *the following facts take place.*

1. $L \subseteq \nabla(L)$.
2. *If* $L_1 \in \mathsf{Jhn}^+$ *and* $L \subseteq L_1$, *then* $\nabla(L) \subseteq \nabla(L_1)$.
3. $\nabla(L) \in \mathsf{Jhn}^+$.
4. $\nabla\nabla(L) = \nabla(L)$.
5. $\nabla(L) = \mathcal{F}$ *if and only if* $L \in \mathsf{Neg} \cup \{\mathcal{F}\}$.

Proof. 1. This is true because $\varphi \to \neg\neg\varphi \in \mathbf{Lj}$.

2. This item trivially follows from the definition.

3. Let formulas φ and $\varphi \to \psi$ belong to $\nabla(L)$. Consider a model $\mathcal{A} \models L$ and take an arbitrary $\mathcal{A}$-valuation v. By definition of $\nabla(L)$ we have $\neg\neg\varphi, \neg\neg(\varphi \to \psi) \in L$, which means that the values of formulas $\varphi \vee \bot$ and $(\varphi \to \psi) \vee \bot$ are dense, $v(\varphi \vee \bot), v((\varphi \to \psi) \vee \bot) \in F_d(\mathcal{A}^\bot)$. Calculate

$$\begin{aligned} v(\varphi \vee \bot) \wedge v((\varphi \to \psi) \vee \bot) &= v((\varphi \wedge (\varphi \to \psi)) \vee \bot) \\ &= v((\varphi \wedge \psi) \vee \bot) \leq v(\psi \vee \bot) \in \nabla(\mathcal{A}). \end{aligned}$$

Thus, $\mathcal{A} \models \neg\neg\psi$ for an arbitrary model $\mathcal{A}$ for L, i.e., $L \vdash \neg\neg\psi$, whence $\psi \in \nabla(L)$. In this way, the set $\nabla(L)$ is closed under *modus ponens.* The fact that it is closed under the substitution rule follows directly from the definition. We have thus proved that $\nabla(L)$ is a logic, the fact that it extends **Lj** follows from Item 1.

4. First, we note that the object $\nabla\nabla(L)$ is well defined in view of the previous item. The inclusion $\nabla(L) \subseteq \nabla\nabla(L)$ follows from Item 1.

Take a formula $\varphi \in \nabla\nabla(L)$, in this case $\neg\neg\varphi \in \nabla(L)$ and $\neg\neg\neg\neg\varphi \in L$. The last formula is equivalent in **Lj** to $\neg\neg\varphi$, and so $\varphi \in \nabla(L)$, which proves the inverse inclusion.

5. If $L \in \mathsf{Neg} \cup \{\mathcal{F}\}$, then $\nabla(L) = \mathcal{F}$, because an arbitrary negative formula belongs to L in this case. Assume $L \in \mathsf{Jhn} \setminus \mathsf{Neg}$. Then $L \subseteq \mathbf{Lk}$ and by Item 2 $\nabla(L) \subseteq \nabla(\mathbf{Lk}) = \mathbf{Lk}$. The last equality is due to the fact that a formula and its double negation are equivalent in **Lk**.

□

The operator ∇ is closely related to the negative equivalence relation, as we can see from the following

Proposition 6.3.2 *1. For any $L \in \mathsf{Jhn}^+$, we have $L \equiv_{neg} \nabla(L)$.*

2. For any $L_1, L_2 \in \mathsf{Jhn}^+$, the following equivalence holds

$$L_1 \equiv_{neg} L_2 \iff \nabla(L_1) = \nabla(L_2).$$

Proof. 1. It follows from Item 3 of Corollary 6.1.5.

2. Let $L_1 \equiv_{neg} L_2$. By definition $\varphi \in \nabla(L_1)$ if and only if $\neg\neg\varphi \in L_1$. In virtue of the negative equivalence of L_1 and L_2, the last fact is equivalent to $\neg\neg\varphi \in L_2$, which is equivalent, in turn, to $\varphi \in \nabla(L_2)$. We have thus proved that $\nabla(L_1) = \nabla(L_2)$.

To prove the inverse implication assume $\nabla(L_1) = \nabla(L_2)$. If $\varphi \in L_1$, then also $\varphi \in \nabla(L_1)$, whence, by assumption $\varphi \in \nabla(L_2)$, and so $\neg\neg\varphi \in L_2$. In the same way, $\varphi \in L_2$ implies $\neg\neg\varphi \in L_1$. Applying Item 3 of Corollary 6.1.5 we conclude that L_1 and L_2 are negatively equivalent.

□

For a logic $L \in \mathsf{Jhn}^+$, we denote by $[L]_{neg}$ its abstract class with respect to negative equivalence,

$$[L]_{neg} := \{L_1 \in \mathsf{Jhn}^+ \mid L_1 \equiv_{neg} L\}.$$

It turns out that each of such abstract classes forms an interval in the lattice Jhn^+, moreover the greatest point of the interval $[L]_{neg}$ can be calculated by ∇.

Proposition 6.3.3 *For any* $L \in \mathsf{Jhn} \setminus \mathsf{Neg}$,

$$[L]_{neg} = [L', \nabla(L)],$$

where $L' \subseteq \mathbf{Lg}$.

Proof. First we state that the set $[L]_{neg}$ is convex. Letting $L_1, L_2 \in [L]_{neg}$ we check that the interval $[L_1, L_2]$ is contained in $[L]_{neg}$. Take an arbitrary $L' \in [L_1, L_2]$, we then have $L_1 \leq_{neg} L' \leq_{neg} L_2$. Taking into account $L_1 \equiv_{neg} L_2$ we immediately obtain $L' \in [L]_{neg}$.

The logic $\nabla(L)$ is the greatest point of $[L]_{neg}$. Indeed, if $L' \equiv_{neg} L$ and $\varphi \in L'$, then $\neg\neg\varphi \in L$, i.e., $\varphi \in \nabla(L)$, and we have the inclusion $L' \subseteq \nabla(L)$.

To state that $[L]_{neg}$ has the least point it is enough to observe that the intersection of an arbitrary family of logics from the class $[L]_{neg}$ again belongs to this class. One can give a more explicit presentation of the least logic from $[L]_{neg}$. Put

$$L' := \mathbf{Lj} + \{\neg\neg\varphi \mid \varphi \in L\}.$$

Due to Proposition 6.1.2 every logic negatively equivalent to L must contain L'. On the other hand, the logic L' itself belongs to $[L]_{neg}$. Indeed, the relation $L' \leq_{neg} L$ follows from an obvious inclusion $L' \subseteq L$, the inverse relation $L \leq_{neg} L'$ follows from Proposition 6.1.2.

According to Corollary 6.2.2, the logic $\mathbf{Lg}$ is the least logic in $[\mathbf{Lk}]_{neg}$, and so it has a presentation

$$\mathbf{Lg} = \mathbf{Lj} + \{\neg\neg\varphi \mid \varphi \in \mathbf{Lk}\}.$$

Using this fact and the inclusion $L \subseteq \mathbf{Lk}$ we immediately obtain $L' \subseteq \mathbf{Lg}$.

□

Logics of the form $\nabla(L)$ admit another interesting characterization independent of the operator ∇ and the notion of negative equivalence.

We define ∇-logics as fixed-points of the operator ∇, i.e., we say that a logic $L \in \mathsf{Jhn}$ is a ∇*-logic* if $\nabla(L) = L$. In view of Item 4 of Proposition 6.3.1, any logic of the form $\nabla(L)$ is a ∇-logic. The ∇-logics have a description, in which again arises the rule $\frac{\varphi \vee \bot}{\varphi}$.

Proposition 6.3.4 *A logic* $L \in \mathsf{Jhn}$ *is a* ∇*-logic if and only if* $L_{int} = \mathbf{Lk}$ *and* L *is closed under the rule*

$$\frac{\varphi \vee \bot}{\varphi}.$$

Proof. Recall that $\mathbf{Lj} \vdash \neg\neg(p \vee \neg p)$. This means that for every $L \in \mathsf{Jhn}$, the formula $p \vee \neg p$ belongs to $\nabla(L)$. It was proved in Item 2 of Corollary 5.3.2 that $\mathbf{Lj} + \{p \vee \neg p\} = \mathbf{Lk} * \mathbf{Ln}$, and so any logic of the form $\nabla(L)$ contains $\mathbf{Lk} * \mathbf{Ln}$. An inclusion of logics implies the inclusion of respective counterparts (see Proposition 4.2.2), therefore, $\mathbf{Lk} \subseteq (\nabla(L))_{int}$. We have thus proved that intuitionistic counterparts of ∇-logics are classical.

We now observe that every logic $\nabla(L)$ is closed under the rule $\frac{\neg\neg\varphi}{\varphi}$. This fact easily follows from the idempotentness of ∇. If $\neg\neg\varphi \in \nabla(L)$, then $\varphi \in \nabla\nabla(L) = \nabla(L)$. According to the lemma below, the double negation $\neg\neg\varphi$ is equivalent to $\varphi\vee\bot$ in $\mathbf{Lk} * \mathbf{Ln}$, and so in any ∇-logic, which completes the proof of the direct implication.

Lemma 6.3.5 $\neg\neg p \leftrightarrow (p \vee \bot) \in \mathbf{Lk} * \mathbf{Ln}$.

Proof. By definition of free combination we have

$$\neg\neg(p \vee \bot) \leftrightarrow (p \vee \bot) = In(\neg\neg p \leftrightarrow p) \in \mathbf{Lk} * \mathbf{Ln}.$$

It remains to note that $\neg(p \vee \bot) \leftrightarrow \neg p \in \mathbf{Lj}$.

□

Prove the inverse implication. The condition $L_{int} = \mathbf{Lk}$ implies the inclusion $\mathbf{Lk}*\mathbf{Ln} \subseteq L$, and we apply Lemma 6.3.5 to conclude that L is closed under the rule $\frac{\neg\neg\varphi}{\varphi}$. If $\varphi \in \nabla(L)$, then by definition $\neg\neg\varphi \in L$, and applying the above rule we obtain $\varphi \in L$.

□

6.4 The Structure of Jhn^+ up to Negative Equivalence

In this section, we give a characterization of the partial ordering

$$\langle \mathsf{Jhn}^+/\equiv_{neg}, \sqsubseteq_{neg}\rangle,$$

where $\sqsubseteq_{neg}:=\leq_{neg} / \equiv_{neg}$. To obtain the main results we apply the technique of Jankov's formulas suggested by V. A. Jankov [37, 38] and modified by H. Ono [83] and A. Wroński [125, 126]. Usually, this technique is used for constructing uncountable families of logics. We are interested first of all for Jankov's formulas themselves. In our considerations, they will have the form of negative formulas, which allows one to prove that different logics are not negatively equivalent. We recall basic elements of Jankov's method adopting it for j-algebras.

A relation $X \models_{\mathcal{A}} \varphi$, where X is a set of formulas, φ a formula, and $\mathcal{A}$ a j-algebra, means that for any $\mathcal{A}$-valuation v, if $v(\psi) = 1$ for all $\psi \in X$, then $v(\varphi) = 1$. If $\mathcal{K}$ is a class of j-algebras, then $X \models_{\mathcal{K}} \varphi$ means that $X \models_{\mathcal{A}} \varphi$ for all $\mathcal{A} \in \mathcal{K}$. Finally, write $X \models_L \varphi$ instead of $X \models_{Mod(L)} \varphi$.

Let $\mathcal{A} = \langle A, \vee, \wedge, \rightarrow, \bot, 1\rangle$ be a not more than countable and subdirectly irreducible j-algebra. For each element $a \in \mathcal{A}$, $a \neq \bot$, we attach a unique propositional variable p_a. Further, for any $a \in \mathcal{A}$, we attach a unique atomic formula Z_a as follows

$$Z_a := \begin{cases} p_a, & \text{if } a \neq \bot \\ \bot, & \text{if } a = \bot. \end{cases}$$

A *diagram* $\mathcal{D}(\mathcal{A})$ of $\mathcal{A}$ is the following set of formulas

$$\begin{aligned} \mathcal{D}(\mathcal{A}) \quad &:= \quad \{Z_{a\vee b} \leftrightarrow (Z_a \vee Z_b) \mid a, b \in A\} \quad \cup \\ &\cup \quad \{Z_{a\wedge b} \leftrightarrow (Z_a \wedge Z_b) \mid a, b \in A\} \quad \cup \\ &\cup \quad \{Z_{a\rightarrow b} \leftrightarrow (Z_a \rightarrow Z_b) \mid a, b \in A\}. \end{aligned}$$

Let $\mathcal{A}$ be a finite subdirectly irreducible j-algebra. Then $\mathcal{D}(\mathcal{A})$ is a finite set of formulas and we can define a *Jankov formula* of $\mathcal{A}$ by

$$J(\mathcal{A}) := (\bigwedge \mathcal{D}(\mathcal{A})) \rightarrow Z_{\star_{\mathcal{A}}},$$

where $(\bigwedge \mathcal{D}(\mathcal{A}))$ is the conjunction of all formulas in $\mathcal{D}(\mathcal{A})$, and $\star_{\mathcal{A}}$ is the opremum of $\mathcal{A}$. It is easy to see that $J(\mathcal{A}) \notin L\mathcal{A}$. Moreover, the following statement holds.

Lemma 6.4.1 *Let $\mathcal{A}$ be a finite and subdirectly irreducible j-algebra. For each j-algebra $\mathcal{B}$, the following two conditions are equivalent.*

1. *$J(\mathcal{A}) \notin L\mathcal{B}$.*
2. *$\mathcal{A}$ is embeddable into a quotient algebra of $\mathcal{B}$.*

Proof. $1 \Rightarrow 2$. Assume $\mathcal{B} \not\models J(\mathcal{A})$. Let v be a $\mathcal{B}$-valuation such that $v(\bigwedge \mathcal{D}(\mathcal{A})) \not\leq v(Z_{\star_{\mathcal{A}}})$. Put $a_0 := v(\bigwedge \mathcal{D}(\mathcal{A}))$ and consider the quotient $\mathcal{B}/\langle a_0\rangle$. Define a mapping $h : \mathcal{A} \rightarrow \mathcal{B}/\langle a_0\rangle$ by the rule $h(a) := v(Z_a)/\langle a_0\rangle$. It follows from $v(\bigwedge \mathcal{D}(\mathcal{A})) \in \langle a_0\rangle$ that h is a homomorphism. Since $a_0 \not\leq v(Z_{\star_{\mathcal{A}}})$, we have $h(\star_{\mathcal{A}}) \neq 1$, i.e., $\star_{\mathcal{A}} \notin Ker(h)$. This means that h is an embedding.

$2 \Rightarrow 1$. Let F be a filter on $\mathcal{B}$ and $h : \mathcal{A} \rightarrow \mathcal{B}/F$ be an embedding. Consider a $\mathcal{B}$-valuation v such that $v(Z_a)/F = h(a)$ for all $a \in \mathcal{A}$. Homomorphism properties of h imply that for all $\psi \in \mathcal{D}(\mathcal{A})$ we have $v(\psi) \in F$, and so $v(\bigwedge \mathcal{D}(\mathcal{A})) \in F$. At the same time, $h(\star_\mathcal{A}) \neq 1/F$ since h is an embedding, which implies $v(Z_{\star_\mathcal{A}}) \notin F$. In this way, $v(\bigwedge \mathcal{D}(\mathcal{A})) \not\leq v(Z_{\star_\mathcal{A}})$, i.e., $\mathcal{B} \not\models J(\mathcal{A})$.

□

In case $\mathcal{A}$ is not finite, we cannot, of course, define a Jankov formula of $\mathcal{A}$. However, one can prove

Lemma 6.4.2 *Let $\mathcal{A}$ be a countable and subdirectly irreducible j-algebra. For each j-algebra $\mathcal{B}$, the following conditions are equivalent.*

1. *$\mathcal{D}(\mathcal{A}) \not\models_\mathcal{B} Z_{\star_\mathcal{A}}$.*
2. *$\mathcal{A}$ is embeddable into $\mathcal{B}$.*

Proof. 1 ⇒ 2. Let v be a $\mathcal{B}$-valuation such that $v(\psi) = 1$ for all $\psi \in \mathcal{D}(\mathcal{A})$ and $v(Z_{\star_\mathcal{A}}) \neq 1_\mathcal{B}$. Consider a mapping $h : \mathcal{A} \to \mathcal{B}$ given by the rule $h(a) = v(Z_a)$. It follows easily from our assumption and the definition of $\mathcal{D}(\mathcal{A})$ that h is a homomorphism. If h is not a monomorphism, then $Ker(h) \neq \{1_\mathcal{A}\}$ and $\star_\mathcal{A} \in Ker(h)$, i.e., $h(\star_\mathcal{A}) = 1_\mathcal{B}$. The latter conflicts with the assumption that $v(Z_{\star_\mathcal{A}}) \neq 1_\mathcal{B}$.

2 ⇒ 1. Assume h embeds $\mathcal{A}$ into $\mathcal{B}$. Consider a $\mathcal{B}$-valuation such that $v(p_a) = h(a)$ for $a \neq \bot_\mathcal{A}$. Naturally, $v(\bot) = h(\bot_\mathcal{A}) = \bot_\mathcal{B}$. It is clear that $v(\psi) = 1$ for all $\psi \in \mathcal{D}(\mathcal{A})$ and $v(Z_{\star_\mathcal{A}}) \neq 1_\mathcal{B}$, i.e., $\mathcal{D}(\mathcal{A}) \not\models_\mathcal{B} Z_{\star_\mathcal{A}}$.

□

A sequence $\{L_i\}_{i<\omega}$ of logics from Jhn is said to be *strongly independent* if $L_i \not\subseteq \Sigma_{j\neq i} L_j$ for each $i < \omega$, where $\Sigma_{j\neq i} L_j$ is the least logic containing all L_j, $j \neq i$.

The following two facts are natural generalizations of Proposition 1.2 and Lemma 1.4 of [106] to the class of extensions of minimal logic.

Proposition 6.4.3 *Let $\{L_i\}_{i<\omega}$ be a strongly independent sequence of logics from Jhn. For any subsets I and J of ω, $\Sigma_{i\in I} L_i = \Sigma_{i\in J} L_i$ if and only if $I = J$.*

Proof. Prove the non-trivial implication. Assume that $I \neq J$ and there is $k \in I \setminus J$. Obviously, $\Sigma_{i\in J} L_i \subseteq \Sigma_{i\neq k} L_i$. Since $\{L_i\}_{i<\omega}$ is strongly independent, we have $L_k \not\subseteq \Sigma_{i\neq k} L_i$, moreover, $L_k \not\subseteq \Sigma_{i\in J} L_i$. Therefore, $\Sigma_{i\in I} L_i \neq \Sigma_{i\in J} L_i$.

□

Proposition 6.4.4 *Let $\{\mathcal{A}_i\}_{i<\omega}$ be a sequence of subdirectly irreducible j-algebras satisfying the following conditions:*

(1) each $\mathcal{A}_i$ is finite;

(2) for every $i, j < \omega$, $i \neq j$ implies that $\mathcal{A}_i$ cannot be embedded into any quotient algebra of $\mathcal{A}_j$.

If a logic L is contained in every $L\mathcal{A}_i$ $(i < \omega)$, the sequence $\{L_i\}_{i<\omega}$ of logics defined by $L_i = L + \{J(\mathcal{A}_i)\}$ $(i < \omega)$ is strongly independent.

Proof. It suffices to show that $J(\mathcal{A}_i) \notin \Sigma_{j\neq i} L_i$ for all i. According to Lemma 6.4.1 we have $J(\mathcal{A}_i) \in L\mathcal{A}_j$ for $j \neq i$. Consequently, $L \cup \{J(\mathcal{A}_j) \mid j \neq i\} \subseteq L\mathcal{A}_i$, and we conclude $\Sigma_{j\neq i} L_j \subseteq L\mathcal{A}_i$. In view of $J(\mathcal{A}_i) \notin L\mathcal{A}_i$ we have $J(\mathcal{A}_i) \notin \Sigma_{j\neq i} L_i$.

□

We now turn to studying the partial ordering $\langle \mathsf{Jhn}^+ / \equiv_{neg}, \sqsubseteq_{neg} \rangle$. First we prove

Proposition 6.4.5 *Let $L \in \mathsf{Neg}$. For any $L_1, L_2 \in Spec(\mathbf{Lk}, L)$, we have the equivalence*

$$L_1 \equiv_{neg} L_2 \iff L_1 = L_2.$$

Proof. We need only prove that different logics from the class $Spec(\mathbf{Lk}, L)$ are not negatively equivalent. Let logics $L_1, L_2 \in Spec(\mathbf{Lk}, L)$ be such that there is a formula $\varphi \in L_1 \setminus L_2$. Since $\varphi \notin L_2$, according to Proposition 2.2.6 it must be refuted at some subdirectly irreducible model of L_2.

Let $\mathcal{A}$ be a subdirectly irreducible j-algebra such that $\mathcal{A} \models L_2$ and $\mathcal{A} \not\models \varphi$. First of all we note that $\mathcal{A}$ is not negative. If $\mathcal{A}$ is negative, then $\mathcal{A}$ is a model of $(L_2)_{neg} = L_2 + \{\bot\} = L$. The logics L_1 and L_2 have a common negative counterpart L and $L_1 \subseteq L$. In this way, we conclude that $\mathcal{A} \models L_1$, which conflicts with our assumption that $\mathcal{A}$ refutes φ.

We now consider the upper algebra $\mathcal{A}^\bot$, which is non-trivial due to the last remark. The opremum of $\mathcal{A}$ equals to the opremum of $\mathcal{A}^\bot$. Indeed, $\bot_\mathcal{A} \neq 1_\mathcal{A}$, and so $\bot_\mathcal{A} \leq \star_\mathcal{A}$. Therefore, Heyting algebra $\mathcal{A}^\bot$ is a subdirectly irreducible model of $(L_1)_{int} = \mathbf{Lk}$, whence, it is a Boolean algebra. It is well known that only two-element Boolean algebra is subdirectly irreducible, and so $\star_\mathcal{A} = \bot_\mathcal{A}$. We have thus proved the following

Lemma 6.4.6 *Let $L \in \mathsf{Neg}$ and $L_1 \in Spec(\mathbf{Lk}, L)$. Every subdirectly irreducible model $\mathcal{A}$ of L_1 is either negative, in which case $\mathcal{A} \models L$, or has the form $\mathcal{A} \cong \mathcal{B} \oplus \mathbf{2}$, where $\mathcal{B}$ is a negative algebra modelling L and $\mathbf{2}$ is a two-element Boolean algebra.*

Assume additionally that $\mathcal{A}$ is finite. Since $\mathcal{A}$ is not a model of L_1, it is not embeddable into a quotient algebra of any j-algebra $\mathcal{B}$ modelling L_1. According to Lemma 6.4.1, this means that for any model $\mathcal{B}$ of L_1, $\mathcal{B} \models J(\mathcal{A})$, and so $J(\mathcal{A}) \in L_1$. It is clear that $J(\mathcal{A}) \notin L_2$ since $J(\mathcal{A}) \notin L\mathcal{A}$.

As noted above, $\star_{\mathcal{A}} = \bot_{\mathcal{A}}$, whence,

$$J(\mathcal{A}) = (\bigwedge \mathcal{D}(\mathcal{A})) \to \bot = \neg(\bigwedge \mathcal{D}(\mathcal{A})).$$

This formula proves that logics L_1 and L_2 are not negatively equivalent.

It remains to consider the case when a j-algebra $\mathcal{A}$ refuting $\varphi \in L_1$ is infinite. We use Lemma 6.4.2 to conclude that for any j-algebra $\mathcal{B}$ with $\mathcal{B} \models L_1$, we have $\mathcal{D}(\mathcal{A}) \models_{\mathcal{B}} \bot$, i.e., that $\mathcal{D}(\mathcal{A}) \models_{L_1} \bot$.

We may consider j-algebras as models of the first-order predicate calculus of the language $\langle \vee, \wedge, \to, \bot, 1 \rangle$ with identity "$=$" and classical negation "$\sim$". Propositional variables will be treated as individual ones, propositional formulas as terms, and valuations as denotations of individual variables. To avoid confusion we will denote the first order satisfiability relation by $\models^1$. In this case, for any j-algebra $\mathcal{B}$, $\mathcal{B}$-valuation v, and formula φ, we have the following equivalences:

$$v(\varphi) = 1 \;\Leftrightarrow\; \mathcal{B} \models^1 \varphi = 1[v],$$

$$\mathcal{B} \models \varphi \;\Leftrightarrow\; \mathcal{B} \models^1 (\varphi = 1)',$$

where $(\psi)'$ denotes the universal closure of a first-order formula ψ.

The relation $\mathcal{D}(\mathcal{A}) \models_{L_1} \bot$ is equivalent to the statement that there is no j-algebra $\mathcal{B}$ and $\mathcal{B}$-valuation v such that $\mathcal{B} \models L_1$, $\bot_{\mathcal{B}} \neq 1_{\mathcal{B}}$, and $v(\varphi) = 1$ for all $\varphi \in \mathcal{D}(\mathcal{A})$. In view of the above equivalences, the latter means exactly that the set of formulas

$$\Sigma := \{(\varphi = 1)' \mid \varphi \in L_1\} \cup \{\varphi = 1 \mid \varphi \in \mathcal{D}(\mathcal{A})\} \cup \{\sim (\bot = 1)\}$$

is not satisfiable. Due to the Malcev local theorem for first-order logic, there exists a finite non-satisfiable subset $X \subseteq \Sigma$.

Further, we argue to state that $\sim(\bot = 1) \in X$. Consider the quotient algebra $\mathcal{A}/\langle\bot\rangle$. The set of formulas $\{\varphi = 1 \mid \varphi \in \mathcal{D}(\mathcal{A})\}$ is satisfiable in $\mathcal{A}$, it will be also satisfiable in the quotient algebra $\mathcal{A}/\langle\bot\rangle$. The algebra $\mathcal{A}/\langle\bot\rangle$ is negative, and so it is a model for $L = (L_2)_{neg}$. Since $L_1 \subseteq L$, the sentences from the set $\{(\varphi = 1)' \mid \varphi \in L_1\}$ will be true in $\mathcal{A}/\langle\bot\rangle$. We have thus proved that the set of formulas $\Sigma \setminus \{\sim(\bot = 1)\}$ is satisfiable, from which we conclude that $\sim(\bot = 1) \in X$.

Denote by X' the following subset of $\mathcal{D}(\mathcal{A})$,

$$X' := \{\varphi \mid \varphi \in \mathcal{D}(\mathcal{A}), \varphi = 1 \in X\}$$

Coming back to identities of j-algebras, we can see that

$$X' \models_{\mathcal{B}} \bot$$

for any $\mathcal{B}$ with $\mathcal{B} \models L_1$, i.e., $X' \models_{L_1} \bot$.

We claim that the formula $(\bigwedge X') \to \bot = \neg(\bigwedge X')$, where $\bigwedge X'$ is the conjunction of all formulas from X', is true in all subdirectly irreducible models $\mathcal{B}$ of L_1, and so belongs to L_1.

Take an arbitrary subdirectly irreducible model $\mathcal{B}$ of L_1. If $\mathcal{B}$ is negative, then any negated formula is true on $\mathcal{B}$, in particular, $\mathcal{B} \models (\bigwedge X') \to \bot$. Assume that $\mathcal{B}$ is not negative and v is a $\mathcal{B}$-valuation. Relation $X' \models_{L_1} \bot$ means that the set of formulas X' is not satisfiable in any non-negative model of L_1, whence $v(\bigwedge X') \neq 1$. Taking into account that $\mathcal{B}$ is subdirectly irreducible and $\bot_{\mathcal{B}} = \star_{\mathcal{B}}$ by Lemma 6.4.6 we obtain $v(\bigwedge X') \leq \bot_{\mathcal{B}}$, consequently, $v((\bigwedge X') \to \bot) = 1$. We have thus proved that $\neg(\bigwedge X') \in L_1$. Recall that X' is a subset of $\mathcal{D}(\mathcal{A})$, the diagram of the non-negative model $\mathcal{A}$ for L_2, therefore, it is satisfiable in $\mathcal{A}$ and $\neg(\bigwedge X')$ does not belong to L_2. In this way, the formula $\neg(\bigwedge X')$ proves that L_1 and L_2 are not negatively equivalent.

□

Proposition 6.4.7 *For any logics* $L \in \mathsf{Jhn} \setminus \mathsf{Neg}$ *and* $L_1 \in \mathsf{Neg}$ *such that* $L_1 \subseteq L_{neg}$, *there exists a logic* $L' \in Spec(\mathbf{Lk}, L_1)$ *such that* $L' \equiv_{neg} L$.

Proof. Consider the logic $\nabla(L)$. By Proposition 6.3.2 $\nabla(L) \equiv_{neg} L$. Due to Proposition 6.3.4 $(\nabla(L))_{int} = \mathbf{Lk}$. At the same time, by Proposition 6.3.1 $L \subseteq \nabla(L)$ and consequently, $L_{neg} \subseteq \nabla(L)_{neg}$. Combining the last two facts yields $\nabla(L) \in Spec(\mathbf{Lk}, L_2)$ for some $L_2 \supseteq L_{neg} \supseteq L_1$, in particular, $\mathbf{Lk} * L_2 \subseteq \nabla(L)$.

Define a logic L' by setting

$$L' := \mathbf{Lk} * L_1 + \{\neg\varphi \mid \neg\varphi \in L\} (= \mathbf{Lk} * L_1 + \{\neg\varphi \mid \neg\varphi \in \nabla(L)\}).$$

It is clear that $\mathbf{Lk} * L_1 \subseteq \nabla(L)$, whence, $L' \subseteq \nabla(L)$, and so $L' \leq_{neg} L$. Further, let $\varphi \in L$. Then also $\neg\neg\varphi \in L$, and $\neg\neg\varphi \in L'$ by definition. We now apply Proposition 6.1.2 to conclude $L \leq_{neg} L'$. We have thus proved that L and L' are negatively equivalent and it remains to check that L' belongs to the desired interval $Spec(\mathbf{Lk}, L_1)$.

Obviously, $L_1 \subseteq L'_{neg}$. If this inclusion is proper, there is a negative algebra $\mathcal{A}$, which is a model of L_1 but not of L'_{neg}. It is clear that $\mathcal{A}$ models $\mathbf{Lk} * L_1$. All negated formulas are true in $\mathcal{A}$, therefore, $\mathcal{A} \models L'$. Moreover, $\mathcal{A}$ is negative, whence, $\mathcal{A} \models L'_{neg}$, a contradiction.

□

Now the ordering $\langle \mathsf{Jhn}^+/\equiv_{neg}, \sqsubseteq_{neg}\rangle$ can be described as follows.

Proposition 6.4.8 *1. The following isomorphism takes place*

$$\langle \mathsf{Jhn}^+/\equiv_{neg}, \sqsubseteq_{neg}\rangle \cong \langle Spec(\mathbf{Lk}, \mathbf{Ln}) \cup \{\mathcal{F}\}, \subseteq\rangle.$$

2. For $L_1 \in \mathsf{Int}$ and $L_2 \in \mathsf{Neg}$, we have

$$\langle Spec(L_1, L_2)/\equiv_{neg}, \sqsubseteq_{neg}\rangle \cong \langle Spec(\mathbf{Lk}, L_2), \subseteq\rangle.$$

Proof. 1. Applying Proposition 6.4.7 for the case $L_1 = \mathbf{Ln}$, we conclude that for any $L \in \mathsf{Jhn} \setminus \mathsf{Neg}$, there is a logic in $Spec(\mathbf{Lk}, \mathbf{Ln})$ negatively equivalent to L. Such logic is unique due to Proposition 6.4.5 and we denote it by $h(L)$. For every $L \in \mathsf{Neg} \cup \{\mathcal{F}\}$, put $h(L) = \mathcal{F}$. We have thus defined a mapping

$$h : \mathsf{Jhn}^+ \to Spec(\mathbf{Lk}, \mathbf{Ln}) \cup \{\mathcal{F}\}.$$

Obviously, $h(L) = L$ for all $L \in Spec(\mathbf{Lk}, \mathbf{Ln})$, therefore, h is a mapping onto. Moreover, it satisfies the property that $L \equiv_{neg} h(L)$ for every $L \in \mathsf{Jhn}^+$.

Take arbitrary $L_1, L_2 \in \mathsf{Jhn} \setminus \mathsf{Neg}$ such that $L_1 \leq_{neg} L_2$. In view of the envisaged property of h we have $h(L_1) \leq_{neg} h(L_2)$. If $\varphi \in h(L_1)$, then $\neg\neg\varphi \in h(L_1)$ and by Proposition 6.1.2 $\neg\neg\varphi \in h(L_2)$. Due to the same proposition we have $h(L_1) \leq_{neg} h(L_1) \cap h(L_2)$. The inverse relation $h(L_1) \cap h(L_2) \leq_{neg} h(L_1)$ follows from the set-theoretical inclusion. Applying Proposition 6.4.5 to $h(L_1) \equiv_{neg} h(L_1) \cap h(L_2)$ we infer $h(L_1) \subseteq h(L_2)$. Note also that for any $L_1 \in \mathsf{Jhn}$ and $L_2 \in \mathsf{Neg} \cup \{\mathcal{F}\}$, $L_1 \leq_{neg} L_2$ and $h(L_1) \subseteq h(L_2) = \mathcal{F}$. We have thus proved that h is an epimorphism of $\langle \mathsf{Jhn}^+, \leq_{neg}\rangle$ onto $\langle Spec(\mathbf{Lk}, \mathbf{Ln}) \cup \{\mathcal{F}\}, \subseteq\rangle$. Passing to the quotient mapping $h/\equiv_{neg}$ of this epimorphism with respect to the negative equivalence we obtain the desired isomorphism of $\langle \mathsf{Jhn}^+/\equiv_{neg}, \sqsubseteq_{neg}\rangle$ and $\langle Spec(\mathbf{Lk}, \mathbf{Ln}) \cup \{\mathcal{F}\}, \subseteq\rangle$.

2. Arguing as in the previous item we construct a mapping

$$h : Spec(L_1, L_2) \to Spec(\mathbf{Lk}, L_2)$$

such that $L \equiv_{neg} h(L)$ for any $L \in Spec(L_1, L_2)$. We need only check that this is a mapping onto. Take an arbitrary $L \in Spec(\mathbf{Lk}, L_2)$ and put

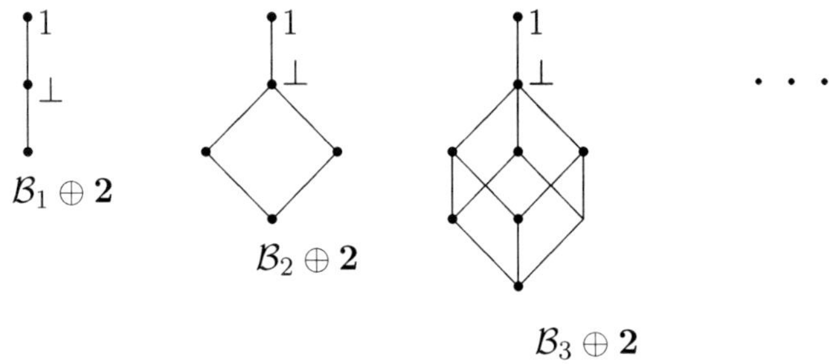

Figure 6.1

$L' := L_1 * L_2 + \{\neg\varphi \mid \neg\varphi \in L\}$. As in Proposition 6.4.7 we prove that $L' \equiv_{neg} L$. Now we conclude the proof as in the previous item.

□

In conclusion, we make some remarks on the cardinality of intervals $Spec(L_1, L_2)$. It turns out that any interval of this form is infinite, moreover, infinite up to negative equivalence, and in many cases such an interval has the power of the continuum. We start with the simplest interval $Spec(\mathbf{Lk}, \mathbf{Lmn})$.

Proposition 6.4.9 *The interval* $Spec(\mathbf{Lk}, \mathbf{Lmn})$ *has the following structure:*

$$\mathbf{Lk} * \mathbf{Lmn} \subseteq \ldots \subseteq L_n \subseteq \ldots \subseteq L_1 \subseteq L_0,$$

where $L_0 = \mathbf{Le} = \mathbf{Lk} \cap \mathbf{Lmn}$*; for* $n > 0$, $L_n = L(\mathcal{B}_n \oplus \mathbf{2}) = \mathbf{Lk} * \mathbf{Lmn} + \{J(\mathcal{B}_{n+1} \oplus \mathbf{2})\}$, *where* $\mathcal{B}_n$ *is a negative Peirce algebra with* n *atoms (see Figure 6.1); finally,* $\mathbf{Lk} * \mathbf{Lmn} = L(\mathcal{B} \oplus \mathbf{2})$, *where* $\mathcal{B}$ *is an arbitrary infinite negative Peirce algebra.*

Proof. All logics from the interval $Spec(\mathbf{Lk}, \mathbf{Lmn})$ have the same negative models, namely, the models of **Lmn**. Therefore, every logic L in $Spec(\mathbf{Lk}, \mathbf{Lmn})$ is determined by the class of its non-negative finitely generated subdirectly irreducible models and we denote this class by $Mod^{+}_{fsi}(L)$. Due to Lemmas 6.4.6 and 6.4.10 every such model model has the form $\mathcal{B} \oplus \mathbf{2}$, where $\mathcal{B}$ is a finitely generated model of **Lmn**, i.e., a finitely generated negative Peirce algebra.

Lemma 6.4.10 *The algebra* $\mathcal{B} \oplus \mathbf{2}$ *is finitely generated if and only if* $\mathcal{B}$ *is finitely generated.*

Proof. Let $\mathcal{B}' := \mathcal{B} \oplus \mathbf{2}$ and $\star := \star_{\mathcal{B}\oplus\mathbf{2}}$. Note that $|\mathcal{B}'| = |\mathcal{B}| \cup \{1_{\mathcal{B}'}\}$ and $\star = 1_{\mathcal{B}} = \bot_{\mathcal{B}}$.

If X generates $\mathcal{B}'$, then $X \setminus \{\star\}$ generates $\mathcal{B}$. Indeed, if $b \in \mathcal{B}$ is a value of some term defined on generators, then replacing $\star$ by $c \to c$, where $c \in X \setminus \{\star\}$, does not change the value of this term.

If X generates $\mathcal{B}$, then $X \cup \{\star\}$ generates $\mathcal{B}'$. It is a consequence of the following obvious relations. For any $a, b \in \mathcal{B}$, we have $a \wedge_{\mathcal{B}} b = a \wedge_{\mathcal{B}'} b$, $a \vee_{\mathcal{B}} b = a \vee_{\mathcal{B}'} b$, if $a \to_{\mathcal{B}} b \neq 1_{\mathcal{B}}$, then $a \to_{\mathcal{B}} b = a \to_{\mathcal{B}'} b$.

□

The only difference between Boolean and negative Peirce algebras is in the interpretation of $\bot$. In Boolean algebras, $\bot$ is interpreted as the least element, whereas in negative Peirce algebras, it is interpreted as the greatest element of an algebra. In this way, all finitely generated negative Peirce algebras are finite. And we have a countable chain of different (up to isomorphism) finitely generated negative Peirce algebras $\{\mathcal{B}_n \mid n \in \omega\}$, where $\mathcal{B}_n$ is a negative Peirce algebra with n-atoms, i.e., with n minimal elements in the set $\mathcal{B}_n \setminus \{f\}$ for f denoting the least element of $\mathcal{B}_n$. Clearly, $\mathcal{B}_n$ is isomorphically embedded into $\mathcal{B}_m$ if and only if $n \leq m$. Therefore, the class $Mod^+_{fsi}(L)$ for $L \in Spec(\mathbf{Lk}, \mathbf{Lmn})$ is of the form

$$M_\alpha = \{\mathcal{B}_n \oplus \mathbf{2} \mid n < \alpha\},$$

where $1 \leq \alpha \leq \omega$. It is not hard to check that each of these sets can be realized as a set of non-negative finitely generated subdirectly irreducible models of a suitable logic L_α from the interval $Spec(\mathbf{Lk}, \mathbf{Lmn})$. Indeed, the class $M_1 = \{\mathcal{B}_0 \oplus \mathbf{2}\} = \{\mathbf{2}\}$ corresponds to the logic $\mathbf{Le} = \mathbf{Lk} \cap \mathbf{Lmn}$, which has a unique non-negative subdirectly irreducible model, $\mathbf{2}$. For $1 < \alpha < \omega$, consider a logic

$$L_\alpha := \mathbf{Lk} * \mathbf{Lmn} + \{J(\mathcal{B}_\alpha \oplus \mathbf{2})\}.$$

According to Lemma 6.4.1 the class $Mod^+_{fsi}(L_\alpha)$ coincides with the class of algebras of the form $\mathcal{B}_n \oplus \mathbf{2}$ such that the algebra $\mathcal{B}_\alpha \oplus \mathbf{2}$ is not embeddable into a quotient algebra of $\mathcal{B}_n \oplus \mathbf{2}$. Every proper quotient algebra of $\mathcal{B}_n \oplus \mathbf{2}$ is negative and non-negative algebra can not be embedded into it. The algebra $\mathcal{B}_m \oplus \mathbf{2}$ is embeddable into $\mathcal{B}_n \oplus \mathbf{2}$ if and only if $\mathcal{B}_m$ is embeddable into $\mathcal{B}_n$. Combining all these facts we obtain

$$Mod^+_{fsi}(L_\alpha) = M_\alpha.$$

From the last equality and the fact that $\mathbf{Lmn} = L(\mathcal{B}_n)$ for any n we have $L_\alpha = L(\mathcal{B}_\alpha \oplus \mathbf{2})$.

The class $Spec(\mathbf{Lk}, \mathbf{Lmn})$ has the least element $L_\omega = \mathbf{Lk} * \mathbf{Lmn}$ and, obviously, $Mod^+_{fsi}(\mathbf{Lk} * \mathbf{Lmn}) = M_\omega$. If we take an arbitrary infinite negative Peirce algebra $\mathcal{B}$, any algebra of the form $\mathcal{B}_n \oplus \mathbf{2}$ will be embeddable into $\mathcal{B} \oplus \mathbf{2}$. Therefore, $L_\omega = L(\mathcal{B} \oplus \mathbf{2})$.

Classes of models $M_\alpha, 1 \leq \alpha \leq \omega$, form an ascending chain of type $\omega + 1$ with respect to inclusion, which means that the corresponding logics $L_\alpha, 1 \leq \alpha \leq \omega$, form a descending chain of type $(\omega + 1)^*$.

□

Further, we obtain a sufficient condition guaranteeing that $Spec(L_1, L_2)$ is of the power of the continuum.

Proposition 6.4.11 *Let $L_1 \in \mathsf{Int}$ and $L_2 \in \mathsf{Neg}$. Assume that there exists a family $\{\mathcal{B}_i \mid i < \omega\}$ of finite negative algebras such that $\mathcal{B}_i \models L_2$ for all $i < \omega$, and $\mathcal{B}_i$ is not embeddable into $\mathcal{B}_j$ for $i \neq j$. Then*

$$|Spec(L_1, L_2)| = 2^\omega.$$

Proof. In view of Proposition 6.4.8, it is enough to consider the case $L_1 = \mathbf{Lk}$. Consider the sequence $\{\mathcal{A}_i\}_{i<\omega}$ of j-algebras defined by $\mathcal{A}_i = \mathcal{B}_i \oplus \mathbf{2}$, $i < \omega$. It is a sequence of subdirectly irreducible j-algebras modelling $\mathbf{Lk} * L_2$. By assumption if $i \neq j$, $\mathcal{B}_i$ is not embeddable into $\mathcal{B}_j$, in which case also $\mathcal{A}_i$ is not embeddable into $\mathcal{A}_j$. Further, note that every proper quotient algebra of $\mathcal{A}_j$ is negative. This means that the non-negative algebra $\mathcal{A}_i$ cannot be embedded into a quotient algebra of $\mathcal{A}_j$ for $i \neq j$.

Define a sequence of logics $\{L_i\}_{i<\omega}$ by $L_i = \mathbf{Lk} * L_2 + \{J(\mathcal{A}_i)\}$. By Proposition 6.4.4 this sequence is strongly independent and we obtain a continuum of different logics of the form $\Sigma_{i\in I} L_i$, $I \subseteq \omega$. Each of these logics belongs to the interval $Spec(\mathbf{Lk}, L_2)$, because all $J(\mathcal{A}_i)$ are negative formulas and belong to L_2.

□

Corollary 6.4.12 $|\mathsf{Jhn}/\equiv_{neg}| = |Spec(\mathbf{Lk}, \mathbf{Ln})| = 2^\omega$.

Proof. It is not hard to check that the family $\{\mathcal{B}_i \mid i \in \omega\}$ (see Figure 6.2) of negative algebras satisfies the condition of the previous proposition.

□

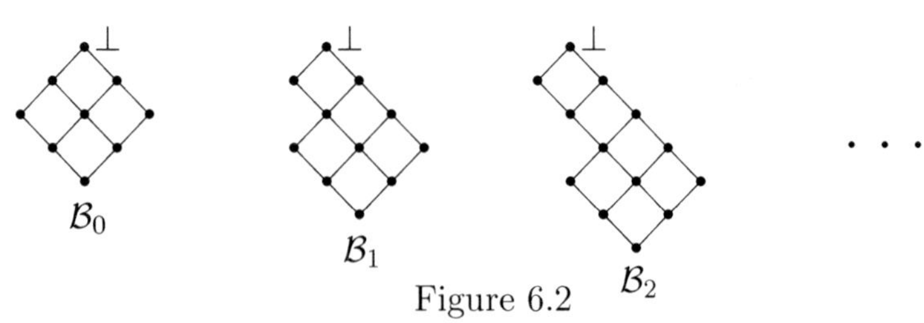

Figure 6.2

Chapter 7

Absurdity as Unary Operator[1]

7.1 Introduction

This chapter finishes the first part of the book devoted to the concept of negation as reduction to absurdity. As was mentioned in Chapter 1, minimal logic lies on the border line of paraconsistency. We have in **Lj** for arbitrary formulas φ and ψ,

$$\{\varphi, \neg\varphi\} \vdash \neg\psi.$$

This means that although inconsistent **Lj**-theories may be non-trivial, they are trivial with respect to negation. Any negated formula is provable in any inconsistent **Lj**-theory.

Due to this "paraconsistent paradox" of minimal logic it looks natural to finish the investigation of the class of **Lj**-extensions with an attempt to overcome this paradox. We try to do it by merging the class of **Lj**-extensions into a more general class of paraconsistent logics and pointing out some special property distinguishing extensions of minimal logic in the latter class. We also suggest that the negation in logics from the above-mentioned class should preserve the most essential property of intuitionistic negation, namely, that the negation must be defined as reduction to absurdity. In intuitionistic logic, the negation is characterized by three important features: 1) we assert $\neg\varphi$ if supposing φ leads to absurdity; 2) absurdity may be explicated as a propositional constant $\bot$; 3) absurdity implies everything,

[1]Parts of this chapter were originally published in [73] (Nicholas Copernicus University Press, Poland) and in [81] (Rodopi, Netherlands). Reprinted here by permission of the publishers.

$\bot \to p$, or $\neg p \to (p \to q)$. Note that Item 3 implies, in fact, Item 2. If any contradiction implies everything, all contradictions are equivalent, and so we may use a propositional constant do denote an arbitrary contradictory or absurd statement. In minimal logic **Lj**, we omit Item 3, which allows one to distinguish non-equivalent contradictions. In an extension L of **Lj**, the contradiction $\varphi \wedge \neg\varphi$ is equivalent to $\bot$ if and only if φ is provable in $L_{neg} = L + \{\bot\}$, the negative counterpart of L (see Chapter 4), moreover, it was stated that L_{neg} may be considered as the logic of contradictions of L. At the same time, Item 2 is preserved in **Lj**, which results, in particular, in the above-mentioned "paraconsistent paradox". It looks natural to make the next step and to resign not only Item 3, but also its consequence, Item 2. This gives rise to a question: how should the absurdity be explicated in a logical system in this case?

We suggest an answer, which is based on the interrelations between logic of classical refutability **Le** and Łukasiewicz's modal logic Ł. In Chapter 3 it was proved that the simplest matrix for **Le** has four elements, and so **Le** can be considered as a four-valued logic as well as Łukasiewicz's modal logic.

In Section 3.2, it was proved that in **Le**, one can define one isomorph of classical logic and two different isomorphs of maximal negative logic. These isomorphs, on the one hand, correspond to translations of **Lk** and **Lmn** into **Le** playing the key role in studying the structure of the class Jhn (see Section 4.1). On the other hand, the isomorphs were induced by mappings, which can be identified in a natural way with the modalities of Ł, which easily implies the fact that **Le** is definition equivalent to the positive fragment of Ł and that **Le** extended by the classical negation is definition equivalent to Ł. Similar interrelations between **Le** and Ł were stated previously by J. Porte [84, 85]. For our goals the most essential is the fact that the necessity operator of Ł may be identified with the contradiction operator $C(\varphi) := \varphi \wedge \neg\varphi$ of logic of classical refutability. This leads to the idea of considering a contradiction operator C as a primary logical connective and defining a negation as a reduction to this operator, $\neg\varphi := \varphi \to C(\varphi)$. In this way, we merge the class Jhn into the class of so-called C-logics and show that logics from Jhn will be distinguished by some property of C, which becomes paradoxical if C is considered as a modal operator. So, more natural modal properties of the contradiction operator result in a more satisfiable consequence relation from the point of view of paraconsistent logic. Further, it turns out that the same negation may be defined via different operators. There is a sense to distinguish, for example, between the contradiction operator C that should satisfy the axiom $C(\varphi) \to \varphi$ and the more general absurdity operator, for which a similar axiom does not obey.

At the end of this chapter, we show that in such well-known paraconsistent logics as the logic **CLuN** by D. Batens [6, 7] and Sette's maximal paraconsistent logic P^1 [88, 102] the negation can be defined via a suitable contradiction or absurdity operator. It is interesting that the negation of **CLuN** may be defined via an absurdity operator about which nothing is postulated. In this way, **CLuN** may be considered as a counterpart of **Le**, which has the same positive fragment and where the negation is defined via the constant $\bot$, about which nothing is postulated.

Let $\mathcal{L}_1$ and $\mathcal{L}_2$ be propositional languages, L_1 and L_2 logics in languages $\mathcal{L}_1$ and $\mathcal{L}_2$ respectively, and let $\theta : \mathcal{L}_1 \to \mathcal{L}_2$ and $\rho : \mathcal{L}_2 \to \mathcal{L}_1$ be translations from the language $\mathcal{L}_1$ into $\mathcal{L}_2$ and *vice versa*, i.e., mappings from the set of $\mathcal{L}_1$-formulas to the set $\mathcal{L}_2$-formulas and *vice versa*. We say that L_1 is *faithfully embedded into* L_2 *via* θ if for any $\mathcal{L}_1$-formula φ, we have the equivalence

$$\varphi \in L_1 \iff \theta(\varphi) \in L_2.$$

Further, the logic L_1 is said to be *definition equivalent to* L_2 *via translations* θ *and* ρ if L_1 is faithfully embedded into L_2 via θ, L_2 is faithfully embedded into L_1 via ρ and moreover, for any $\mathcal{L}_1$-formula φ and $\mathcal{L}_2$-formula ψ we have

$$L_1 \vdash \varphi \leftrightarrow \rho\theta(\varphi) \text{ and } L_2 \vdash \psi \leftrightarrow \theta\rho(\psi).$$

A translation α from $\mathcal{L}_1$ into $\mathcal{L}_2$ *preserves propositional variables* if $\alpha(p) = p$ for any propositional variable p and it *preserves an* n*-ary connective* $*$ if $* \in \mathcal{L}_1 \cap \mathcal{L}_2$ and

$$\alpha(*(\varphi_1, \dots, \varphi_n)) = *(\alpha\varphi_1, \dots \alpha\varphi_n)$$

for all $\mathcal{L}_1$-formulas $\varphi_1, \dots, \varphi_n$.

Remark. If L_1 is faithfully embedded into L_2 via θ, θ preserves the implication connective and logics L_1 and L_2 satisfy the Deduction Theorem, then it is not hard to prove that for any set of $\mathcal{L}_1$-formulas X and $\mathcal{L}_1$-formula φ, the equivalence holds:

$$X \vdash_{L_1} \varphi \iff \theta(X) \vdash_{L_2} \theta(\varphi),$$

where $\theta(X) = \{\theta(\psi) \mid \psi \in X\}$. In other words, θ is a strong translation.

Taking into account this remark and the above definitions, Proposition 2.1.1 can be re-worded as follows.

Proposition 7.1.1 *Logics* $\mathbf{Lj}^{\neg}$ *and* $\mathbf{Lj}^{\bot}$ *are definition equivalent via translations* θ *and* ρ.

□

7.2 Le and Łukasiewicz's Modal Logic

Recall that logic of classical refutability **Le** is characterized by the class of Peirce-Johansson (or *pj*-)algebras. A *pj*-algebra is an implicative lattice satisfying the Peirce law and with $\bot$ interpreted as an arbitrary element of its universe. This characterization easily implies that any characteristic model of **Le** must contain at least four elements: unit element 1, $0(=\bot=\neg 1)$, some element -1 under 0, because **Le** is paraconsistent and finally, element $a = 0 \to -1$ incomparable with 0. It turns out that the four-element lattice $\mathbf{4}'$ with universe $\{1, 0, a, -1\}$ and $\bot$ interpreted as 0 is really a characteristic model for **Le** (see Corollary 3.1.6).

Łukasiewicz's modal logic Ł was introduced in [52] (see also [53]). The intention of J. Łukasiewicz was to construct a system of modal logic, which is an extension of classical propositional logic with two interdefinable modal operators of "necessity" L and of "possibility" M, $Lp \leftrightarrow \neg M \neg p$ and $Mp \leftrightarrow \neg L \neg p$. These operators, in turn, should satisfy the conditions

$$\vdash Lp \to p, \vdash p \to Mp,$$

and

$$\nvdash p \to Lp, \nvdash Mp \to p.$$

He used the matrix approach to define his system. Since the ordinary two-element matrix **2** for classical logic does not allow one to define modal operators satisfying the above-mentioned conditions, Łukasiewicz took the four-element matrix $\mathbf{2} \times \mathbf{2}$, which also defines classical logic.

Note that if we identify the elements of $\mathbf{2} \times \mathbf{2}$ and the elements of **Le**-model $\mathbf{4}'$ in the following way $(1, 1) \to 1$, $(1, 0) \to a$, $(0, 1) \to 0$, and $(0, 0) \to -1$, the binary operations of classical logic and of **Le** will coincide on $\mathbf{4}'$, only negations will act differently on the set of truth values $\{1, a, 0, -1\}$.

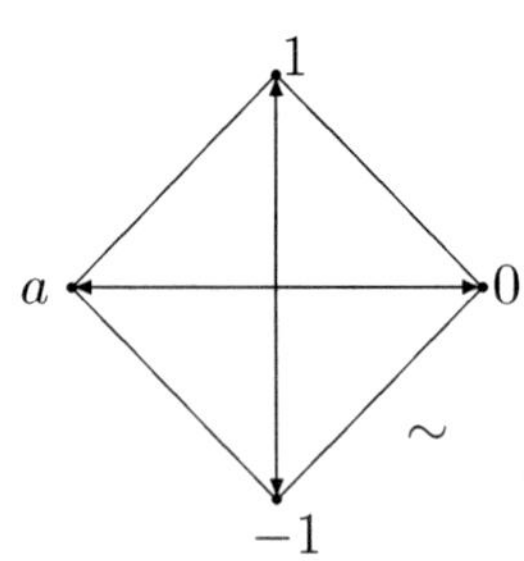

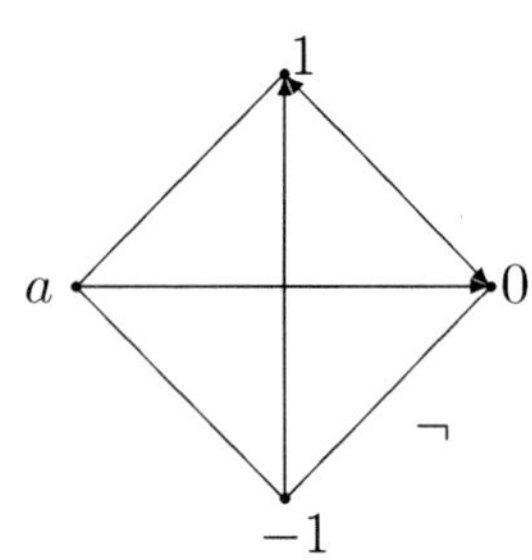

It these diagrams the classical negation $\sim$ is on the left and the negation $\neg$ of **Le** on the right.

Modalities L and M and dual modalities W and V are defined on $\mathbf{2} \times \mathbf{2}$ via the following truth-tables:

x	Lx	Mx	Wx	Vx
1	0	1	a	1
a	-1	a	a	1
0	0	1	-1	0
-1	-1	a	-1	0

It is easy to check that the modalities L and M are interdefinable via classical negation and satisfy Łukasiewicz's conditions. We define Łukasiewicz's modal logic Ł in the language $\mathcal{L}^+ + \{L, \sim\}$ as the set of tautologies of the following algebra

$$\mathbf{4}^{\text{Ł}} := \langle \{-1, a, 0, 1\}, \vee, \wedge, \rightarrow, \sim, L, 1 \rangle.$$

Denote by Ł^+ its positive fragment in the language $\mathcal{L}^+ + \{L\}$. Note that the second modality M is definable through L in positive fragment Ł^+ as follows

$$Mp \leftrightarrow L(p \rightarrow p) \rightarrow p.$$

The pairs of modalities L, M and W, V are interdefinable, e.g., as follows

$$Wp \leftrightarrow p \wedge M({\sim}(p \rightarrow p)), \;\; Lp \leftrightarrow p \wedge V({\sim}(p \rightarrow p)).$$

If we use modalities W and V instead of L and M, respectively, we obtain the logic with the same set of tautologies. Taking into account the above identification of truth-values of logics Ł and **Le** we can easily see that the actions of modalities L, M, and V on the set of truth-values coincide, respectively with the mappings τ, δ, and ε, which are definable in the logic of classical refutability and which were used in Section 3.2 to construct isomorphs of classical (ε) and maximal negative logic (τ, δ) in **Le**. On the other hand, the modality W cannot be defined in **Le**, because algebra $\mathbf{4}'$ has a subalgebra with the universe $\{1, 0\}$, i.e., the set $\{1, 0\}$ is closed under all operators definable in **Le**.

Thus, we can see that there is a close interrelation between Ł and **Le**. J. Porte [84, 85] was the first to pay attention to the above-mentioned interrelation but he had Łukasiewicz's modal logic as a starting point for his considerations. More exactly, in [84], it was proved that the modalities of

Ł can be defined in a system obtained from classical logic by the addition of a constant, Ω in Porte's denotation, about which nothing is postulated. And conversely, that the modalities of Ł can be used to define the constant Ω. In [85], J. Porte observed that his constant Ω is close to the negation of logic of classical refutability and stated that **Le** is definition equivalent to the positive part of Ł.

Extend the language $\mathcal{L}^{\perp}$ of **Le** via a new negation symbol $\sim$, $\mathcal{L}^{\perp,\sim} := \mathcal{L}^{\perp} \cup \{\sim\}$, and define the logic $\mathbf{Le}^{\sim}$ as a logic in the language $\mathcal{L}^{\perp,\sim}$ with axiom schemes of **Le** and, additionally, axiom schemes of classical negation for $\sim$:

$1^{\sim}$. $(p \to q) \to ((p \to\sim q) \to\sim p)$

$2^{\sim}$. $\sim p \to (p \to q)$

$3^{\sim}$. $p\vee \sim p$

Define a translation $\theta : \mathcal{L}^{\perp,\sim} \to \mathcal{L}^{+} + \{L, \sim\}$ from the language of $\mathbf{Le}^{\sim}$ to the language of Łukasiewicz's modal logic and an inverse translation ρ from the language of Ł to the language of $\mathbf{Le}^{\sim}$ in such a way that they preserve propositional variables and all classical connectives and moreover,

$$\theta(\perp) = L(p \to p), \;\; \rho(L\varphi) = \rho\varphi \wedge \neg\rho\varphi,$$

where p is a propositional variable and φ is an arbitrary formula in the language $\mathcal{L}^{+} + \{L, \sim\}$.

Defining an algebra $\mathbf{4}^{\sim}$ as an expansion of $\mathbf{4}'$ via the classical negation, i.e.,

$$\mathbf{4}^{\sim} := \langle\{-1, a, 0, 1\}, \vee, \wedge, \to, \perp, \sim, 1\rangle,$$

we state the following facts.

Proposition 7.2.1 $\mathbf{4}^{\sim}$ *is a characteristic model for* $\mathbf{Le}^{\sim}$.

Proof. Obviously, $\mathbf{Le}^{\sim}$ has only two subdirectly irreducible models, $\mathbf{2}^{\sim}$ and $\mathbf{2}'^{\sim}$, which are expansions of Boolean algebra $\mathbf{2}$ and negative algebra $\mathbf{2}'$ via an additional Boolean negation $\sim$. Therefore, $\mathbf{Le}^{\sim} = L\mathbf{2}^{\sim} \cap L\mathbf{2}'^{\sim}$. Taking into account $\mathbf{4}^{\sim} \cong \mathbf{2}^{\sim} \times \mathbf{2}'^{\sim}$ we immediately arrive at $\mathbf{L4}^{\sim} = \mathbf{Le}^{\sim}$.

□

Proposition 7.2.2 $\mathbf{Le}^{\sim}$ *is a conservative extension of* **Le**.

Proof. If φ is a formula in the language $\mathcal{L}^{\perp}$ and $\mathbf{Le}^{\sim} \vdash \varphi$, then $\mathbf{4}^{\sim} \models \varphi$, and also $\mathbf{4}' \models \varphi$, because the connective $\sim$ does not occur in φ. Since $\mathbf{L4}' = \mathbf{Le}$, we have $\varphi \in \mathbf{Le}$, which completes the proof.

□

We now give in a slightly modified version of Porte's results [84, 85].

Proposition 7.2.3 *The logics* $\mathbf{Le}^{\sim}$ *and* Ł *are definition equivalent via* θ *and* ρ*, the logic* **Le** *is definition equivalent to the positive fragment of* Ł *via the same translations.*

The *proof* is by direct verification, because all logics involved in this statement are four-valued.

□

The only difference in the above translations from Porte's translations is the item $\rho(L\varphi) = \rho\varphi \wedge \neg\rho\varphi$. The original version was $\rho(L\varphi) = \rho\varphi \wedge \perp$, but, obviously, $\mathbf{Lj} \vdash (\varphi \wedge \neg\varphi) \leftrightarrow (\varphi \wedge \perp)$.

Proposition 7.2.3 means, first of all, that there is a close connection between Łukasiewicz's modalities and the paraconsistent negation of logic of classical refutability. From this fact Porte [85, pp. 87–88.] inferred a rather categorical conclusion that "the modalities of the Ł-system are very far from what everybody calls "possibility" and "necessity" and/or that the weak negation of CR (the logic of classical refutability) is very far from what everybody calls "negation"." As noted above, Łukasiewicz constructed his system so as to satisfy the minimal list of requirements for modal operators, which gives rise to a long history of critics of Łukasiewicz's modal logic *L*, but special discussion of this logic lies outside the scope of the present research. The resent work by Font and Hajek [32] may be recommended to become acquainted with the topic. Regarding the critics of the negation in the logic of classical refutability based on its similarity to modal operators, in recent years such similarity was not considered as something "negative" and was intensively studied. For example, K. Dosen in a series of works [23, 24, 25] treated the negation namely as a modal operator. In fact, the definition equivalence of Łukasiewicz's modal logic and logic of classical refutability means that the list of properties of modal operators given in the beginning of this section is not broad enough to distinguish the necessity operator and the operator of contradiction. The interrelation between **Le** and Ł stated in Proposition 7.2.3 will be used in the next section to suggest the way to generalize the notion of negation, which allows one to overcome the paradox of minimal logic.

7.3 Paradox of Minimal Logic and Generalized Absurdity

Recall that by the paradox of minimal logic we mean the following property of **Lj**: from any contradiction one can infer in **Lj** an arbitrary negative formula, i.e., for any φ and ψ, we have

$$\varphi, \ \neg\varphi \vdash_{\mathbf{Lj}} \neg\psi.$$

Due to this property, the negation makes no sense in inconsistent **Lj**-theories, because all negative formulas are provable in them. This property is conditioned, on the one hand, by the axiom for implication

$$p \to (q \to p)$$

(which is sometimes called "the positive implication paradox") and, on the other hand, by the unrestricted law *reductio ad absurdum*,

$$(p \to q) \to ((p \to \neg q) \to \neg p),$$

saying that if a formula implies a contradiction, one can negate this formula without any restriction on the nature of this contradiction. Indeed, let T be some inconsistent **Lj**-theory and $\varphi, \neg\varphi \in T$. Then for an arbitrary formula ψ, we can infer in T the implications $\psi \to \varphi$ and $\psi \to \neg\varphi$ using the positive paradox, and we then infer $\neg\psi$ applying *reductio ad absurdum.*

Of course, one can try to overcome the above mentioned paradox via rejecting the positive implication paradox and passing in this way into the field of relevant logic. But we choose another way, leave intuitionistic implication unchanged and consider possible ways of restricting *reductio ad absurdum.* It is worth noting that this idea has been exploited many times in investigations in the field of paraconsistency. For example, such well-known paraconsistent logics as Sette's P^1 [88, 102] and Da Costa's C_1 [19] have *reductio ad absurdum* restricted to complex formulas, in case of P^1, and to formulas "behaving consistently", in case of C_1.

Our approach is based on the correspondence between the necessity operator L of Łukasiewicz's modal logic and the contradiction operator C, $C(\varphi) := \varphi \wedge \neg\varphi$, in the logic of classical refutability, which was stated in the previous section. In **Le** as well as in an arbitrary extension of minimal logic, the negation can be defined via the constant "absurdity", $\neg\varphi := \varphi \to \bot$, but it can also be defined via the contradiction operator $C(\varphi)$, $\neg\varphi := \varphi \to C(\varphi)$. It is clear that

$$\mathbf{Lj} \vdash \neg\varphi \leftrightarrow (\varphi \to C(\varphi)).$$

This leads to the idea of defining a negation via the contradiction operator C considered as a primary logical connective, $\neg\varphi := \varphi \to C(\varphi)$. It will be shown below that the unrestricted *reductio ad absurdum* for negation defined as above exactly corresponds to some paradoxical properties of $C(\varphi)$ considered as a modal operator.

The results of Section 7.2 allow us to identify **Le** with a subsystem of Ł and after such identification, we have

$$Ł \vdash C(\varphi) \leftrightarrow L(\varphi) \text{ and } Ł \vdash \neg\varphi \leftrightarrow (\varphi \to L(\varphi)).$$

Notice that modal operators L and M have the following properties:

$$Ł \vdash L(\varphi \wedge \psi) \leftrightarrow L(\varphi) \wedge \psi \text{ and } Ł \vdash M(\varphi \vee \psi) \leftrightarrow M(\varphi) \vee \psi.$$

Each of these properties can be inferred from the other modulo classical logic and the relation defining M through L, and both of these properties have a paradoxical nature. Indeed, if we accept the conjunction of two conditions as necessary, it means from the intuitive point of view more than stating that one of these conditions is necessary and the second just takes place. In a similar way, assuming that it is possible that one of the two conditions takes place should be weaker than the alternative of one of the conditions and the possibility of the other. It is interesting that the numerous authors who criticized Łukasiewicz's modalities did not pay any attention to these paradoxes.

For any formula φ, we have

$$\varphi \leftrightarrow \varphi \wedge (p \to p) \text{ and } \varphi \leftrightarrow \varphi \vee \sim (p \to p),$$

from which we infer, using the above paradoxical properties,

$$Ł \vdash L(\varphi) \leftrightarrow \varphi \wedge L(p \to p) \text{ and } Ł \vdash M(\varphi) \leftrightarrow \varphi \vee M({\sim}(p \to p)).$$

The first of these equivalences corresponds to the relation $C(\varphi) \leftrightarrow \varphi \wedge \bot \leftrightarrow \varphi \wedge \neg(p \to p)$ for the contradiction operator in **Le**. Thus, the paradoxical properties envisaged above allow us to define the negation via the constant "absurdity".

Now we consider the language $\mathcal{L}^C = \langle \vee, \wedge, \to, C \rangle$, i.e., $\mathcal{L}^+$ extended by the contradiction operator C, and define a C-*logic* as a logic in this language containing axioms of **Lp** and the formula $C(p) \to p$. We say that C is *extensional* in a C-logic L if L is closed under the rule

$$\frac{\varphi \leftrightarrow \psi}{C(\varphi) \leftrightarrow C(\psi)}.$$

Lemma 7.3.1 *Let L be a C-logic and*

$$C(p \wedge q) \leftrightarrow C(p) \wedge q,\ \ C(p \wedge q) \leftrightarrow p \wedge C(q) \in L,$$

then C is extensional in L.

Proof. Let L satisfy the conditions of the lemma. Assume $\varphi \leftrightarrow \psi \in L$. Using the first equivalence and the axiom $C(p) \rightarrow p$ we have

$$C(\varphi \wedge \psi) \ \leftrightarrow\ C(\varphi) \wedge \psi \ \leftrightarrow\ C(\varphi) \wedge \varphi \ \leftrightarrow\ C(\varphi).$$

In a similar way, we infer from the equivalence $C(p \wedge q) \leftrightarrow p \wedge C(q)$ that $C(\varphi \wedge \psi) \leftrightarrow C(\psi)$, and, finally, $C(\varphi) \leftrightarrow C(\psi)$.

□

Proposition 7.3.2 *Let L be a C-logic and $\neg\varphi := \varphi \rightarrow\ C(\varphi)$. Then*

$$C(p) \leftrightarrow p \wedge \neg p \in L, \tag{7.1}$$

moreover,

$$(p \rightarrow q) \rightarrow ((p \rightarrow \neg q) \rightarrow \neg p) \in L$$

iff

$$C(p \wedge q) \leftrightarrow C(p) \wedge q,\ \ C(p \wedge q) \leftrightarrow p \wedge C(q) \in L.$$

Proof. From the well-known property $p \wedge (p \rightarrow q) \leftrightarrow p \wedge q$ of the intuitionistic implication and the axiom $C(p) \rightarrow p$, we immediately obtain the equivalence $C(p) \leftrightarrow p \wedge \neg p$, which justifies the name "contradiction operator" for C.

Assume that $(p \rightarrow q) \rightarrow ((p \rightarrow \neg q) \rightarrow \neg p) \in L$ and show that in this case the negation may be defined via a constant, $\neg\varphi \leftrightarrow \varphi \rightarrow \bot$, where $\bot := C(p_0 \rightarrow p_0)$ for some fixed variable p_0. Substituting the definition of negation in *reductio ad absurdum* and exporting the second premise we obtain

$$((p \rightarrow q) \wedge (p \rightarrow (q \rightarrow C(q)))) \rightarrow (p \rightarrow C(p))$$

Taking into account $\mathbf{Lp} \vdash p \rightarrow (q \wedge r) \leftrightarrow (p \rightarrow q) \wedge (p \rightarrow q)$ we arrive at

$$(p \rightarrow (q \wedge (q \rightarrow C(q))) \rightarrow (p \rightarrow C(p)).$$

Due to (7.1), the latter is equivalent to

$$(p \rightarrow C(q)) \rightarrow (p \rightarrow C(p)). \tag{7.2}$$

Substituting the tautology $p_0 \to p_0$ for q, we obtain

$$L \vdash (p \to C(p_0 \to p_0)) \to \neg p.$$

To state the inverse implication, we substitute $p_0 \to p_0$ for p and p for q in (7.2)

$$((p_0 \to p_0) \to C(p)) \to ((p_0 \to p_0) \to C(p_0 \to p_0)),$$

from which we have $C(p) \to C(p_0 \to p_0)$. Taking into account that the intuitionistic implication is increasing in the second argument, we arrive at

$$\neg p \to (p \to C(p_0 \to p_0))$$

and, finally, at

$$\neg p \leftrightarrow p \to \bot.$$

Further, from this fact and the equivalence $C(p) \leftrightarrow p \wedge \neg p$ we infer as in **Lj** that $L \vdash C(p) \leftrightarrow p \wedge \bot$. This equivalence implies in a trivial fashion the extensionality of C in L, moreover,

$$C(p \wedge q) \leftrightarrow (p \wedge q) \wedge \bot \leftrightarrow (p \wedge \bot) \wedge q \leftrightarrow C(p) \wedge q.$$

The second desired equivalence easily follows from the one just proved by extensionality.

We now assume that the equivalences $C(p \wedge q) \leftrightarrow C(p) \wedge q$ and $C(p \wedge q) \leftrightarrow p \wedge C(q)$ hold in L. By Lemma 7.3.1 the operator C will be extensional in L. Defining $\bot := C(p_0 \to p_0)$ and using the extensionality of C we obtain

$$C(p) \leftrightarrow C((p_0 \to p_0) \wedge p) \leftrightarrow C(p_0 \to p_0) \wedge p \leftrightarrow p \wedge \bot$$

and, further,

$$\neg p \leftrightarrow p \to C(p) \leftrightarrow p \to (p \wedge \bot) \leftrightarrow p \to \bot.$$

The equivalence $\neg p \leftrightarrow p \to \bot$ easily implies the unrestricted version of *reductio ad absurdum.*

□

As noted at the beginning of this section, the paradox of minimal logic is conditioned by the "positive implication paradox" and the unrestricted version of *reductio ad absurdum.* In the definition of C-logics we leave the intuitionistic implication unchanged, therefore, a C-logic meets the paradox of minimal logic if and only if the negation defined via its operator C satisfies the unrestricted version of *reductio ad absurdum.* In this way, we infer from the last proposition

Corollary 7.3.3 *A C-logic L satisfy for all formulas φ and ψ the condition*

$$\varphi,\ \neg\varphi \vdash_L \neg\psi$$

iff

$$C(p \wedge q) \leftrightarrow C(p) \wedge q,\ C(p \wedge q) \leftrightarrow p \wedge C(q) \in L.$$

□

As we can see from this corollary, deleting the paradoxical property of C allows one to avoid the paradox of minimal logic. In other words, if the operator C of contradiction has more natural properties from the viewpoint of modal logic, then the negation defined via C generates more a satisfiable inference relation from the viewpoint of paraconsistent logic.

It was noted above that in minimal logic the negation may be defined in two equivalent ways: via the constant $\bot$ and via the non-constant operator $C(\varphi)$ of contradiction. This simple observation leads naturally to the distinction between the contradiction operator and what we will call the *absurdity operator*, which we denote $A(\varphi)$. In case of **Lj**, the absurdity operator is constant ($A(\varphi) \leftrightarrow \bot$). There are different intuitions behind these operators. By $C(\varphi)$ we mean a contradiction expressed in terms of φ, i.e., a simultaneous stating of φ and its negation $\neg\varphi$, from which follows that $C(\varphi)$ should imply φ. By $A(\varphi)$ we mean such a statement that reducing φ to it, i.e., proving the implication $\varphi \to A(\varphi)$, is enough to negate φ. This understanding of $A(\varphi)$ does not assume that $A(\varphi)$ implies φ. The formulas $A(\varphi)$ and φ may be incomparable, as takes place in case of **Lj**, where the formulas φ and $\bot$ are incomparable in the general case. In fact, the contradiction operator C can be considered as a special case of an absurdity operator satisfying an additional assumption that $C(\varphi) \to \varphi$ for any φ. Moreover, if a negation of some system can be defined in terms of an absurdity operator, such a negation itself can be taken as an absurdity operator, which will produce the same negation. To present these considerations in a precise form, we define an A-language as the positive language $\mathcal{L}^+$ with an additional unary operator A, $\mathcal{L}^A := \mathcal{L}^+ \cap \{A\}$, and an A-*logic* as a logic in the language $\mathcal{L}^A$ containing axioms of positive logic **Lp**.

Proposition 7.3.4 *Let L be an A-logic. Define the contradiction operator C as $C(\varphi) := \varphi \wedge A(\varphi)$ and the negation $\neg$ as $\neg\varphi := \varphi \to A(\varphi)$. The following formulas are provable in L:*

$$\neg p \leftrightarrow p \to \neg p \quad \textit{and} \quad \neg p \leftrightarrow p \to C(p);$$
$$C(p) \leftrightarrow p \wedge \neg p \quad \textit{and} \quad C(p) \to p;$$
$$C(p) \to A(p) \quad \textit{and} \quad A(p) \to \neg p.$$

Proof. All the formulas listed in the proposition can easily be inferred from the given definitions and axioms of **Lp**. For example, the equivalence $\neg p \leftrightarrow p \to \neg p$ is an abbreviation for

$$p \to A(p) \leftrightarrow p \to (p \to A(p))$$

and the latter is just a particular case of the contraction law.

□

This proposition shows that C defined in terms of the absurdity operator A really can be considered as a contradiction operator corresponding to the negation defined via A and that there is the whole "interval" of operators, from C to $\neg$ defining the same negation.

7.4 *A*- and *C*-Presentations

7.4.1 Definitions and First Results

Now we want to define more exactly what a negation in one or another logic can be defined via an absurdity operator means. We restrict ourselves to the class of $\neg$-*logics*, which are sets of formulas in the language $\mathcal{L}^\neg$ containing axioms of positive logic and closed under the rules of substitution and *modus ponens*. Further, we need several translations between languages $\mathcal{L}^\neg$, $\mathcal{L}^C$, and $\mathcal{L}^A$.

Translations $\theta : \mathcal{L}^\neg \to \mathcal{L}^C$, $\rho : \mathcal{L}^C \to \mathcal{L}^\neg$, $\xi : \mathcal{L}^\neg \to \mathcal{L}^A$, and $\zeta : \mathcal{L}^C \to \mathcal{L}^A$ preserve propositional variables and all positive connectives and act on other connectives as follows:

$$\theta(\neg\varphi) := \theta\varphi \to C(\theta\varphi); \;\; \rho(C(\varphi)) := \rho\varphi \wedge \neg\rho\varphi;$$

$$\xi(\neg\varphi) := \xi\varphi \to A(\xi\varphi); \;\; \zeta(C(\varphi)) := \zeta\varphi \wedge A(\zeta\varphi).$$

Let L be a $\neg$-logic. An A-logic L' such that ξ defines a faithful embedding of L into L' and is called an *A-presentation* of L. A *C-presentation* of L is a C-logic L' such that θ faithfully embeds L into L'.

By an *exact A-(C-)presentation* of a $\neg$-logic L we mean its A-(C-) presentation, which is the least among all A-(C-)presentations of L.

A *strong C-presentation* of L is a C-logic L' definition equivalent to L via θ and ρ.

If L is a C-logic, then its *A-presentation* is an A-logic L' such that ζ faithfully embeds L into L', and its *exact A-presentation* is the least A-presentation.

If L is a C-logic, then its *¬-presentation* is a ¬-logic L' such that ρ faithfully embeds L into L', and its *exact ¬-presentation* is the least ¬-presentation.

First we state some simple properties of the notions introduced.

Proposition 7.4.1 *1. Let L be an A-(C-)logic. There exists a unique ¬-logic L' such that L is an A-(C-)presentation of L'.*

2. Let L be an A-logic. There exists a unique C-logic L' such that L is an A-presentation of L'.

Proof. 1. It is enough to consider the case of A-logics. We define by induction the subset of *¬-formulas* in the set of all $\mathcal{L}^A$-formulas:

1. a propositional variable is a ¬-formula;
2. if φ and ψ are ¬-formulas, then $\varphi \vee \psi$, $\varphi \wedge \psi$, and $\varphi \to \psi$ are ¬-formulas;
3. if φ is a ¬-formula, then $\varphi \to A(\varphi)$ is a ¬-formula.

For a ¬-formula φ, define a formula $\varphi^{\neg}$ as a result of replacing all subformulas of the form $\psi \to A(\psi)$ by subformulas $\neg\psi$. Note that $\varphi = \xi(\varphi^{\neg})$. Consider the intersection $\mathcal{S}$ of the logic L with the set of all ¬-formulas. It is clear that $\mathcal{S}$ is closed under the substitution of ¬-formulas and the restricted version of *modus ponens*: if $\varphi, \varphi \to \psi \in \mathcal{S}$ and ψ does not have the form $A(\varphi)$, then $\psi \in \mathcal{S}$. Therefore, the set

$$L' := \{\varphi^{\neg} \mid \varphi \in \mathcal{S}\} \tag{7.3}$$

is a ¬-logic. Check that L is an A-presentation of L'.

If $L' \vdash \varphi$, then $\varphi = \psi^{\neg}$ for a suitable formula $\psi \in \mathcal{S} \subseteq L$. Consequently, $\xi\varphi = \xi(\psi^{\neg}) = \psi \in L$. Conversely, if $L \vdash \xi\varphi$, then $\xi\varphi \in \mathcal{S}$ and $(\xi\varphi)^{\neg} = \varphi$, whence, $L' \vdash \varphi$.

The uniqueness of a ¬-logic L', for which L is an A-presentation, easily follows from the above considerations. Indeed, every such logic L' must have the form (7.3).

2. Define a set of *C-formulas* by replacing Item 3 in the definition of ¬-formulas by

3′. if φ is a C-formula, then $\varphi \wedge A(\varphi)$ is a C-formula.

The proof of this item is similar to the previous one. The desired C-logic has the form

$$L' := \{\varphi^C \mid \varphi \in L \text{ and } \varphi \text{ is a } C\text{-formula}\},$$

where φ^C is obtained from φ by replacing every subformula $\psi \wedge A(\psi)$ by $C(\psi)$.

□

In view of this statement Proposition 7.3.2 can be re-worded as

Corollary 7.4.2 *Let L be a C-logic. Then*

$$C(p \wedge q) \leftrightarrow C(p) \wedge q,\ C(p \wedge q) \leftrightarrow p \wedge C(q) \in L$$

iff L is a C-presentation for some extension of **Lj**.

□

Proposition 7.4.3 *1. If L is a $\neg$-logic and L has an A-(C-)presentation, then L has also an exact A-(C-)presentation.*

2. If L is a C-logic having an A-presentation, it also has an exact A-presentation.

Proof. This statement follows from the next simple lemma.

Lemma 7.4.4 *Let $\mathcal{L}_0$ and $\mathcal{L}_1$ be propositional languages and $\tau : \mathcal{L}_0 \to \mathcal{L}_1$ a translation. Further, let L be a logic in the language $\mathcal{L}_0$ and $\{L_i \mid i \in I\}$ a family of logics in the language $\mathcal{L}_1$ such that τ faithfully embeds L into L_i for all $i \in I$. Then τ faithfully embeds L into $\bigcap_{i\in I} L_i$.*

In this way, if the set of A-(C-)presentations of some logic is non-empty, the intersection of all such presentations gives an exact A-(C-)presentation of that logic.

□

To prove further results we impose the restriction of extensionality on the considered logics. Define the extensionality of negation and absurdity operators in the same way as for the operator C above. Let L be a $\neg$-(A-) logic. We say that the negation $\neg$ (respectively, the absurdity operator A) is *extensional* in L if the logic L is closed under the rule

$$\frac{\varphi \leftrightarrow \psi}{\neg\varphi \leftrightarrow \neg\psi}\ \text{(respectively } \frac{\varphi \leftrightarrow \psi}{A(\varphi) \leftrightarrow A(\psi)}\text{)}.$$

If L is a $\neg$-(C-, A-)logic and the operator $\neg$ (C, A) is extensional in L, we say that L is an *extensional* $\neg$-(C-, A-) logic.

Remark. Note that in all logics under consideration the positive connectives $\vee$, $\wedge$, and $\rightarrow$ are extensional in a natural sense, which easily follows from the fact that $\neg$-(C-, and A-) logics was defined so as to satisfy all axioms of the positive logic **Lp**.

Proposition 7.4.5 *If a $\neg$-logic L has an A-presentation, then we have*

$$L \vdash \neg p \leftrightarrow p \rightarrow \neg p.$$

Conversely, if the equivalence $\neg p \leftrightarrow p \rightarrow \neg p$ can be proved in a $\neg$-logic L and additionally, L is an extensional $\neg$-logic, then L has an A-presentation.

Proof. Let a $\neg$-logic L has an A-presentation L'. By the contraction law for $\rightarrow$, we have

$$L' \vdash (p \rightarrow A(p)) \leftrightarrow (p \rightarrow (p \rightarrow A(p))).$$

The last equivalence graphically equals

$$\theta(\neg p) \leftrightarrow \theta(p \rightarrow \neg p) = \theta(\neg p \leftrightarrow p \rightarrow \neg p).$$

By assumption θ is a faithful embedding of L into L', consequently,

$$L \vdash \neg p \leftrightarrow p \rightarrow \neg p.$$

We now assume that L is an extensional $\neg$-logic and $L \vdash \neg p \leftrightarrow p \rightarrow \neg p$.

For a formula φ in the language $\mathcal{L}^{\neg}$, define a formula φ^A as a result of replacing every subformula of the form $\neg\psi$ by $A(\psi)$. Put

$$L^A := \{\varphi^A \mid \varphi \in L\}.$$

It is clear that L^A is an A-logic and the extensionality of $\neg$ in L immediately implies that A is extensional in L^A. We check that ξ is a faithful embedding of L into L^A. First of all, we note that for any formula φ,

$$L \vdash \varphi \text{ iff } L^A \vdash \varphi^A.$$

Therefore, it is enough to prove that $L^A \vdash \varphi^A \leftrightarrow \xi(\varphi)$ for any φ. We use the induction on the structure of formulas.

This statement trivially holds for propositional variables. The case of positive connectives easily follows by extensionality. It remains to consider the case $\varphi = \neg\psi$. By the induction hypothesis we have $\psi^A \leftrightarrow \xi(\psi)$ in L^A.

This fact and the extensionality of A imply $L^A \vdash A(\psi^A) \leftrightarrow A(\xi(\psi))$. On the other hand, we have $A(\xi\psi) \leftrightarrow (\xi(\psi) \rightarrow A(\xi(\psi)))$ since the equivalence $\neg p \leftrightarrow (p \rightarrow \neg p)$ holds in L by assumption. The last two facts together imply

$$L^A \vdash A(\psi^A) \leftrightarrow (\xi(\psi) \rightarrow A(\xi(\psi))),$$

i.e., $L^A \vdash (\neg\psi)^A \leftrightarrow \xi(\neg\psi)$.

□

Proposition 7.4.6 *1. If an extensional $\neg$-logic L has an A-presentation, it has also a C-presentation.*

2. If an extensional $\neg$-logic L has a C-presentation, this C-presentation is unique and it is also an exact and strong C-presentation.

Proof. 1. The set $\mathcal{C}$ of formulas in the language $\mathcal{L}^\neg$ we define as the least set of formulas satisfying the following conditions:

1) propositional variables belong to $\mathcal{C}$;

2) if $\varphi \in \mathcal{C}$ and $\psi \in \mathcal{C}$, then the formulas $\varphi \vee \psi$, $\varphi \wedge \psi$, and $\varphi \rightarrow \psi$ are in $\mathcal{C}$;

3) if $\varphi \in \mathcal{C}$, then $\varphi \wedge \neg\varphi \in \mathcal{C}$.

Let $L^C := L \cap \mathcal{C}$. The set L^C is closed under the rules of *modus ponens* and substitution of formulas from $\mathcal{C}$, moreover, it contains all formulas of the form $(\varphi \wedge \neg\varphi) \rightarrow \varphi$ for φ in $\mathcal{C}$. Therefore, the set

$$L' := \{\varphi^C \mid \varphi \in L^C\},$$

where φ^C is obtained from φ by substituting a subformula $C(\psi)$ for any subformula of the form $\psi \wedge \neg\psi$, forms a C-logic.

Check that L' is the desired C-presentation of L. First we note that the equivalence

$$L \vdash \varphi \text{ iff } L' \vdash \varphi^C$$

holds for any $\varphi \in \mathcal{C}$. The logic L has an A-presentation by assumption, whence, by Proposition 7.4.5 we have in L the equivalences

$$\neg p \leftrightarrow (p \rightarrow \neg p) \leftrightarrow (p \rightarrow (p \wedge \neg p)).$$

Therefore, in L any formula φ is equivalent to a formula φ' obtained by replacing every subformula $\neg\psi$ of φ by $\psi \rightarrow (\psi \wedge \neg\psi)$. Note that φ' belongs to $\mathcal{C}$, moreover, $\theta(\varphi) = \varphi'^{C}$. In this way, we arrive at the following chain of equivalences

$$L \vdash \varphi \iff L \vdash \varphi' \iff L' \vdash \varphi'^{C} \iff L' \vdash \theta\varphi,$$

which completes the proof of this item.

2. It follows easily from the considerations of the previous item that any C-presentation of an extensional $\neg$-logic L coincides with the set $\{\varphi^C \mid \varphi \in L \cap \mathcal{C}\}$. Therefore, the logic L may have at most one C-presentation. The exact C-presentation equals the intersection of all C-presentations, it therefore coincides with the unique C-presentation of L.

It remains to prove that a C-presentation L' of L is strong, i.e., that ρ is a faithful embedding of L' into L and that the translations θ and ρ are mutually inverse. It is clear that for any formula φ in the language $\mathcal{L}^C$, the formula $\rho(\varphi)$ belongs to $\mathcal{C}$ and, conversely, any formula in $\mathcal{C}$ has the form $\rho(\varphi)$. Therefore, the above presentation of L' immediately implies that ρ is a faithful embedding.

Let φ be a formula in the language $\mathcal{L}^{\neg}$. The equivalence $\varphi \leftrightarrow \rho\theta(\varphi)$ follows from the fact $\rho\theta(\varphi)$ graphically equals the formula φ' defined in the previous item. For a formula φ in the language $\mathcal{L}^C$, the equivalence $\varphi \leftrightarrow \theta\rho(\varphi)$ easily follows from the extensionality of L' and the equivalence $C(p) \leftrightarrow (p \wedge (p \rightarrow C(p)))$ which holds in any C-logic.

□

Proposition 7.4.7 *Let L be an extensional $\neg$-logic, L_1 an extensional A-presentation of L, and L_2 a C-presentation of L. Then L_1 is an A-presentation of L_2.*

Proof. 1. By the previous proposition L_2 is a strong C-presentation of L. Therefore, we have for any formula φ,

$$L_2 \vdash \varphi \iff L \vdash \rho\varphi \iff L_1 \vdash \xi\rho\varphi.$$

Moreover, $\xi\rho C(\varphi) = \xi(\rho\varphi \wedge \neg\rho\varphi) = \xi\rho\varphi \wedge (\xi\rho\varphi \rightarrow A(\xi\rho\varphi))$. The latter formula is equivalent in L_1 to $\xi\rho\varphi \wedge A(\xi\rho\varphi)$. From this fact and the extensionality of L_1, we obtain by induction

$$L_1 \vdash \xi\rho\varphi \leftrightarrow \zeta\varphi$$

for any φ.

□

The above results concerning the existence of A- and C-presentations, uniqueness of C-presentations, etc., used essentially the extensionality property. Most of the well-known systems of paraconsistent logic are not extensional and need, therefore, special treatment. In the next sections, we take two basic systems of paraconsistent logic and consider the problem of presenting the negation via absurdity and contradiction operators for these systems.

7.4.2 Logic CLuN

All logics considered in this and the following subsections have the same positive fragment, namely, $\mathbf{Lk}^{+}$, and admit semantic characterization in terms of $\{0,1\}$-interpretations non-truth-functional for operators of negation, absurdity or contradiction. The latter means that the value of an interpretation ν on a formula of the form $\neg\varphi$ ($A(\varphi)$, $C(\varphi)$) is not determined uniquely by the value $\nu(\varphi)$.

For the language $\mathcal{L}^{\diamond}$, where $\diamond \in \{\neg, A, C\}$, we define an *interpretation* ν as a mapping from the set of $\mathcal{L}^{\diamond}$-formulas to the set $\{0,1\}$ satisfying for arbitrary φ and ψ the following properties:

1^{ν}) $\nu(\varphi \wedge \psi) = 1$ iff $\nu(\varphi) = 1$ and $\nu(\psi) = 1$;

2^{ν}) $\nu(\varphi \vee \psi) = 1$ iff $\nu(\varphi) = 1$ or $\nu(\psi) = 1$;

3^{ν}. $\nu(\varphi \rightarrow \psi) = 1$ iff $\nu(\varphi) = 0$ or $\nu(\psi) = 1$.

The logic **CLuN** was introduced in [6] (under the name PI). The name **CLuN** we can understand as classical logic admitting gl*u*ts (see also [7]). It is a fairly weak paraconsistent logics, which is in a sense basic for the class of logics considered by D. Batens in [6]. The logic **CLuN** can be axiomatized modulo classical positive logic via the law of excluded middle, i.e., we have

$$\mathbf{CLuN} = \mathbf{Lk}^{+\neg} + \{p \vee \neg p\},$$

where $\mathbf{Lk}^{+\neg}$ is an expansion of $\mathbf{Lk}^{+}$ to the language $\mathcal{L}^{\neg}$.

We denote by $\mathcal{U}$ the class of interpretations ν satisfying for any φ the property:

U. If $\nu(\varphi) = 0$, then $\nu(\neg\varphi) = 1$.

Proposition 7.4.8 [6, p.205] *For any formula* φ*,* **CLuN** $\vdash \varphi$ *iff* $\nu(\varphi) = 1$ *for all* $\nu \in \mathcal{U}$*.*

We start to consider A- and C-presentations of **CLuN** with the following observation.

Proposition 7.4.9 *The logic* **CLuN** *has no strong* C*-presentation.*

Proof. If **CLuN** has a strong C-presentation, the equivalence $\varphi \leftrightarrow \rho\theta\varphi$ holds in **CLuN** for any φ. Take a formula $\neg\neg p$ and compute

$$\rho\theta(\neg\neg p) = (p \to (p \wedge \neg p)) \to ((p \to (p \wedge \neg p)) \wedge \neg(p \to (p \wedge \neg p))) \leftrightarrow$$

$$\leftrightarrow (p \to \neg p) \to \neg(p \to (p \wedge \neg p)).$$

Let ν be an interpretation from $\mathcal{U}$ such that $\nu(p) = 0$, $\nu(\neg p) = \nu(\neg\neg p) = 1$, and $\nu(\neg(p \to (p \wedge \neg p))) = 0$. One can find out such interpretation in $\mathcal{U}$, because $\nu(p) = 0$ implies $\nu(p \to (p \wedge \neg p)) = 1$. As we can see, $\nu(\rho\theta(\neg\neg p)) = 0$, and so $\nu(\neg\neg p \leftrightarrow \rho\theta(\neg\neg p)) = 0$. By the previous proposition we have

$$\mathbf{CLuN} \nvdash \neg\neg p \leftrightarrow \rho\theta(\neg\neg p).$$

□

However, one can find exact A- and C-presentations of **CLuN**.

Define an A-logic $\mathbf{CLuN}^A$ as an expansion of $\mathbf{Lk}^+$ to the language $\mathcal{L}^A$, and let $\mathcal{U}^A$ be the set of all interpretations of the language $\mathcal{L}^A$. We have the following completeness theorem.

Proposition 7.4.10 *For any* φ*,* $\mathbf{CLuN}^A \vdash \varphi$ *if and only if* $\nu(\varphi) = 1$ *for all* $\nu \in \mathcal{U}^A$*.*

Proof. The soundness can be checked directly and to prove the completeness we use an easy modification of Henkin's construction. As usual, define a *prime* $\mathbf{CLuN}^A$*-theory* as a set of formulas T, which contains $\mathbf{CLuN}^A$, is closed under *modus ponens* and has the disjunction property, i.e.,

$$\varphi \vee \psi \in T \text{ implies } \varphi \in T \text{ or } \psi \in T.$$

In a standard way, one can prove the extension lemma.

Lemma 7.4.11 *Let a set of formulas* S *and a formula* φ *be such that* $S \nvdash_{\mathbf{CLuN}^A} \varphi$*. There is then a prime* $\mathbf{CLuN}^A$*-theory* T *such that* $S \subseteq T$ *and* $\varphi \notin T$*.*

□

Let $\mathbf{CLuN}^A \nvdash \varphi_0$ and T be a prime $\mathbf{CLuN}^A$-theory such that $\varphi_0 \notin T$. Define an interpretation ν_T on propositional variables and formulas of the form $A(\varphi)$ as follows:

$$\nu_T(p) = 1 \iff p \in T, \quad \nu_T(A(\varphi)) = 1 \iff A(\varphi) \in T.$$

By definition of interpretation, the mapping ν_T extends uniquely to the set of all formulas. For any prime $\mathbf{CLuN}^A$-theory, the following equivalences hold:

$$\begin{array}{lcl} \varphi \vee \psi \in T & \Leftrightarrow & \varphi \in T \text{ or } \psi \in T; \\ \varphi \wedge \psi \in T & \Leftrightarrow & \varphi \in T \text{ and } \psi \in T; \\ \varphi \to \psi \in T & \Leftrightarrow & \varphi \notin T \text{ or } \psi \in T. \end{array}$$

The first equivalence follows from the disjunction property, the second is obvious, check the last one. If $\varphi \to \psi \in T$ and $\varphi \in T$, then $\psi \in T$ by *modus ponens*. Conversely, if $\varphi \notin T$, then $\varphi \to \psi \in T$ in view of $\varphi \vee (\varphi \to \psi) \in \mathbf{CLuN}^A$. If $\psi \in T$, then $\varphi \to \psi \in T$ by the positive axiom $\psi \to (\varphi \to \psi)$.

From the definition of ν_T and the above equivalences we obtain

$$\nu_T(\varphi) = 1 \Leftrightarrow \varphi \in T,$$

consequently, $\nu_T(\varphi_0) = 0$.

□

Proposition 7.4.12 $\mathbf{CLuN}^A$ *is an exact A-presentation of* $\mathbf{CLuN}$.

Proof. First we check that for any φ, if $\mathbf{CLuN} \vdash \varphi$, then $\mathbf{CLuN}^A \vdash \xi\varphi$. The logics $\mathbf{CLuN}$ and $\mathbf{CLuN}^A$ have the same positive fragment, therefore, it is enough to check that the formula $\xi(p \vee \neg p) = p \vee (p \to A(p))$ is provable in $\mathbf{CLuN}^A$, which is obviously true.

Conversely, assume that for some φ_0 we have $\mathbf{CLuN} \nvdash \varphi_0$. In this case, $\nu(\varphi_0) = 0$ for suitable $\nu \in \mathcal{U}$. Consider an interpretation ν' of the language $\mathcal{L}^A$ satisfying the following conditions: $\nu'(p) = \nu(p)$ for any propositional variable p; $\nu'(A(\xi\psi)) = \nu(\psi \wedge \neg\psi)$ for any ψ in the language $L^{\neg}$. Obviously, we can construct such an interpretation and it belongs to the class $\mathcal{U}^A$, which contains all interpretations of the language $\mathcal{L}^A$.

Prove by induction on the structure of formulas that $\nu(\psi) = \nu'(\xi\psi)$ for all ψ. This is obviously true for propositional variables. The case of positive connectives is trivial, so it remains to consider the case of negation. Assume that $\nu(\psi) = \nu'(\xi\psi)$ and calculate

$$\nu'(\xi(\neg\psi)) = \nu'(\xi\psi \rightarrow A(\xi\psi)) = \nu'(\xi\psi) \rightarrow \nu'(A(\xi\psi)) =$$

$$\nu\psi \rightarrow \nu(\psi \wedge \neg\psi) = \nu(\psi \rightarrow (\psi \wedge \neg\psi)) = \nu(\neg\psi).$$

The last equation is due to the fact that $\mathbf{CLuN} \vdash \neg p \leftrightarrow p \rightarrow (p \wedge \neg p)$. According to the proven equality $\nu'(\xi\varphi_0)) = \nu(\varphi_0) = 0$, i.e., the formula $\xi\varphi_0$ is not provable in $\mathbf{CLuN}^A$. We have thus proved that $\mathbf{CLuN}^A$ is an A-presentation of **CLuN**. The exactness of this presentation immediately follows from the fact that $\mathbf{CLuN}^A$ is the least A-logic with classical positive fragment. Obviously, any A-presentation of **CLuN** must have $\mathbf{Lk}^+$ as its positive fragment.

□

Similarly to the case of A-presentation we can take the least C-logic with classical positive fragment as a C-presentation of **CLuN**. Put

$$\mathbf{CLuN}^C := \mathbf{Lk}^{+C} + \{C(p) \rightarrow p\},$$

where $\mathbf{Lk}^{+C}$ is an expansion of $\mathbf{Lk}^+$ to the language $\mathcal{L}^C$. We denote by $\mathcal{U}^C$ the class of all interpretations ν of the language $\mathcal{L}^C$, satisfying for any φ the condition:

U^C. If $\nu(\varphi) = 0$, then $\nu(C(\varphi)) = 0$.

Proposition 7.4.13 *For any* φ, $\mathbf{CLuN}^C \vdash \varphi$ *if and only if* $\nu(\varphi) = 1$ *for all* $\nu \in \mathcal{U}^C$.

Proof is analogous to Proposition 7.4.10.

□

Proposition 7.4.14 $\mathbf{CLuN}^C$ *is an exact* C*-presentation of* **CLuN**.

Proof of this statement can be obtained, in fact, by replacing A by C and ξ by θ in the proof of Proposition 7.4.12.

□

Due to Proposition 7.4.9 the logic $\mathbf{CLuN}^C$ cannot be a strong C-presentation, but we can prove that the inverse translation also embeds faithfully $\mathbf{CLuN}^C$ into **CLuN**.

Proposition 7.4.15 **CLuN** *is an exact* $\neg$*-presentation of* $\mathbf{CLuN}^C$.

Proof. The fact that $\mathbf{CLuN}^C \vdash \varphi$ implies $\mathbf{CLuN} \vdash \rho\varphi$ for any φ can be checked directly.

Assume that for some formula φ_0 we have $\mathbf{CLuN}^C \nvdash \varphi_0$ and consider an interpretation $\nu \in \mathcal{U}^C$ with $\nu(\varphi_0) = 0$. In $\mathcal{U}$, one can find an interpretation ν' of the language $\mathcal{L}^\neg$ such that $\nu'(p) = \nu(p)$ for all propositional variables p and $\nu'(\neg\rho\psi) = \nu(\psi \to C\psi)$ for any ψ. Using induction on the structure of formulas, we check that $\nu'\rho\psi = \nu\psi$ for any ψ. Clearly, it is enough to consider the case of C-operation. Let $\nu'\rho\psi = \nu\psi$ for some ψ. Then

$$\nu'(\rho C\psi) = \nu'(\rho\psi \wedge \neg\rho\psi) = \nu'\rho\psi \wedge \nu'(\neg\rho\psi) = \nu\psi \wedge \nu(\psi \to C\psi) = \nu(C\psi).$$

The desired equality is proved, in particular, we have $\nu'(\rho\varphi_0) = 0$. Consequently, **CLuN** is a $\neg$-presentation of $\mathbf{CLuN}^C$.

Further, any $\neg$-presentation of $\mathbf{CLuN}^C$ has positive fragment $\mathbf{Lk}^+$ and contains the formula $p \vee \neg p$, because this formula is equivalent modulo $\mathbf{Lk}^+$ to $p \vee (p \to (p \wedge \neg p)) = \rho(p \vee (p \to Cp))$. Therefore, **CLuN** is the least C-presentation of $\mathbf{CLuN}^C$ and so it is exact.

□

7.4.3 Sette's Logic P^1

We now turn to the first example of maximal paraconsistent logic, namely, to the system P^1 suggested by A. Sette in [102]. This is a sublogic of classical logic maximal in a sense that adjoining to P^1 any new classical tautology yields classical logic. The negation of P^1 is not extensional, however, as we will see, P^1 has even a strong C-presentation. Originally P^1 was defined as a three-valued logic with the set of truth-values $\{T^0, T^1, F\}$, where T^0 and F are, in fact, the classical truth and falsehood, and T^1 is in a sense a second truth, whose negation is also true. Both truths T^0 and T^1 are distinguished values. The matrix for P^1 is $\langle\{T^0, T^1, F\}, \to, \neg, \{T^0, T^1\}\rangle$, where operations $\to$ and $\neg$ are defined via the following truth tables:

$\to$	T^0	T^1	F
T^0	T^0	T^0	F
T^1	T^0	T^0	F
F	T^0	T^0	T^0

$\neg$	
T^0	F
T^1	T^0
F	T^0

The connectives $\vee$ and $\wedge$ are introduced via the definitions:

$$\varphi \vee \psi := (\varphi \to \neg\neg\varphi) \to (\neg\varphi \to \psi),$$

$$\varphi \wedge \psi := (((\varphi \to \varphi) \to \varphi) \to \neg((\psi \to \psi) \to \psi)) \to \neg(\varphi \to \neg\psi).$$

A. Sette [102] gives the following axiomatics for P^1:

1. $p \to (q \to p)$
2. $(p \to (q \to r)) \to ((p \to q) \to (p \to r))$
3. $(\neg p \to \neg q) \to ((\neg p \to \neg\neg q) \to p))$
4. $(p \to q) \to \neg\neg(p \to q)$

The only inference rules are, as usual, substitution and *modus ponens*. It is clear that P^1 is a sublogic of **Lk**, moreover, the following theorem holds.

Theorem 7.4.16 [102] *For any* $\varphi \in \mathbf{Lk}$, *if* φ *is not provable in* P^1, *then* $P^1 + \{\varphi\} = \mathbf{Lk}$.

□

In [2], it was stated that P^1 is equivalent to the system β_2 introduced in [50]. Axiomatics of β_2 includes traditional axioms of classical positive logic and the following axiom schema for negation:

1^β. $(\neg\varphi \to \psi) \to ((\neg\varphi \to \neg\psi) \to \varphi)$, where ψ is molecular, i.e., different from a propositional variable.

From β_2 we can easily pass to an axiomatization of P^1 convenient for our goals.

Proposition 7.4.17 *The logic* P^1 *is axiomatizable modulo the classical positive logic via the following axiom schemes:*

1^S) $\neg\neg\varphi \to \varphi$;

2^S) $(\varphi \to \psi) \to ((\varphi \to \neg\psi) \to \neg\varphi)$, *where* ψ *is molecular.*

Proof. On the one hand, we can use truth-tables to check that schemes 1^S and 2^S hold in P^1. On the other, the schema 1^β easily follows from $(\neg\varphi \to \psi) \to ((\neg\varphi \to \neg\psi) \to \neg\neg\varphi)$, a substitution case of 2^S, and 1^S.

□

Using the maximal property we can easily obtain a semantic characterization of P^1 in terms of $\{0,1\}$-interpretations. Denote by $\mathcal{P}$ the class of all interpretations ν of the language $\mathcal{L}^\neg$ satisfying the properties:

1^P. If $\nu\varphi = 0$, then $\nu\neg\varphi = 1$.

2^P. If φ is molecular and $\nu\varphi = 1$, then $\nu\neg\varphi = 0$.

Proposition 7.4.18 *For any formula φ, $P^1 \vdash \varphi$ if and only if $\nu\varphi = 1$ for all $\nu \in \mathcal{P}$.*

Proof. Put $L^* := \{\varphi \mid \nu\varphi = 1 \text{ for all } \nu \in \mathcal{P}\}$. We have $P^1 \subseteq L^* \subseteq \mathbf{Lk}$. Indeed, the former inclusion can be verified directly, the latter follows from the fact that all classical interpretations belong to $\mathcal{P}$. Due to Theorem 7.4.16 one of these inclusions is not proper. In $\mathcal{P}$, there is an interpretation ν such that $\nu p = \nu\neg p = 1$ and $\nu\neg\neg p = 0$, which means that $p \to \neg\neg p \notin L^*$, i.e., $L^* \neq \mathbf{Lk}$. Therefore, $L^* = P^1$.

□

Having convenient axiomatics and semantics for P^1, we may turn to its presentations.

Lemma 7.4.19 *If L is an A-presentation of P^1, then L contains the following schemes:*

1^A. $A(\varphi \to A\varphi) \to \varphi$;

2^A. $(\psi \wedge A\psi) \to \varphi$, *where ψ is molecular.*

Proof. Consider a ξ-translation of 2^S,

$$(\varphi \to \psi) \to ((\varphi \to (\psi \to A\psi)) \to (\varphi \to A\varphi).$$

Modulo $\mathbf{Lk}^+$ it is equivalent to

$$(\varphi \to (\psi \wedge A\psi)) \to (\varphi \to A\varphi) \leftrightarrow$$

$$\leftrightarrow (\varphi \wedge (\varphi \to (\psi \wedge A\psi))) \to A\varphi \leftrightarrow (\varphi \wedge (\psi \wedge A\psi)) \to A\varphi \leftrightarrow$$
$$\leftrightarrow (\psi \wedge A\psi) \to (\varphi \to A\varphi).$$

In particular, we have

$$L \vdash (\psi \wedge A\psi) \to ((\varphi \to A\varphi) \to A(\varphi \to A\varphi)).$$

At the same time, the formula

$$((\varphi \to A\varphi) \to A(\varphi \to A\varphi)) \to \varphi, \tag{7.4}$$

a ξ-translation of $\neg\neg\varphi \to \varphi$, is provable in L. From the last two formulas we immediately have

$$L \vdash (\psi \wedge A\psi) \to \varphi,$$

where ψ is molecular.

Further, the positive axiom

$$A(\varphi \to A\varphi) \to ((\varphi \to A\varphi) \to A(\varphi \to A\varphi))$$

together with (7.4) yields the scheme 1^A.

□

Denote by P^A the A-logic axiomatized modulo $\mathbf{Lk}^+$ by schemes 1^A and 2^A and by $\mathcal{P}^A$ the class of interpretations ν of the language $\mathcal{L}^A$ satisfying the condition

$1^{A\nu}$. If φ is molecular and $\nu\varphi = 1$, then $\nu(A\varphi) = 0$.

Lemma 7.4.20 *If $P^A \vdash \varphi$, then $\nu\varphi = 1$ for all $\nu \in \mathcal{P}^A$.*

The *proof* is by direct verification.

□

Proposition 7.4.21 *The logic P^A is an exact A-presentation of P^1.*

Proof. Let $L^* := \{\varphi \mid \xi\varphi \in P^A\}$. As it was stated in the proof of Lemma 7.4.19, the ξ-translation of 2^S is equivalent in P^A to a particular case of 2^A. The formula $\xi(\neg\neg\varphi \to \varphi)$ has the following quasi-proof in P^A:

1) $A(\varphi \to A\varphi) \to \varphi$, an axiom;
2) $((\varphi \to A\varphi) \to A(\varphi \to A\varphi)) \to ((\varphi \to A\varphi) \to \varphi)$, by monotonicity of $\to$ in the second argument from 1);
3) $((\varphi \to A\varphi) \to \varphi) \to \varphi$, the Peirce law;
4) $((\varphi \to A\varphi) \to A(\varphi \to A\varphi)) \to \varphi$ $(= \xi(\neg\neg\varphi \to \varphi))$, by transitivity from 2) and 3).

We have thus proved the inclusion $P^1 \subseteq L^*$.

Let $P^A \vdash \xi\varphi_0$ and ν is an arbitrary classical interpretation. Define an interpretation ν' of $\mathcal{L}^A$ by the conditions $\nu'p = \nu p$ for any p and $\nu'(A\varphi) = |1 - \nu\varphi|$. It is not hard to see that $\nu' \in \mathcal{P}^A$ and $\nu'\xi\varphi = \nu\varphi$ for any φ. By Lemma 7.4.20, $\nu\varphi_0 = 1$. And we have in this way $L^* \subseteq \mathbf{Lk}$. This inclusion is proper, because the formula $\xi(p \to \neg\neg p)$ is not in P^A. The last fact can be easily checked using Lemma 7.4.20. Only one of the inclusions $P^1 \subseteq L^* \subseteq \mathbf{Lk}$ is proper, whence $P^1 = L^*$.

We have thus proved that P^A is an A-presentation of P^1, the exactness immediately follows from Lemma 7.4.19.

□

We now turn to C-presentations of P^1.

Lemma 7.4.22 *If L is a C-presentation of P^1, then it contains the schema:*

1^C) $C\psi \to \varphi$, *where ψ is molecular.*

Proof. By Lemma 7.4.19 $L \vdash (\psi \wedge C\psi) \to \varphi$ for molecular ψ, which is equivalent to 1^C in view of the axiom $Cp \to p$ of C-logics.

□

Let P^C be a C-logic axiomatized modulo $\mathbf{Lk}^+$ by the axiom scheme 1^C. We denote by $\mathcal{P}^C$ the class of all interpretations ν of $\mathcal{L}^C$ satisfying the conditions:

$1^{C\nu}$. If $\nu\varphi = 0$, then $\nu C\varphi = 0$.

$2^{C\nu}$. If φ is molecular and $\nu\varphi = 1$, then $\nu C\varphi = 0$.

Lemma 7.4.23 *If $P^C \vdash \varphi$, then $\nu\varphi = 1$ for all $\nu \in \mathcal{P}^C$.*

□

Proposition 7.4.24 *The logic P^C is an exact and strong C-presentation of P^1.*

Proof. The fact that P^C is an exact presentation can be proved similarly to Proposition 7.4.21. We can directly check that P^1 is a $\neg$-presentation of P^C. It remains to prove that any formula φ is equivalent to $\rho\theta\varphi$ in P^1 and that ψ is equivalent to $\theta\rho\psi$ in P^C. This can be easily stated using induction on the structure of formulas and the facts that $P^1 \vdash (\varphi \leftrightarrow \psi) \to (\neg\varphi \leftrightarrow \neg\psi)$ for molecular φ and ψ and that $P^C \vdash (\varphi \leftrightarrow \psi) \to (C(\varphi) \leftrightarrow C(\psi))$ for molecular φ and ψ.

□

Remark. Note that P^1 is not an exact $\neg$-presentation of P^C. The least $\neg$-presentation of P^C can be defined via the class of interpretations ν satisfying the condition $\nu\varphi = 1 \;\Rightarrow\; \nu\neg\varphi = 0$ for molecular φ. And this logic has no C-presentation.

Part II

Strong Negation

Chapter 8

Semantical Study of Paraconsistent Nelson's Logic[1]

This chapter starts the second part of the book devoted to paraconsistent Nelson's logic **N4** and to the class of its extensions.

The natural first step in the investigation of the class of **N4**-extensions is to provide an adequate algebraic semantics for the logic **N4**, i.e., characterizing **N4** via a variety of algebraic systems $\mathcal{V}$ such that there exists a natural dual isomorphism between the lattice of **N4**-extensions and the lattice of subvarieties of $\mathcal{V}$.

We consider two variants of Nelson's paraconsistent logic. The logic **N4** is determined in the language $\langle \vee, \wedge, \to, \sim \rangle$ and its positive fragment coincides with positive logic **Lp**. The logic, which we denote $\mathbf{N4}^{\perp}$, is a conservative extension of **N4** in the language with an additional symbol $\perp$ for the constant "absurdity" and $\langle \vee, \wedge, \to, \perp \rangle$-fragment of $\mathbf{N4}^{\perp}$ coincides with intuitionistic logic **Li**. In the course of semantic investigations, we work mainly with the logic **N4** and we then show how the obtained results can be adapted to $\mathbf{N4}^{\perp}$.

For the logic **N3**, an explosive variant of Nelson's logic, the algebraic semantics was introduced in [90], where **N3** was characterized via the variety of N-lattices. Later D. Vakarelov [110] and independently M.M. Fidel [28] found a very convenient representation of N-lattices in the form of so-called twist-structures. Finally, M.M. Fidel [29] suggested a characterization of **N3**

[1]Parts of this chapter were originally published in [72].

via Heyting algebras with a special family of unary predicates distinguishing classes of counterexamples for elements of a Heyting algebra. Structures of this kind are called Fidel-structures, or **F**-structures.

In our semantic studies of **N4**, we go in the opposite direction compared to **N3** and start with Fidel-structures (Section 8.2). A semantics of this kind for **N4** can be obtained as an immediate generalization of the semantics from [29] and the completeness theorem has an easy and short proof. Then (Section 8.3) we introduce twist-structures for **N4** and show that Fidel-structures and twist-structures are mutually definable, which leads to a characterization of **N4** in terms of twist-structures. In Section 8.4, generalizing the notion of N-lattices we define **N4**-lattices and state that all twist-structures are **N4**-lattices and conversely, that any **N4**-lattice is isomorphic to a twist-structure. With any **N4**-lattice $\mathcal{A}$, we associate the set $D^{\mathcal{A}}$ of its distinguished elements and prove the completeness result with respect to the obtained class of logical matrices. It is essential that the set $D^{\mathcal{A}}$ is not one-element in the general case. This fact has the following explanation. In **N4**, as well as in **N3**, the usual equivalence $\varphi \leftrightarrow \psi := (\varphi \to \psi) \wedge (\psi \to \varphi)$ does not have the congruence property with respect to negation. It is possible that the negations of equivalent formulas are not equivalent. Only the strong equivalence $\varphi \Leftrightarrow \psi := (\varphi \leftrightarrow \psi) \wedge (\sim \psi \leftrightarrow \sim \varphi)$ is a congruence wrt negation. Fortunately, due to the axiom schema $\sim \varphi \to (\varphi \to \psi)$, all negations of tautologies are equivalent in **N3**. Namely this fact allows one to characterize **N3** by the class of N-lattices, where the unit element (the greatest element wrt the lattice ordering) is the only distinguished element. Another situation is in case of **N4**. For example, $\sim (p \to p)$ and $\sim (p \to (q \to p))$ are not equivalent in this logic. The first formula is equivalent to $p \wedge \sim p$, the second to $(p \wedge \sim p) \wedge q$, and these two contradictions are not equivalent in the paraconsistent logic **N4**. This complication can be overcome due to the fact that sets of distinguished elements in **N4**-matrices are definable. In Section 8.5, we prove that **N4**-lattices form a variety and establish a dual lattice isomorphism between the lattice of **N4**-extensions and the lattice of subvarieties of the variety of **N4**-lattices. To obtain an axiomatization of the variety of **N4**-lattices the results of A. Pynko [89] are essentially used. The desired axiomatization is obtained as appropriate weakening of an axiomatization of the class of implicative De Morgan lattices introduced in [89].

In the last section of this chapter we show how to define twist-structures for the logic $\mathbf{N4}^{\perp}$, and then define $\mathbf{N4}^{\perp}$-lattices and establish the dual isomorphism between the lattice of $\mathbf{N4}^{\perp}$-extensions and the lattice of subvarieties of the variety of $\mathbf{N4}^{\perp}$-lattices.

8.1 Preliminaries

In this part of the book, we deal with the propositional languages $\mathcal{L} := \{\vee, \wedge, \to, \sim\}$, where $\sim$ is a symbol for strong negation, and $\mathcal{L}^\perp := \mathcal{L} \cup \{\perp\}$ with an additional symbol for the constant "absurdity". By For ($For^\perp$) we denote the trivial logic in the language $\mathcal{L}$ ($\mathcal{L}^\perp$), i.e., the set of formulas of this language. The connectives of equivalence $\leftrightarrow$ and of strong equivalence $\Leftrightarrow$ are defined as follows:

$$\varphi \leftrightarrow \psi := (\varphi \to \psi) \wedge (\psi \to \varphi), \ \ \varphi \Leftrightarrow \psi := (\varphi \leftrightarrow \psi) \wedge (\sim \psi \leftrightarrow \sim \varphi).$$

As above, logics will be defined via Hilbert-style deductive systems with only the rules of substitution and *modus ponens*. In this way, to define a logic it is enough to give its axioms. Paraconsistent Nelson's logic **N4** is a logic in the language $\mathcal{L}$ characterized by the following list of axioms:

A1) $p \to (q \to p)$;

A2) $(p \to (q \to r)) \to ((p \to q) \to (p \to r))$;

A3) $(p \wedge q) \to p$;

A4) $(p \wedge q) \to q$;

A5) $(p \to q) \to ((p \to r) \to (p \to (q \wedge r)))$;

A6) $p \to (p \vee q)$;

A7) $q \to (p \vee q)$;

A8) $(p \to r) \to ((q \to r) \to ((p \vee q) \to r))$;

A9) $\sim\sim p \leftrightarrow p$;

A10) $\sim (p \vee q) \leftrightarrow (\sim p \wedge \sim q)$;

A11) $\sim (p \wedge q) \leftrightarrow (\sim p \vee \sim q)$;

A12) $\sim (p \to q) \leftrightarrow (p \wedge \sim q)$.

To obtain explosive Nelson's logic **N3** we add to the list of **N4**-axioms, the explosion axiom for strong negation:

A13) $\sim p \to (p \to q)$.

The logic $\mathbf{N4}^{\perp}$ is a logic in the language $\mathcal{L}^{\perp}$ determined by the axioms A1–A13 and the two additional axioms for constant $\perp$:

A14) $\perp \to p$ and A15) $p \to \sim \perp$.

Due to axiom A14 the intuitionistic negation can be defined in $\mathbf{N4}^{\perp}$ as $\neg\varphi := \varphi \to \perp$. If we put $\perp := \sim (p_0 \to p_0)$, one can prove

$$\mathbf{N3} \vdash \perp \to p, \ p \to \sim \perp.$$

The first formula is equivalent to a particular case of axiom A13 modulo positive logic, the second follows from the equivalence $\sim \perp \leftrightarrow (p_0 \to p_0)$ obtained in turn from A9. In particular, the intuitionistic negation is definable in **N3**. This is why we do not consider two different logics **N3** and $\mathbf{N3}^{\perp}$.

We say that a formula φ of the language $\mathcal{L}$ or $\mathcal{L}^{\perp}$ is a *negation normal form (nnf)* if it contains the symbol $\sim$ only before atomic formulas. The following translation $\overline{(\cdot)}$ sends every formula φ to a negation normal form, where $p \in Prop$ and $\diamond \in \{\vee, \wedge, \to\}$:

$$\begin{array}{rclrcl}
\overline{p} & = & p & \overline{\sim p} & = & \sim p \\
\overline{\sim\sim \varphi} & = & \overline{\varphi} & \overline{\varphi \diamond \psi} & = & \overline{\varphi} \diamond \overline{\psi} \\
\overline{\sim (\varphi \vee \psi)} & = & \overline{\sim \varphi} \wedge \overline{\sim \psi} & \overline{\sim (\varphi \wedge \psi)} & = & \overline{\sim \varphi} \vee \overline{\sim \psi} \\
\overline{\sim (\varphi \to \psi)} & = & \overline{\varphi} \wedge \overline{\sim \psi} & \overline{\perp} & = & \perp
\end{array}$$

Proposition 8.1.1 *For any formula φ in the language $\mathcal{L}$ ($\mathcal{L}^{\perp}$),*

$$\mathbf{N4}(\mathbf{N4}^{\perp}) \vdash \varphi \leftrightarrow \overline{\varphi}.$$

The *proof* easily follows from strong negation axioms A9–A12.

□

Proposition 8.1.2 *The logic* $\mathbf{N4}^{\perp}$ *is a conservative extension of* **N4**.

Proof. Let φ be a formula in the language $\mathcal{L}$ and $\varphi_0, \ldots, \varphi_n = \varphi$ is a proof of φ in $\mathbf{N4}^{\perp}$. Let P be the set of all variables occurring in this proof. Put $\phi := \bigwedge_{p \in P} p \wedge \sim p$. Replacing in the above proof any occurrence of $\sim \perp$ by $p_i \to p_i$, $p_i \notin P$, and then any occurrence of $\perp$ by ϕ we obtain a quasi-proof in **N4**. Indeed, every axiom of the form $\psi \to \sim \perp$ is obviously replaced by a provable formula. That every axiom $\perp \to \psi$ is replaced by a provable formula follows from Proposition 8.1.1 and the fact that in positive logic **Lp**, every formula follows from the conjunction of its variables.

□

An important peculiarity of Nelson's logics **N4** and $\mathbf{N4}^{\perp}$ is that the provable equivalence is not a congruence relation. However, axioms A1-A8 constitute the axiomatization of positive logic **Lp** (see Chapter 2). This means that the provable equivalence is a congruence relation with respect to positive connectives. More exactly, for any formulas φ_0, φ_1 and a positive formula $\psi(p)$ with a propositional parameter, the provability of $\varphi_0 \leftrightarrow \varphi_1$ in **N4** (or $\mathbf{N4}^{\perp}$) implies the provability of $\psi(\varphi_0) \leftrightarrow \psi(\varphi_1)$ in that logic. The provable strong equivalence $\Leftrightarrow$ will be a congruence relation in **N4** as well as in $\mathbf{N4}^{\perp}$.

Proposition 8.1.3 *The logics* **N4** *and* $\mathbf{N4}^{\perp}$ *are closed under the weak replacement rule*

$$\frac{\varphi_0 \leftrightarrow \varphi_1 \quad \sim \varphi_0 \leftrightarrow \sim \varphi_1}{\psi(\varphi_0) \leftrightarrow \psi(\varphi_1)}.$$

Proof. It follows easily from Proposition 8.1.1.

□

8.2 Fidel's Semantics

First, we present a semantics for **N4** in terms of **F**-structures (Fidel-structures), which is an immediate generalization of the semantics for **N3** developed by M.M. Fidel in [29].

Definition 8.2.1 *An* **F**-structure *is a tuple*

$$\mathcal{A} = \langle A, \vee, \wedge, \rightarrow, 1, \{N_a\}_{a\in A}\rangle,$$

where $\langle A, \vee, \wedge, \rightarrow, 1\rangle$ *is an implicative lattice and* $\{N_a\}_{a\in A}$ *is a family of sets satisfying the following properties:*

1) *for any* $a \in A$, $\varnothing \neq N_a \subseteq A$;

2) *for any* $a, b \in A$, $a' \in N_a$, *and* $b' \in N_b$, *the following relations hold*

$$a' \vee b' \in N_{a\wedge b}, \;\; a' \wedge b' \in N_{a\vee b}, \;\; a \in N_{a'};$$

3) *for any* $a, b \in A$ *and* $b' \in N_b$, *we have* $a \wedge b' \in N_{a\rightarrow b}$.

We say that an **F**-structure $\mathcal{A}$ is *saturated* if the equality $N_a = A$ holds for all $a \in A$. For **F**-structures we use the denotation $\langle \mathcal{A}, \{N_a\}_{a\in|\mathcal{A}|}\rangle$, where $\mathcal{A}$ is an implicative lattice and $\{N_a\}_{a\in|\mathcal{A}|}$ a family of subsets of $|\mathcal{A}|$.

With an arbitrary set of formulas Γ non-trivial with respect to **N4** we associate an equivalence relation $\equiv_\Gamma$ on the set of formulas in a standard way. For any formulas φ and ψ, we set

$$\varphi \equiv_\Gamma \psi \iff \Gamma \vdash_{\mathbf{N4}} \varphi \leftrightarrow \psi.$$

The coset of an element φ with respect to $\equiv_\Gamma$ is denoted by $[\varphi]_\Gamma$ and the family of all cosets by L_Γ. In case of **N4**, the relation $\equiv_\Gamma$ is a congruence with respect to all positive connectives, which allows one to define on the set L_Γ operations $\vee$, $\wedge$ and $\rightarrow$ as follows:

$$[\varphi]_\Gamma * [\psi]_\Gamma := [\varphi * \psi]_\Gamma, \text{ where } * \in \{\vee, \wedge, \rightarrow\}.$$

Due to the fact that **N4** contains all axioms of positive logic, the structure $\langle L_\Gamma, \vee, \wedge, \rightarrow, 1\rangle$, where 1 denotes the coset of formulas deducible from Γ, $1 = [p \rightarrow p]_\Gamma$, is an implicative lattice. We extend it to an **F**-structure putting

$$N_{[\varphi]_\Gamma} := \{[\sim \psi]_\Gamma \mid \psi \in [\varphi]_\Gamma\} \text{ for any } [\varphi]_\Gamma \in L_\Gamma.$$

Later on, we omit the lower index Γ when it does not lead to confusion. Let

$$\mathcal{L}_\Gamma := \langle L_\Gamma, \vee, \wedge, \rightarrow, 1, \{N_{[\varphi]}\}_{[\varphi] \in L_\Gamma}\rangle.$$

Lemma 8.2.2 *For every set of formulas Γ non-trivial wrt* **N4**, *the structure $\mathcal{L}_\Gamma$ is an* **F**-*structure.*

Proof. It is enough to check Items 2, 3 of Definition 8.2.1, which can easily be done with the help of the axioms for strong negation.

□

We say that an **F**-structure $\langle \mathcal{A}, \{N_a\}_{a \in |\mathcal{A}|}\rangle$ is a *substructure* of an **F**-structure $\langle \mathcal{B}, \{N'_b\}_{b \in |\mathcal{B}|}\rangle$ if (1) $\mathcal{A}$ is an implicative sublattice of $\mathcal{B}$ and (2) for any $a \in |\mathcal{A}|$, $N_a \subseteq N'_a$.

Lemma 8.2.3 *Any* **F**-*structure is a substructure of a saturated* **F**-*structure.*

Proof. Any structure $\langle \mathcal{A}, \{N_a\}_{a \in |\mathcal{A}|}\rangle$ can be embedded into a saturated structure $\langle \mathcal{A}, \{N'_a\}_{a \in |\mathcal{A}|}\rangle$, where $\mathcal{A}$ is the same implicative lattice and $N'_a = |\mathcal{A}|$ for all $a \in A$.

□

Definition 8.2.4 *Let $\langle \mathcal{A}, \{N_a\}_{a \in |\mathcal{A}|}\rangle$ be an* **F***-structure. The mapping v from For into $|\mathcal{A}|$ is said to be a* valuation *into this structure if the following conditions hold for arbitrary formulas φ and ψ:*

1) $v(\sim p) \in N_{v(p)}$ for a propositional variable p;

2) $v(\varphi * \psi) = v(\varphi) * v(\psi)$ for $* \in \{\vee, \wedge, \rightarrow\}$;

3) $v(\sim (\varphi \vee \psi)) = v(\sim \varphi) \wedge v(\sim \psi)$ and $v(\sim (\varphi \wedge \psi)) = v(\sim \varphi) \vee v(\sim \psi)$;

4) $v(\sim (\varphi \rightarrow \psi)) = v(\varphi) \wedge v(\sim \psi)$;

5) $v(\sim\sim \varphi) = v(\varphi)$.

Remark. It can easily be seen from this definition that every valuation into an **F**-structure is uniquely determined by its restriction to the set of propositional variables and their negations.

The following fact can be checked by induction on the structure of formulas.

Lemma 8.2.5 *For any* **F***-structure $\langle \mathcal{A}, \{N_a\}_{a \in |\mathcal{A}|}\rangle$, any valuation v and formula φ, we have $v(\sim \varphi) \in N_{v(\varphi)}$.*

□

Now we define a semantic consequence relation $\models_{\mathbf{F}}$ between sets of formulas and formulas. Let Γ be a set of formulas and φ be a formula. For an **F**-structure $\mathcal{A}$, the relation $\Gamma \models_{\mathbf{F}}^{\mathcal{A}} \varphi$ holds if and only if for any valuation v into $\mathcal{A}$ the condition $v(\psi) = 1$ for all $\psi \in \Gamma$ implies $v(\varphi) = 1$. The relation $\Gamma \models_{\mathbf{F}} \varphi$ means that $\Gamma \models_{\mathbf{F}}^{\mathcal{A}} \varphi$ for any **F**-structure $\mathcal{A}$.

Lemma 8.2.6 *Let $\mathcal{B}$ be a substructure of an* **F***-structure $\mathcal{A}$. Then $\Gamma \models_{\mathbf{F}}^{\mathcal{A}} \varphi$ implies $\Gamma \models_{\mathbf{F}}^{\mathcal{B}} \varphi$.*

Proof. Indeed, if v is a valuation into $\mathcal{B}$, it is also a valuation into $\mathcal{A}$.

□

It is not hard to check that the introduced consequence relation is closed under substitution.

Lemma 8.2.7 *Let $\Gamma(p_1, \ldots, p_n)$ be a set of formulas, $\varphi(p_1, \ldots, p_n), \psi_1, \ldots, \psi_n$ be formulas, and $\mathcal{A}$ an* **F***-structure. If $\Gamma \models_{\mathbf{F}}^{\mathcal{A}} \varphi$, then*

$$\Gamma(\psi_1, \ldots, \psi_n) \models_{\mathbf{F}}^{\mathcal{A}} \varphi(\psi_1, \ldots, \psi_n).$$

□

Theorem 8.2.8 *Let Γ be a set of formulas non-trivial with respect to* **N4** *and φ a formula. The following conditions are equivalent:*

1) $\Gamma \vdash_{\mathbf{N4}} \varphi$*;*

2) $\Gamma \models_{\mathbf{F}} \varphi$*;*

3) $\Gamma \models_{\mathbf{F}}^{\mathcal{A}} \varphi$ *for any saturated* **F***-structure* $\mathcal{A}$*;*

4) $\Gamma \models_{\mathbf{F}}^{\mathcal{L}_\Gamma} \varphi$*.*

Proof. 1) $\Rightarrow$ 2) The soundness of consequence relation with respect to the introduced semantics can be proved in a standard way via induction on the length of proof.

Implications 2) $\Rightarrow$ 3) and 3) $\Rightarrow$ 4) are obvious.

4) $\Rightarrow$ 1) Assume $\Gamma \models_{\mathbf{F}}^{\mathcal{L}_\Gamma} \varphi$ and consider a mapping v' from the set of propositional variables and their negations to L_Γ such that for a variable p, we have $v'(p) = [p]$ and $v'(\sim p) = [\sim p]$. This mapping extends in a unique way to a valuation v into $\mathcal{L}_\Gamma$. Using Items 2–5 of Definition 8.2.4 one can easily check that $v(\varphi) = [\varphi]$ for any formula φ.

If $\psi \in \Gamma$, then clearly $\Gamma \vdash_{\mathbf{N4}} \psi$, i.e., $[\psi] = 1$. In view of the above property of v, we have $v(\psi) = 1$ for all $\psi \in \Gamma$. This implies by assumption that $v(\varphi) = [\varphi] = 1$, i.e., $\Gamma \vdash_{\mathbf{N4}} \varphi$.

□

If φ is a positive formula, the validity of this formula on an **F**-structure $\langle \mathcal{A}, \{N_a\}_{a \in |\mathcal{A}|} \rangle$ is equivalent to its validity on the implicative lattice $\mathcal{A}$. This fact and Theorem 8.2.8 imply

Corollary 8.2.9 **N4** *is a conservative extension of positive logic.*

8.3 Twist-structures

Following M.M. Fidel [28] and D. Vakarelov [110] we introduce the notion of twist-structures. Because we need a semantics for the logic **N4**, where the intuitionistic negation is not definable, as underlying algebraic systems for twist-structures we take implicative lattices unlike the works [28, 110], where twist-structures were defined over Heyting algebras. The apt term "twist-structure" was introduced in [47]. We define an algebraic structure on the direct square of the universe of an underlying algebra in such a way that operations are not componentwise, but they are "twisted" in some sense.

Definition 8.3.1 *Let* $\mathcal{A} = \langle A, \vee, \wedge, \to, 1\rangle$ *be an implicative lattice.*

1. *The* full twist-structure *over* $\mathcal{A}$ *is the algebra*
$$\mathcal{A}^{\bowtie} = \langle A \times A, \vee, \wedge, \to, \sim\rangle$$
with the twist-operations *defined for* $(a, b), (c, d) \in A \times A$ *as follows:*
$$(a, b) \vee (c, d) := (a \vee c, b \wedge d), \;\; (a, b) \wedge (c, d) := (a \wedge c, b \vee d)$$
$$(a, b) \to (c, d) := (a \to c, a \wedge d), \;\; \sim (a, b) := (b, a).$$
2. *A* twist-structure *over* $\mathcal{A}$ *is an arbitrary subalgebra* $\mathcal{B}$ *of the full twist-structure* $\mathcal{A}^{\bowtie}$ *such that* $\pi_1(\mathcal{B}) = \mathcal{A}$ *(in which case also* $\pi_2(\mathcal{B}) = \mathcal{A}$*).*
3. *The class of all twist-structures over* $\mathcal{A}$ *is denoted* $S^{\bowtie}(\mathcal{A})$*.*

It should be noted that our terminology differs from that of [47], where the term "twist-structure" is used for twist-structures over Heyting algebras (see Section 8.6) and with the unverse of the form $\{(a, b) \mid a \wedge b = 0\}$, where 0 is the least element of the underlying algebra.

A valuation into a twist-structure $\mathcal{B}$ is defined in the usual way as a homomorphism of the algebra of formulas into $\mathcal{B}$. The semantic consequence relation $\models_{\bowtie}$ over twist-structures is defined as follows. Let Γ be a set of formulas, φ a formula, and $\mathcal{B}$ a twist-structure. The relation $\Gamma \models^{\mathcal{B}}_{\bowtie} \varphi$ holds if and only if for any valuation v in $\mathcal{B}$ the condition that $\pi_1 v(\psi) = 1$ for all $\psi \in \Gamma$ implies $\pi_1 v(\varphi) = 1$. The relation $\Gamma \models_{\bowtie} \varphi$ means that $\Gamma \models^{\mathcal{B}}_{\bowtie} \varphi$ for all twist-structures $\mathcal{B}$.

One can easily pass from **F**-structures to twist-structures and back. Let $\mathcal{A}^* = \langle \mathcal{A}, \{N_a\}_{a \in |\mathcal{A}|}\rangle$ be an **F**-structure. We associate with $\mathcal{A}^*$ a twist-structure $\mathcal{A}^*_{\bowtie}$ over $\mathcal{A}$ with the following universe
$$|\mathcal{A}^*_{\bowtie}| = \{(a, b) \in |\mathcal{A}| \times |\mathcal{A}| \mid b \in N_a\}.$$
Using the definitions of twist-operations and of **F**-structures one can check that for any **F**-structure $\mathcal{A}^*$ the set $|\mathcal{A}^*_{\bowtie}|$ is closed under all twist-operations, and so the structure $\mathcal{A}^*_{\bowtie}$ is well defined.

On the other hand, given a twist-structure $\mathcal{B} \in S^{\bowtie}(\mathcal{A})$ we define an **F**-structure $\mathcal{B}^{\mathbf{F}} = \langle \mathcal{A}, \{N_a\}_{a \in |\mathcal{A}|}\rangle$, where
$$N_a = \{b \in A \mid (a, b) \in \mathcal{B}\}$$
for any $a \in |\mathcal{A}|$. The correctness of this definition can be verified directly.

Further, let $\mathcal{A}$ be an **F**-structure and v a valuation into $\mathcal{A}$. We define a valuation $v_{\bowtie}$ into $\mathcal{A}_{\bowtie}$ for a propositional variable p as

$$v_{\bowtie}(p) := (v(p), v(\sim p)).$$

By induction on the structure of formulas we obtain then the following statement.

Lemma 8.3.2 *For any formula* φ, $v_{\bowtie}(\varphi) = (v(\varphi), v(\sim \varphi))$.

□

Now, let $\mathcal{B}$ be a twist-structure and v a valuation. We define a valuation $v_{\mathbf{F}}$ into $\mathcal{B}^{\mathbf{F}}$ as a unique valuation satisfying the conditions

$$v_{\mathbf{F}}(p) = \pi_1 v(p) \text{ and } v_{\mathbf{F}}(\sim p) = \pi_2 v(p)$$

for all propositional variables p. Again, by induction on the structure of formulas one can prove the following lemma.

Lemma 8.3.3 *For any formula* φ, $v_{\mathbf{F}}(\varphi) = \pi_1 v(\varphi)$.

□

Proposition 8.3.4 *1. Let* $\mathcal{A}^* = \langle \mathcal{A}, \{N_a\}_{a\in|\mathcal{A}|}\rangle$ *be an* **F***-structure and* v *a valuation into* $\mathcal{A}^*$*. Then*

$$(\mathcal{A}^*_{\bowtie})^{\mathbf{F}} = \mathcal{A}^* \text{ and } (v_{\bowtie})_{\mathbf{F}} = v.$$

2. Let $\mathcal{B}$ *be a twist-structure and* v *a valuation into* $\mathcal{B}$*. Then*

$$(\mathcal{B}^{\mathbf{F}})_{\bowtie} = \mathcal{B} \text{ and } (v_{\mathbf{F}})_{\bowtie} = v.$$

Proof. 1. Let $(\mathcal{A}^*_{\bowtie})^{\mathbf{F}} = \langle \mathcal{A}, \{N'_a\}_{a\in|\mathcal{A}|}\rangle$. For any pair of elements $a, b \in |\mathcal{A}|$, we have the equivalences

$$b \in N'_a \iff (a, b) \in \mathcal{A}^*_{\bowtie} \iff b \in N_a.$$

This proves the equality $(\mathcal{A}^*_{\bowtie})^{\mathbf{F}} = \mathcal{A}^*$.

Take an arbitrary formula φ and calculate

$$(v_{\bowtie})_{\mathbf{F}}(\varphi) = \pi_1 v_{\bowtie}(\varphi) = \pi_1((v(\varphi), v(\sim \varphi))) = v(\varphi).$$

The second equality in this chain is due to Lemma 8.3.2.

2. Let $\mathcal{B}$ be a twist-structure over $\mathcal{A}$, $\mathcal{B}^{\mathbf{F}} = \langle \mathcal{A}, \{N_a\}_{a \in |\mathcal{A}|} \rangle$ and let $a, b \in |\mathcal{A}|$. The first equality of this item follows from the equivalences:

$$(a,b) \in (\mathcal{B}^{\mathbf{F}})_{\bowtie} \iff b \in N_a \iff (a,b) \in \mathcal{B}.$$

For a formula φ, we calculate applying Lemma 8.3.3

$$(v_{\mathbf{F}})_{\bowtie}(\varphi) = (v_{\mathbf{F}}(\varphi), v_{\mathbf{F}}(\sim \varphi)) =$$

$$= (\pi_1 v(\varphi), \pi_1 v(\sim \varphi)) = (\pi_1 v(\varphi), \pi_2 v(\varphi)) = v(\varphi).$$

□

Proposition 8.3.5 *Let Γ be a set of formulas and φ a formula.*

1. For any **F***-structure $\mathcal{A}$,*

$$\Gamma \models^{\mathcal{A}}_{\mathbf{F}} \varphi \iff \Gamma \models^{\mathcal{A}_{\bowtie}}_{\bowtie} \varphi.$$

2. For any twist-structure $\mathcal{A}$,

$$\Gamma \models^{\mathcal{A}}_{\bowtie} \varphi \iff \Gamma \models^{\mathcal{A}^{\mathbf{F}}}_{\mathbf{F}} \varphi.$$

Proof. 1. In view of the previous proposition, any valuation v into $\mathcal{A}$ has the form $v'_{\mathbf{F}}$ for a suitable $\mathcal{A}_{\bowtie}$-valuation v' and any valuation v into $\mathcal{A}_{\bowtie}$ can be represented as $v'_{\bowtie}$ for some $\mathcal{A}$-valuation v'. These observations easily imply the desired equivalence.

2. Again, in view of the previous proposition $(\mathcal{A}^{\mathbf{F}})_{\bowtie} = \mathcal{A}$, which allows one to reduce this item to the previous one.

□

For a set of formulas Γ non-trivial with respect to **N4**, we define a Lindenbaum twist-structure $\mathcal{L}^{\bowtie}_{\Gamma} := (\mathcal{L}_{\Gamma})_{\bowtie}$, i.e.,

$$\mathcal{L}^{\bowtie}_{\Gamma} = \langle \{([\varphi]_{\Gamma}, [\sim \psi]_{\Gamma}) \mid \psi \in [\varphi]_{\Gamma}\}, \vee, \wedge, \rightarrow, \sim \rangle.$$

The next completeness result easily follows from the completeness theorem for Fidel structures, Proposition 8.3.5 and an obvious equivalence:

(an **F**-structure $\mathcal{B}$ is saturated) $\iff$ (the twist-structure $\mathcal{B}_{\bowtie}$ is full).

Theorem 8.3.6 *Let* Γ *be a set of formulas, non-trivial with respect to* **N4**, *and* φ *a formula. The following conditions are equivalent:*

1) $\Gamma \vdash_{\mathbf{N4}} \varphi$;

2) $\Gamma \models_{\bowtie} \varphi$;

3) $\Gamma \models_{\bowtie}^{\mathcal{B}} \varphi$ *for any full twist-structure* $\mathcal{B}$;

4) $\Gamma \models_{\bowtie}^{\mathcal{L}_{\Gamma}^{\bowtie}} \varphi$.

Remark. It was already mentioned that the provable equivalence $\Leftrightarrow$ is a congruence in **N4**, therefore, on the set

$$L_{\mathbf{N4}^{\perp}}^{\Leftrightarrow} := \{[\varphi]_{\Leftrightarrow} \mid \varphi \in For\},$$

where $[\varphi]_{\Leftrightarrow} := \{\psi \mid \mathbf{N4} \vdash \varphi \Leftrightarrow \psi\}$, one can define a structure of Lindenbaum algebra in a standard way. Denote this algebra by $\mathcal{L}_{\mathbf{N4}}^{\Leftrightarrow}$. Note that the mapping $[\varphi]_{\Leftrightarrow} \mapsto ([\varphi], [\sim \varphi])$ determines an isomorphism of Lindenbaum algebra $\mathcal{L}_{\mathbf{N4}}^{\Leftrightarrow}$ and Lindenbaum twist-structure $\mathcal{L}_{\mathbf{N4}}^{\bowtie}$.

8.3.1 Embedding of N3 into N4[2]

Having the semantics for **N4** in terms of twist-structures we are ready to prove that explosive **N3** is faithfully embedded into paraconsistent **N4**. First we describe twist-structures, which are models of **N3**.

Lemma 8.3.7 *Let* $\mathcal{A}$ *be an implicative lattice and* $\mathcal{B} \in S^{\bowtie}(\mathcal{A})$. *We have* $\mathcal{B} \models$ **N3** *if and only if* $\mathcal{A}$ *has the least element* 0 *and for any* $(a, b) \in \mathcal{B}$, $a \wedge b = 0$.

Proof. As follows from the definition of twist-operations, the truth of axiom A13 on $\mathcal{B}$ is equivalent to the fact that for any $(a, b), (c, d) \in \mathcal{B}$, $a \wedge b \leq c$. By Definition 8.3.1 c is an arbitrary element of $\mathcal{A}$, whence for any $(a, b) \in \mathcal{B}$, $a \wedge b$ is the least element of $\mathcal{A}$.

□

Definition 8.3.8 *Let* $\mho :=\sim (p_0 \to p_0)$, *where* p_0 *is a fixed propositional variable. We define a transformation* $(\cdot)^*$ *of formulas as follows.*

1. For a propositional variable p,

$$(p)^* := p \vee \mho, \ \ (\sim p)^* := (\sim p \wedge (p \to \mho)) \vee \mho.$$

[2]This section was originally published in [71].

2. $(\varphi \diamond \psi)^* := \varphi^* \diamond \psi^*$, *where* φ *and* ψ *are* nnf *and* $\diamond \in \{\vee, \wedge, \rightarrow\}$.

3. $(\varphi)^* \rightleftharpoons (\overline{\varphi})^*$ *for any formula* φ *not a* nnf.

Theorem 8.3.9 *For any formula* φ,

1. $\varphi \in \mathbf{N3}$ *if and only if* $\varphi^* \in \mathbf{N4}$.

2. $\mathbf{N3} \vdash \varphi \leftrightarrow \varphi^*$.

Proof. 1. First, we prove the direct implication. For any substitution instance φ of one or another **N4**-axiom, $\overline{\varphi}$ is provable in **N4** due to Proposition 8.1.1. The transformation $(\cdot)^*$ preserves all positive connectives for *nnf*, therefore, $(\overline{\varphi})^*$ is also provable in **N4** as a substitution instance of $\overline{\varphi}$. Let $\varphi, \varphi \rightarrow \psi \in \mathbf{N3}$ and $\varphi^*, (\varphi \rightarrow \psi)^* \in \mathbf{N4}$. Note that $(\varphi \rightarrow \psi)^* = (\overline{\varphi \rightarrow \psi})^* = (\overline{\varphi} \rightarrow \overline{\psi})^* = \overline{\varphi}^* \rightarrow \overline{\psi}^*$, which immediately implies that $\psi^* = (\overline{\psi})^* \in \mathbf{N4}$. We have thus proved that the set of formulas, for which the considered implication holds, is closed under *modus ponens*. In this way, it remains to check that formulas of the form

$$(\sim \varphi \rightarrow (\varphi \rightarrow \psi))^* = (\sim \varphi)^* \rightarrow (\varphi^* \rightarrow \psi^*)$$

are provable in **N4**. The latter formula is equivalent in **N4** to $((\sim \varphi)^* \wedge \varphi^*) \rightarrow \psi^*$, therefore, the desired result follows from the next lemma.

Lemma 8.3.10 *For any formula* φ, *the following holds:*

a) $\mathbf{N4} \vdash \mho \rightarrow \varphi^*$;

b) $\mathbf{N4} \vdash \mho \leftrightarrow \varphi^* \wedge (\sim \varphi)^*$.

Proof. a) This item is true by definition for propositional variables and their negations. Any formula φ^* can be obtained from formulas of the form p^* and $(\sim p)^*$ with the help of positive connectives, which allows one to complete the proof by an easy induction on the structure of formulas.

b) We use again an induction on the structure of formulas. For a propositional variable p, we have

$$p^* \wedge (\sim p)^* \leftrightarrow (p \wedge \sim p \wedge (p \rightarrow \mho)) \vee \mho \leftrightarrow (\sim p \wedge p \wedge \mho) \vee \mho \leftrightarrow \mho.$$

In case of the conjunction of formulas, we apply the induction hypotheses and Item a) to obtain the following chain of equivalences:

$$(\varphi \wedge \psi)^* \wedge (\sim(\varphi \wedge \psi))^* \leftrightarrow \varphi^* \wedge \psi^* \wedge (\overline{\sim(\varphi \wedge \psi)})^* \leftrightarrow$$

$$\leftrightarrow \varphi^* \wedge \psi^* \wedge (\overline{\sim\varphi} \vee \overline{\sim\psi})^* \leftrightarrow \varphi^* \wedge \psi^* \wedge ((\sim\varphi)^* \vee (\sim\psi)^*) \leftrightarrow$$

$$\leftrightarrow ((\varphi^* \wedge (\sim\varphi)^*) \wedge \psi^*) \vee (\varphi^* \wedge (\psi^* \wedge (\sim\psi)^*)) \leftrightarrow (\mho \wedge \psi^*) \vee (\varphi^* \wedge \mho) \leftrightarrow \mho.$$

The case of disjunction can be considered similarly. For the implication, we have

$$(\varphi \to \psi)^* \wedge (\sim(\varphi \to \psi))^* \leftrightarrow (\varphi^* \to \psi^*) \wedge (\overline{\sim(\varphi \to \psi)})^* \leftrightarrow$$

$$\leftrightarrow (\varphi^* \to \psi^*) \wedge (\overline{\varphi} \wedge \overline{\sim\psi})^* \leftrightarrow (\varphi^* \to \psi^*) \wedge \varphi^* \wedge (\sim\psi)^* \leftrightarrow$$

$$\leftrightarrow \varphi^* \wedge \psi^* \wedge (\sim\psi)^* \leftrightarrow \varphi^* \wedge \mho \leftrightarrow \mho.$$

Thus, it remains to consider the case of negation, which is trivial in view of the graphical equality $(\sim\varphi)^* \wedge (\sim\sim\varphi)^* = (\sim\varphi)^* \wedge \varphi^*$.

□

We now turn to the inverse implication. Assume $\mathbf{N3} \nvdash \varphi$. Then there is a twist-structure $\mathcal{B}$, $\mathcal{B} \models \mathbf{N3}$, and its valuation v such that $\pi_1 v\varphi \neq 1$. By Lemma 8.3.7, $\mathcal{B} \in S^{\bowtie}(\mathcal{A})$, where $\mathcal{A}$ is an implicative lattice with the least element 0. Since $p_0 \to p_0 \in \mathbf{N4}$, we have $\pi_1 v(p_0 \to p_0) = 1$. In view of Lemma 8.3.7 $\pi_2 v(p_0 \to p_0) = 0$. In this way, $v(\mho) = (0, 1)$. For any propositional variable p, $v(p) = (a, b)$, we have

$$v(p^*) = v(p) \vee v(\mho) = v(p) \vee (0, 1) = v(p),$$

$$v((\sim p)^*) = (\sim v(p) \wedge (v(p) \to (0, 1))) \vee (0, 1) =$$

$$= (b, a) \wedge ((a, b) \to (0, 1)) = (b \wedge (a \to 0), a) = (b, a) = v(\sim p).$$

Here we have taken into account the fact that $a \wedge b = 0$ is equivalent to $b \leq a \to 0$. Thus, if ψ is a propositional variable or its negation, we have $v(\psi^*) = v(\psi)$. Due to the fact that $(\cdot)^*$ preserve all logical connectives for *nnf*, the relation $v(\psi^*) = v(\psi)$ holds for any *nnf* ψ. So for any formula ψ, by Item 3 of Definition 8.3.8 we have $v(\psi^*) = v(\overline{\psi})$. By Proposition 8.1.1 we have $\mathbf{N4} \vdash \psi \leftrightarrow \overline{\psi}$, whence $\pi_1 v(\psi^*) = \pi_1 v(\psi)$ for any ψ. In particular, $\pi_1 v(\varphi^*) \neq 1$, i.e., $\varphi^* \notin \mathbf{N4}$.

2. In the proof of the previous item, it was stated that for any formula φ, for any twist-structure $\mathcal{B}$, $\mathcal{B} \models \mathbf{N3}$, and any $\mathcal{B}$-valuation v, $\pi_1 v(\varphi^*) = \pi_1 v(\varphi)$, which implies $\mathbf{N3} \vdash \varphi \leftrightarrow \varphi^*$.

□

We have proved that as well as in case of intuitionistic and minimal logics deleting the explosion axiom from the logic **N3** does not decrease the expressive power of the logic. For this reason, we conclude this section with the following problem: to find a natural example of an explosive logic such that it is not faithfully embeddable into its paraconsistent analog. An example should be natural in a sense that the logic must be described in the literature and not constructed especially to solve this problem.

8.4 N4-Lattices

In this section, we give an algebraic definition of twist-structures. Namely, we describe the class of algebraic systems isomorphic to twist-structures. The definition below is closely related to that of N-lattices (see [92, 33]). Recall that in case of semantic investigations for **N3**, the notion of N-lattices was first introduced [90] and only later was it proved that N-lattices can be represented as twist-structures [28, 110].

Definition 8.4.1 *An algebra* $\mathcal{A} = \langle A, \vee, \wedge, \to, \sim \rangle$ *is said to be an* **N4**-lattice *if the following hold.*

1. *The reduct* $\langle A, \vee, \wedge, \sim \rangle$ *is a De Morgan algebra, i.e.,* $\langle A, \vee, \wedge \rangle$ *is a distributive lattice (non-bounded in general case) and the following identities hold:* $\sim (p \vee q) = \sim p \wedge \sim q$ *and* $\sim\sim p = p$.
2. *The relation* $\preceq$*, where* $a \preceq b$ *denotes* $(a \to b) \to (a \to b) = a \to b$*, is a preordering on* $\mathcal{A}$.
3. *The relation* $\approx$*, where* $a \approx b$ *if and only if* $a \preceq b$ *and* $b \preceq a$*, is a congruence relation with respect to* $\vee, \wedge, \to$ *and the quotient-algebra* $\mathcal{A}_{\bowtie} := \langle A, \vee, \wedge, \to \rangle / \approx$ *is an implicative lattice.*
4. *For any* $a, b \in A$*,* $\sim (a \to b) \approx a \wedge \sim b$.
5. *For any* $a, b \in A$*,* $a \leq b$ *if and only if* $a \preceq b$ *and* $\sim b \preceq \sim a$*, where* $\leq$ *is a lattice ordering on* $\mathcal{A}$.

First we check that the twist-structures belong to the just defined class of lattices.

Proposition 8.4.2 *Every twist-structure $\mathcal{B} \in S^{\bowtie}(\mathcal{A})$ is an* **N4**-*lattice, moreover, the following facts are true for all $(a,b),(c,d) \in |\mathcal{B}|$:*

a) $(a,b) \preceq (c,d)$ if and only if $a \leq c$;

b) $(a,b) \approx (c,d)$ if and only if $a = c$;

c) $(a,b) \leq (c,d)$ if and only if $a \leq c$ and $d \leq b$;

d) The mapping $[(a,b)]_{\approx} \mapsto a$ determines an isomorphism of implicative lattices $\mathcal{B}_{\bowtie}$ and $\mathcal{A}$.

Proof. Check statements a)–d).

a) We calculate $(a,b) \to (c,d) = (a \to c, a \wedge d)$ and $(a \to c, a \wedge d) \to (a \to c, a \wedge d) = (1, (a \to c) \wedge a \wedge d) = (1, a \wedge c \wedge d)$. Thus, the inequality $(a,b) \preceq (c,d)$ is equivalent to the conditions $a \to c = 1$ and $a \wedge d = a \wedge c \wedge d$. The first condition means $a \leq c$ and implies the second.

Item b) immediately follows from a) and Item c) follows from the definition of twist-operations.

d) In view of b) we have $[(a,b)]_{\approx} = \{(a,c) \mid (a,c) \in |\mathcal{B}|\}$. Moreover, twist-operations are componentwise with respect to the first component and $\pi_1(\mathcal{B}) = \mathcal{A}$. These facts easily imply that $[(a,b)]_{\approx} \mapsto a$ determines the desired isomorphism.

We now turn to the definition of **N4**-lattices. The fact that $\langle |\mathcal{B}|, \vee, \wedge, \sim \rangle$ is a De Morgan algebra easily follows from the definition of twist-operations. Items 2 and 3 of the definition follows from b) and d), respectively. Item 4 can be verified directly, and 5 follows from a) and c).

□

Proposition 8.4.3 *Every* **N4**-*lattice $\mathcal{A}$ is isomorphic to a twist-structure over $\mathcal{A}_{\bowtie}$.*

Proof. Consider a mapping h from $\mathcal{A}$ to $(\mathcal{A}_{\bowtie})^{\bowtie}$ defined by the rule

$$h(a) := ([a]_{\approx}, [\sim a]_{\approx}).$$

We claim that h is an isomorphic embedding. Indeed, for any $a, b \in \mathcal{A}$, we have

$$h(a \wedge b) = ([a \wedge b]_{\approx}, [\sim (a \wedge b)]_{\approx}) = ([a]_{\approx} \wedge [b]_{\approx}, [\sim a]_{\approx} \vee [\sim b]_{\approx}) = h(a) \wedge h(b).$$

We have used De Morgan's law and the congruence properties of $\approx$. That h preserves $\vee$ can be checked similarly. Consider the case of implication:

$$h(a \to b) = ([a]_\approx \to [b]_\approx, [a]_\approx \wedge [\sim b]_\approx) = h(a) \to h(b).$$

We have again used the congruence properties of $\approx$ and Item 4 of Definition 8.4.1. Finally, by identity $\sim\sim p = p$ we have

$$h(\sim a) = ([\sim a]_\approx, [\sim\sim a]_\approx) = ([\sim a]_\approx, [a]_\approx) = \sim h(a).$$

Now assume $h(a) = h(b)$, which means that $a \approx b$ and $\sim a \approx \sim b$. By Item 5 of Definition 8.4.1 we immediately obtain $a = b$, which completes the proof.

□

For an **N4**-lattice $\mathcal{A}$, define a set $D^{\mathcal{A}} := \{a \in |\mathcal{A}| \mid a \to a = a\}$. Let $\mathcal{N}$ denote the class of all matrices of the form $\langle \mathcal{A}, D^{\mathcal{A}} \rangle$, where $\mathcal{A}$ is an **N4**-lattice, and let $\models^{\mathcal{N}}$ be a consequence relation defined in a natural way by the class $\mathcal{N}$.

Lemma 8.4.4 *Let* $\mathcal{A} \in S^{\bowtie}(\mathcal{B})$*. Then*

$$D^{\mathcal{A}} = \{(1, b) \mid (1, b) \in \mathcal{A}\}.$$

Proof. Calculate $(a, b) \to (a, b) = (1, a \wedge b)$. Obviously, the equality $(1, a \wedge b) = (a, b)$ is equivalent to $a = 1$.

□

From this lemma and Theorem 8.3.6 we immediately obtain that **N4** is characterized by the class of matrices $\mathcal{N}$.

Theorem 8.4.5 *Let* Γ *be a set of formulas and* φ *a formula. The following equivalence holds:*

$$\Gamma \vdash_{\mathbf{N4}} \varphi \iff \Gamma \models^{\mathcal{N}} \varphi.$$

8.5 The Variety of N4-Lattices

We have thus obtained a characterization of the logic **N4** via the class $\mathcal{N}$ of matrices. We now show that the underlying algebras of these matrices, namely, **N4**-lattices, form a variety. The fact that sets of distinguished elements of matrices in $\mathcal{N}$ are definable allows one to state the dual isomorphism between the lattice $\mathcal{E}$**N4** of logics extending **N4** and the lattice of subvarieties of the variety of **N4**-lattices. In this section, we will use essentially the results of [89], where the class of implicative De Morgan lattices was introduced and it was stated that this class provides an algebraic semantics for Belnap's four-valued logic in the language $\langle \vee, \wedge, \rightarrow, \sim \rangle$. Our axiomatization of the class of **N4**-lattices is obtained by weakening the axiomatization of the class of implicative De Morgan lattices.

Definition 8.5.1 *Let $\mathcal{V}_{\mathbf{N4}}$ be a variety of algebras in the language $\mathcal{L}$ satisfying the identities of De Morgan algebras and the following identities:*

1^N. $(p \rightarrow p) \rightarrow q = q$

2^N. $(p \wedge q) \rightarrow r = p \rightarrow (q \rightarrow r)$

3^N. $p \rightarrow (q \wedge r) = (p \rightarrow q) \wedge (p \rightarrow r)$

4^N. $(p \vee q) \rightarrow r = (p \rightarrow r) \wedge (q \rightarrow r)$

5^N. $q \leq p \rightarrow q$

6^N. $(p \rightarrow q) \wedge (q \rightarrow r) \preceq p \rightarrow r$

7^N. $p \rightarrow p \leq \sim (p \rightarrow q) \rightarrow p$

8^N. $q \rightarrow q \leq p \rightarrow (\sim q \rightarrow \sim (p \rightarrow q))$

9^N. $q \rightarrow q \leq \sim (p \rightarrow q) \rightarrow \sim q$

10^N. $p \wedge (p \rightarrow q) \leq q \vee \sim (\sim q \rightarrow \sim p)$

11^N. $p \leq (q \rightarrow q) \vee \sim (\sim q \rightarrow \sim p)$

Here, we consider $p \leq q$ as an abbreviation of the identity $p \vee q = q$ and $p \preceq q$ as an abbreviation for $(p \rightarrow q) \rightarrow (p \rightarrow q) = p \rightarrow q$.

Remark. Since algebras from $\mathcal{V}_{\mathbf{N4}}$ are De Morgan algebras, for any $\mathcal{A} \in \mathcal{V}_{\mathbf{N4}}$ and $a, b \in \mathcal{A}$, holds

$$a \leq b \Rightarrow \sim b \leq \sim a. \tag{8.1}$$

Indeed, $\sim a \wedge \sim b = \sim (a \vee b) = \sim b$.

As in the case of **N4**-lattices for an arbitrary algebra $\mathcal{A} \in \mathcal{V}_{\mathbf{N4}}$, we define $D^{\mathcal{A}} := \{a \in |\mathcal{A}| \mid a \rightarrow a = a\}$. First we state some properties of sets $D^{\mathcal{A}}$ in algebras of $\mathcal{V}_{\mathbf{N4}}$.

Lemma 8.5.2 *Let* $\mathcal{A} \in \mathcal{V}_{\mathbf{N4}}$. *For any* $a, b \in \mathcal{A}$, *the following holds:*

a) if $a \in D^{\mathcal{A}}$, *then* $a \to b = b$;

b) $D^{\mathcal{A}} = \{a \to a \mid a \in |\mathcal{A}|\}$;

c) if $a \in D^{\mathcal{A}}$ *and* $a \preceq b$, *then* $b \in D^{\mathcal{A}}$;

d) if $a \leq b$, *then* $a \preceq b$;

e) if $a \in D^{\mathcal{A}}$ *and* $a \leq b$, *then* $b \in D^{\mathcal{A}}$;

f) if $a, b \in D^{\mathcal{A}}$, *then* $a \wedge b \in D^{\mathcal{A}}$.

g) $a \preceq a \wedge b$ *if and only if* $a \preceq b$.

Proof. a) We have $a \to b = (a \to a) \to b = b$ by identity 1^N.

b) If $a \in D^{\mathcal{A}}$, then by definition $a = a \to a$. On the other hand, by 1^N we have $(a \to a) \to (a \to a) = a \to a$, whence $a \to a \in D^{\mathcal{A}}$.

c) By assumption $(a \to b) \to (a \to b) = a \to b$. On the other hand, $a \in D^{\mathcal{A}}$ implies by a) $a \to b = b$, whence $b \to b = b$.

d) By 2^N we have $(a \to b) \to (a \to b) = ((a \to b) \wedge a) \to b$. The latter is equal to $a \to b$, because $a \leq b$ by assumption and $b \leq a \to b$ by identity 5^N.

e) It follows immediately from c) and d).

f) Applying identity 2^N and Item a) we calculate $(a \wedge b) \to (a \wedge b) = a \to (b \to (a \wedge b)) = b \to (a \wedge b) = a \wedge b$.

g) By definition $a \preceq a \wedge b$ means $a \to (a \wedge b) \in D^{\mathcal{A}}$, which is equivalent to $(a \to a) \wedge (a \to b) \in D^{\mathcal{A}}$. By Items e) and f) the latter is equivalent to $a \to a \in D^{\mathcal{A}}$ and $a \to b \in D^{\mathcal{A}}$. By b) $a \to a \in D^{\mathcal{A}}$, which completes the proof.

□

Theorem 8.5.3 *For any algebra* $\mathcal{A} = \langle A, \vee, \wedge, \to, \sim \rangle$, *the following holds:*

$$\mathcal{A} \in \mathcal{V}_{\mathbf{N4}} \iff \mathcal{A} \text{ is an } \mathbf{N4}\text{-lattice.}$$

Proof. If $\mathcal{A}$ is an **N4**-lattice, then by Proposition 8.4.3 $\mathcal{A}$ can be isomorphically embedded into $\mathcal{B}^{\bowtie}$ for a suitable implicative lattice $\mathcal{B}$. Checking that the defining identities of $\mathcal{V}_{\mathbf{N4}}$ are satisfied in $\mathcal{A}$ is reduced in this way to a routine calculation with the help of twist-operations and the application of well-known identities of implicative lattices. In case of identity 6^N, we will need also Proposition 8.4.2.a).

We now take an algebra $\mathcal{A} \in \mathcal{V}_{\mathbf{N4}}$ and check successively Items 1–5 of Definition 8.4.1.

1. $\mathcal{A}$ is a De Morgan algebra by definition.

2. Let us check that the relation $\preceq$ on $\mathcal{A}$ is reflexive and transitive. Note that for $a, b \in \mathcal{A}$, $a \preceq b$ if and only if $a \to b \in D^{\mathcal{A}}$. By Lemma 8.5.2 the relation $a \preceq a$ holds for any $a \in \mathcal{A}$.

Take elements $a, b, c \in \mathcal{A}$ such that $a \preceq b$ and $b \preceq c$. We then have $(a \to b) \wedge (b \to c) \in D^{\mathcal{A}}$ by Lemma 8.5.2.f). This fact and identity 6^N imply then $a \preceq c$. In this way, $\preceq$ is transitive.

3. Check first the congruence properties of the equivalence $\approx$. Let $a, b, c, d \in \mathcal{A}$, $a \approx b$ and $c \approx d$.

We have by Lemma 8.5.2.d) $b \to (b \vee d) \in D^{\mathcal{A}}$, moreover, $a \to b \in D^{\mathcal{A}}$ by assumption. By Lemma 8.5.2.f) and identity 6^N we obtain $a \to (b \vee d) \in D^{\mathcal{A}}$. In a similar way, from $c \to d \in D^{\mathcal{A}}$ we infer $c \to (b \vee d) \in D^{\mathcal{A}}$. Further, applying Lemma 8.5.2.f) and identity 4^N yields

$$(a \vee c) \to (b \vee d) = (a \to (b \vee d)) \wedge (c \to (b \vee d)) \in D^{\mathcal{A}},$$

i.e., $a \vee c \preceq b \vee d$. We omit the verification of the inverse inequality.

From the relations $a \to b, c \to d \in D^{\mathcal{A}}$ one can conclude by 6^N and Lemma 8.5.2.f) that the implications $(a \wedge c) \to b$ and $(a \wedge c) \to d$ are in $D^{\mathcal{A}}$. Now we need identity 3^N to obtain $(a \wedge c) \to (b \wedge d) \in D^{\mathcal{A}}$.

Finally, consider the case of implication. We have $(b \to a) \wedge (a \to c) \preceq (b \to c)$ by identity 6^N, whence, $(b \to a) \to ((a \to c) \to (b \to c)) \in D^{\mathcal{A}}$ by 2^N. By assumption, $b \to a \in D^{\mathcal{A}}$, from which we infer by Lemma 8.5.2.c) $(a \to c) \preceq (b \to c)$. Similarly, the inequality $(b \to c) \preceq (b \to d)$ can be obtained from $(b \to c) \wedge (c \to d) \preceq (b \to d)$ and the assumption $c \to d \in D^{\mathcal{A}}$. By transitivity of $\preceq$ we have $(a \to c) \preceq (b \to d)$. Analogously, one can check the inverse inequality.

We now show that $\mathcal{A}_{\bowtie} := \langle A, \vee, \wedge, \to \rangle / \approx$ is an implicative lattice. $\mathcal{A}_{\bowtie}$ is a distributive lattice as a quotient of a distributive lattice. Therefore, it remains to prove that $\to$ is a relative pseudo-complement operation on $\mathcal{A}_{\bowtie}$, i.e., that for any $a, b, c \in \mathcal{A}$, the following condition is met:

$$[c]_{\approx} \leq [a \to b]_{\approx} \text{ if and only if } [a]_{\approx} \wedge [c]_{\approx} \leq [b]_{\approx}.$$

We can replace it by the condition:

$$c \approx c \wedge (a \to b) \text{ if and only if } a \wedge c \wedge b \approx a \wedge c,$$

which is equivalent by Lemma 8.5.2.d) to:

$$c \preceq c \wedge (a \to b) \text{ if and only if } a \wedge c \preceq a \wedge c \wedge b,$$

and by Lemma 8.5.2.g) to:

$$c \preceq (a \to b) \text{ if and only if } a \wedge c \preceq b.$$

The latter immediately follows from the equality $c \to (a \to b) = (a \wedge c) \to b$, which is true by 2^N and the commutativity of $\wedge$.

4. In view of Items b) and e) of Lemma 8.5.2 and 2^N, identities 7^N–9^N can be rewritten as $\sim (p \to q) \preceq p$, $\sim (p \to q) \preceq \sim q$, and $p \wedge \sim q \preceq \sim (p \to q)$ respectively. These facts together mean that the identity $\sim (p \to q) \approx p \wedge \sim q$ holds on $\mathcal{A}_{\bowtie}$.

5. Let $a, b \in \mathcal{A}$. Assume $a \leq b$, then the inequality $\sim b \leq \sim a$ follows by (8.1). By Lemma 8.5.2.d) we have $a \preceq b$ and $\sim b \preceq \sim a$.

We now assume that $a \preceq b$ and $\sim b \preceq \sim a$ and try to prove $a \leq b$. The idea of this proof is taken from the work [89] by A. Pynko, where it was realized in another setting. But as we will see, it works also for $\mathcal{V}_{\mathbf{N4}}$. First we state some new identities in essentially the same way as was done in [89] for the class of implicative De Morgan lattices.

Lemma 8.5.4 *The following identities hold in* $\mathcal{V}_{\mathbf{N4}}$*:*

a) $\sim p \to \sim p = p \to p$;

b) $\sim (p \to p) \leq q \to q$;

c) $q \to q \leq (p \to q) \to (p \to q)$.

Proof. a) Take an algebra $\mathcal{A} \in \mathcal{V}_{\mathbf{N4}}$ and any $a, b \in \mathcal{A}$. From 1^N and 8^N we obtain

$$b \to b \leq (a \to a) \to (\sim b \to \sim ((a \to a) \to b)) = \sim b \to \sim b.$$

Similarly, one can state the inverse inequality $\sim b \to \sim b \leq b \to b$.

b) By Lemma 8.5.2 we have $(a \to a) \wedge (b \to b) \in D^{\mathcal{A}}$. At the same time, by 5^N and a) for any $c \in D^{\mathcal{A}}$, holds $\sim c \leq \sim c \to \sim c = c \to c = c$. Hence, $\sim (a \to a) \vee \sim (b \to b) \leq (a \to a) \wedge (b \to b)$. Finally, $\sim (a \to a) \leq b \to b$.

c) We calculate

$$(a \to b) \to (a \to b) = (a \to b) \to ((a \wedge a) \to b) = [\text{by } 2^N]$$

$$= (a \to b) \to (a \to (a \to b)) = [\text{by } 2^N] \; (a \wedge (a \to b)) \to (a \to b) = [\text{by } 2^N]$$

$$
\begin{aligned}
&= a \to ((a \to b) \to (a \to b)) = [\text{by a)}]\\
&= a \to ({\sim} (a \to b) \to {\sim} (a \to b)) = [\text{by } 5^N]\\
&= a \to (({\sim} b \wedge {\sim} (a \to b)) \to {\sim} (a \to b)) = [\text{by } 2^N]\\
&= {\sim} (a \to b) \to (a \to ({\sim} b \to {\sim} (a \to b))) \geq [\text{by } 5^N]\\
&\geq a \to ({\sim} b \to {\sim} (a \to b)) \geq [\text{by } 8^N]\ b \to b.
\end{aligned}
$$

□

Due to Item c) of the previous lemma conditions $a \preceq b$ and ${\sim} b \preceq {\sim} a$ imply $b \to b \leq a \to b$ and ${\sim} a \to {\sim} a \leq {\sim} b \to {\sim} a$. These inequalities together with 10^N and (8.1) yield

$$a \wedge (b \to b) \leq a \wedge (a \to b) \leq b \vee {\sim} ({\sim} b \to {\sim} a) \leq b \vee {\sim} ({\sim} a \to {\sim} a).$$

The inequality ${\sim} a \to {\sim} a \leq {\sim} b \to {\sim} a$, identity 11^N and (8.1) allow us to obtain

$$a \leq (b \to b) \vee {\sim} ({\sim} a \to {\sim} a).$$

From this fact and item b) of Lemma 8.5.4, we infer $a \leq b \to b$. Substituting in 11^N ${\sim} b$ for p and ${\sim} a$ for q yields ${\sim} b \leq {\sim} a \to {\sim} a$, and so ${\sim} ({\sim} a \to {\sim} a) \leq b$ by (8.1). Comparing this with the chain of inequalities above and taking into account $a \leq b \to b$ we obtain $a \leq b$.

□

This theorem allows one to state one more completeness result for **N4**.

Theorem 8.5.5 *For any formula φ, the following equivalence holds:*

$$\vdash_{\mathbf{N4}} \varphi \iff \varphi \to \varphi = \varphi \in Eq(\mathcal{V}_{\mathbf{N4}}).$$

Proof. Due to Theorem 8.4.5 the fact $\vdash_{\mathbf{N4}} \varphi$ means that for any **N4**-lattice $\mathcal{A}$ and valuation v, $v(\varphi) \in D^{\mathcal{A}}$. By definition of $D^{\mathcal{A}}$ the latter means exactly that for any $\mathcal{A} \in \mathcal{V}_{\mathbf{N4}}$, the identity $\varphi \to \varphi = \varphi$ holds in $\mathcal{A}$.

□

The latter characterization of the logic **N4** is the most attractive, because it can be extended to any extension of **N4**, moreover, we can establish a dual isomorphism between the lattice $\mathcal{E}\mathbf{N4}$ and the lattice $Sub(\mathcal{V}_{\mathbf{N4}})$ of all subvarieties of $\mathcal{V}_{\mathbf{N4}}$.

For any $L \in \mathcal{E}\mathbf{N4}$, define a variety

$$Var(L) := \{\mathcal{A} \mid \varphi \to \varphi = \varphi \in Eq(\mathcal{A}) \text{ for all } \varphi \in L\}.$$

Clearly, $Var(L) \in Sub(\mathcal{V}_{\mathbf{N4}})$. Moreover, the logic L is characterized by the variety $Var(L)$.

Proposition 8.5.6 *For any logic $L \in \mathcal{E}\mathbf{N4}$ and a formula φ, we have the equivalence:*

$$\varphi \in L \iff \varphi \to \varphi = \varphi \in Eq(Var(L)).$$

Proof. The direct implication is by definition of $Var(L)$. To state the inverse implication consider the canonical twist-structure $\mathcal{L}_L^{\bowtie}$ for the logic L (see Section 8.3). Prove that $\mathcal{L}_L^{\bowtie} \in Var(L)$. By definition

$$|\mathcal{L}_L^{\bowtie}| = \{([\varphi], [\sim \psi]) \mid \psi \in [\varphi]\},$$

where $[\varphi]$ is a coset of φ wrt the equivalence $\equiv_L$. Taking into account that $[\varphi] = [\psi]$ for $\psi \in [\varphi]$ we obtain

$$|\mathcal{L}_L^{\bowtie}| = \{([\varphi], [\sim \varphi]) \mid \varphi \in For\}.$$

Let $\varphi = \varphi(p_1, \dots, p_n)$ and an $\mathcal{L}_L^{\bowtie}$-valuation v be such that $v(p_i) = ([\psi_i], [\sim \psi_i])$, $i = 1, \dots, n$. Using induction on the structure of formulas, we state that

$$v(\varphi) = ([\varphi(\psi_1, \dots, \psi_n)], [\sim \varphi(\psi_1, \dots, \psi_n)]).$$

In this way, if $\varphi \in L$, then for any $\mathcal{L}_L^{\bowtie}$-valuation v we have $\pi_1 v(\varphi) = 1$, i.e., $v(\varphi) \in D^{\mathcal{L}_L^{\bowtie}}$. Consequently, $\varphi \to \varphi = \varphi \in Eq(\mathcal{L}_L^{\bowtie})$ and $\mathcal{L}_L^{\bowtie} \in Var(L)$.

Take a formula $\varphi \notin L$ and $\mathcal{L}_L^{\bowtie}$-valuation v such that $v(p) := ([p], [\sim p])$. Then

$$v(\varphi) = ([\varphi], [\sim \varphi]).$$

By assumption $L \nvdash \varphi$, consequently, $[\varphi] \neq 1$. This means that $v(\varphi) \notin D^{\mathcal{L}_L^{\bowtie}}$, i.e., $\varphi \to \varphi = \varphi \notin Eq(\mathcal{L}_L^{\bowtie})$.

□

Corollary 8.5.7 *1. If $L_1, L_2 \in \mathcal{E}\mathbf{N4}$, then*

$$L_1 \subseteq L_2 \iff Var(L_2) \subseteq Var(L_1).$$

2. The mapping $Var : \mathcal{E}\mathbf{N4} \to Sub(\mathcal{V}_{\mathbf{N4}})$ is one-to-one.

Proof. 1. The direct implication immediately follows from the definition. Assume that $Eq(Var(L_1)) \subseteq Eq(Var(L_2))$ and take $\varphi \in L_1$. Then $\varphi \to \varphi = \varphi \in Eq(Var(L_2))$. By the previous proposition we have $\varphi \in L_2$.

2. This item follows from the previous one.

□

Now for any $V \in Sub(\mathcal{V}_{\mathbf{N4}})$, define a set of formulas

$$L(V) := \{\varphi \mid \varphi \to \varphi = \varphi \in Eq(V)\}.$$

Proposition 8.5.8 *1. For any $V \in Sub(\mathcal{V}_{\mathbf{N4}})$, the set $L(V)$ is a logic and $L(V) \in \mathcal{E}\mathbf{N4}$.*

2. The mappings $Var : \mathcal{E}\mathbf{N4} \to Sub(\mathcal{V}_{\mathbf{N4}})$ and $L : Sub(\mathcal{V}_{\mathbf{N4}}) \to \mathcal{E}\mathbf{N4}$ are mutually inverse.

Proof. 1. It is clear that $\mathbf{N4} \subseteq L(V)$ and that the set $L(V)$ is closed under substitution. Assume $\varphi, \varphi \to \psi \in L(V)$ and take an arbitrary **N4**-lattice $\mathcal{A} \in V$ and a valuation v into $\mathcal{A}$. By assumption $v(\varphi \to \varphi) = v(\varphi)$ and $v((\varphi \to \psi) \to (\varphi \to \psi)) = v(\varphi \to \psi)$, i.e., $v(\varphi), v(\varphi) \to v(\psi) \in D^{\mathcal{A}}$. Hence, due to Lemma 8.5.2 we have $v(\psi) \in D^{\mathcal{A}}$. The latter means that $\psi \to \psi = \psi \in Eq(V)$. Thus, $L(V)$ is closed under *modus ponens* and $L(V) \in \mathcal{E}\mathbf{N4}$.

2. It follows from Proposition 8.5.6 that $L(Var(L)) = L$ for any $L \in \mathcal{E}\mathbf{N4}$. Let us compare the varieties V and $Var(L(V))$. It follows from the definition that $Var(L(V)) \subseteq V$. To state the inverse inclusion we need the following lemma.

Lemma 8.5.9 *For any **N4**-lattice $\mathcal{A}$ and formulas φ and ψ, we have $\varphi = \psi \in Eq(\mathcal{A})$ if and only if*

$$(\varphi \Leftrightarrow \psi) \to (\varphi \Leftrightarrow \psi) = \varphi \Leftrightarrow \psi \in Eq(\mathcal{A}).$$

Proof. Without loss of generality we may assume that $\mathcal{A}$ is a twist-structure over a suitable implicative lattice $\mathcal{B}$. It is enough to check the equivalence: $\varphi = \psi \in Eq(\mathcal{A})$ if and only if $v(\varphi \Leftrightarrow \psi) \in D^{\mathcal{A}}$ for any $\mathcal{A}$-valuation v. Due to Lemma 8.4.4 the condition $v(\varphi \Leftrightarrow \psi) \in D^{\mathcal{A}}$ is equivalent to $\pi_1 v(\varphi \Leftrightarrow \psi) = 1$.

Let v be an arbitrary $\mathcal{A}$-valuation. Assume $v(\varphi) = (a, b)$ and $v(\psi) = (c, d)$ for some $a, b, c, d \in B$. The direct computation shows that

$$\pi_1 v(\varphi \Leftrightarrow \psi) = (a \leftrightarrow c) \wedge (b \leftrightarrow d).$$

It is obvious that $(a \leftrightarrow c) \wedge (b \leftrightarrow d) = 1$ if and only if $a = c$ and $b = d$, i.e., $v(\varphi) = v(\psi)$.

□

For brevity we denote the identity $\varphi \to \varphi = \varphi$ as φ^N.

We now turn to the inclusion $V \subseteq Var(L(V))$. Let $\varphi = \psi \in Eq(V)$. By the previous lemma we have $(\varphi \Leftrightarrow \psi)^N \in Eq(V)$, then $\varphi \Leftrightarrow \psi \in L(V)$ and $(\varphi \Leftrightarrow \psi)^N \in Eq(Var(L(V)))$. Again by Lemma 8.5.9 $\varphi = \psi \in Eq(Var(L(V)))$.

□

Theorem 8.5.10 *The mapping Var is a dual lattice isomorphism between $\mathcal{E}\mathbf{N4}$ and $Sub(\mathcal{V}_{\mathbf{N4}})$.*

Proof. By Item 2 of Proposition 8.5.8 the mapping Var is a bijection. Therefore, the fact that Var is a dual lattice isomorphism follows from Corollary 8.5.7.

8.6 The Logic N4$^{\perp}$ and N4$^{\perp}$-Lattices

First, we consider when twist-structures over implicative lattices are bounded.

Lemma 8.6.1 *Let $\mathcal{A}$ be an implicative lattice and $\mathcal{B} \in S^{\bowtie}(\mathcal{A})$.*

1. *If $\mathcal{A}$ has the least element 0, then $(0,1), (1,0) \in \mathcal{B}$ and $\mathcal{B}$ is a bounded lattice with zero element $(0,1)$ and unit element $(1,0)$.*
2. *If $\mathcal{B}$ is bounded, then $\mathcal{A}$ has the least element 0, zero element of $\mathcal{B}$ is $(0,1)$ and unit element is $(1,0)$.*

Proof. 1. By definition of twist-structures there are $a, b \in \mathcal{A}$ such that $(0,a), (1,b) \in \mathcal{B}$, then $(0,1) = (0,a) \wedge (b,1) \in \mathcal{B}$ and $(1,0) = \sim (0,1) \in \mathcal{B}$. That $(0,1)$ is the least and $(1,0)$ the greatest element of $\mathcal{B}$ follows from the definition of twist-operations.

2. Assume that (a,b) is the least element of $\mathcal{B}$ and $c \in \mathcal{A}$. There is $d \in \mathcal{A}$ such that $(c,d) \in \mathcal{B}$. Then $(a,b) \leq (c,d)$, whence $a \leq c$. Thus, a is the least element of $\mathcal{A}$.

□

This allows us to define twist-structures over Heyting algebras as follows.

Definition 8.6.2 *Let $\mathcal{A} = \langle A, \vee, \wedge, \to, 0, 1 \rangle$ be a Heyting algebra.*

1. *A* full twist-structure *over $\mathcal{A}$ is an algebra*

$$\mathcal{A}^{\bowtie} = \langle A \times A, \vee, \wedge, \to, \sim, \perp \rangle$$

such that $\langle A \times A, \vee, \wedge, \to, \sim \rangle$ is the full twist-structure over the implicative lattice $\langle A, \vee, \wedge, \to, 1 \rangle$ and $\perp = (0,1)$.

2. *A* twist-structure *over $\mathcal{A}$ is an arbitrary subalgebra $\mathcal{B}$ of the full twist-structure $\mathcal{A}^{\bowtie}$ such that $\pi_1(\mathcal{B}) = \mathcal{A}$.*
3. *The class of all twist-structures over $\mathcal{A}$ is denoted $S^{\bowtie}(\mathcal{A})$.*

Definition 8.6.3 *An algebra* $\mathcal{A} = \langle A, \vee, \wedge, \to, \sim, \bot \rangle$ *is an* $\mathbf{N4}^{\bot}$-lattice *if* $\langle A, \vee, \wedge, \to, \sim \rangle$ *is an* **N4***-lattice and* $\bot$ *is the least element of* $\mathcal{A}$. *In this case* $\mathcal{A}_{\bowtie} := \langle A, \vee, \wedge, \to, \bot \rangle / \approx$ *is a Heyting algebra, which we call a* basic Heyting algebra *of* $\mathcal{A}$.

We can easily prove the equivalence of the last two definitions.

Proposition 8.6.4 *Every twist-structure* $\mathcal{B} \in S^{\bowtie}(\mathcal{A})$ *over a Heyting algebra* $\mathcal{A}$ *is an* $\mathbf{N4}^{\bot}$*-lattice. Every* $\mathbf{N4}^{\bot}$*-lattice* $\mathcal{A}$ *is isomorphic to a twist-structure over* $\mathcal{A}_{\bowtie}$ *and an isomorphism is given by the rule*

$$a \mapsto ([a]_{\approx}, [\sim a]_{\approx}).$$

□

The mapping $i^{\bowtie} : \mathcal{A} \hookrightarrow (\mathcal{A}_{\bowtie})^{\bowtie}$, $i^{\bowtie}(a) = ([a]_{\approx}, [\sim a]_{\approx})$, is called a *canonical embedding* of an $\mathbf{N4}^{\bot}$-lattice into the full twist-structure over the basic Heyting algebra.

The validity of formulas in twist-structures over Heyting algebras and in $\mathbf{N4}^{\bot}$-lattices is defined in the same way as for twist-structures over implicative lattices and for **N4**-lattices. Note that although an $\mathbf{N4}^{\bot}$-lattice $\mathcal{A}$ has the greatest element, its set of distinguished values is $D^{\mathcal{A}} = \{a \to a \mid a \in \mathcal{A}\}$ and it is not one-element in the general case.

To prove that $\mathbf{N4}^{\bot}$ is complete with respect to the class of twist-structures over Heyting algebras we have only to note that $\mathcal{L}_{\mathbf{N4}^{\bot}}$ is a Heyting algebra due to axiom A14. Consequently, the Lindenbaum twist-structure $\mathcal{L}^{\bowtie}_{\mathbf{N4}}$ is a twist-structure over a Heyting algebra.

Theorem 8.6.5 *For any formula* φ, $\mathbf{N4}^{\bot} \vdash \varphi$ *if and only if* $\mathcal{A} \models \varphi$ *for any* $\mathbf{N4}^{\bot}$*-lattice* $\mathcal{A}$.

□

Taking into account that the action of twist-operations on the first component agreed with the operations of the basic Heyting algebra, we easily obtain

Corollary 8.6.6 $\mathbf{N4}^{\bot}$ *is a conservative extension of intuitionistic logic.*

□

Since $\mathbf{N4}^{\bot}$-lattices are exactly bounded **N4**-lattices, they also form a variety determined by the identities of de Morgan algebras, the set of identities 1^N–11^N and the new identity:

12^N. $\bot \leq p$

The mappings $Var : \mathcal{E}\mathbf{N4}^{\perp} \to Sub(\mathcal{V}_{\mathbf{N4}^{\perp}})$ and $L : Sub(\mathcal{V}_{\mathbf{N4}^{\perp}}) \to \mathcal{E}\mathbf{N4}^{\perp}$ are defined exactly as for the logic **N4** and the variety of **N4**-lattices. Repeating the reasoning from the previous section we prove

Theorem 8.6.7 *The mappings Var and L are mutually inverse dual lattice isomorphisms between $\mathcal{E}\mathbf{N4}^{\perp}$ and $Sub(\mathcal{V}_{\mathbf{N4}^{\perp}})$.*

□

Chapter 9

$\mathbf{N4}^{\perp}$-Lattices[1]

In the previous chapter, we considered two variants of Nelson's paraconsistent logic, **N4** and $\mathbf{N4}^{\perp}$, which were defined in different languages. $\mathbf{N4}^{\perp}$ is a conservative extension of **N4** in the language with the additional constant $\perp$ allowing us to define in this logic the intuitionistic negation. The addition of this constant results in an easy modification of the semantics. Models of **N4** are isomorphic to twist-structures over implicative lattices and due to the fact that implicative lattices do not necessarily have the least element and the lattices modelling **N4** do not have the greatest element either. In fact, no constant can be naturally defined in **N4**. Extending the language with $\perp$ we obtain the class of models isomorphic to twist-structures over Heyting algebras, i.e., implicative lattices with the least element 0. A twist-structure over bounded lattice is also bounded, it contains necessarily elements $(0, 1)$ and $(1, 0)$, which are the least and the greatest elements. Thus, the semantics for $\mathbf{N4}^{\perp}$ is given by the class of bounded lattices. It turns out that the introduction of $\perp$ has essential consequences for the class of extensions $\mathcal{E}\mathbf{N4}^{\perp}$. As we will see in Chapter 10, adding the constant $\perp$ enriches the class of $\mathbf{N4}^{\perp}$-extensions as compared to $\mathcal{E}\mathbf{N4}$, and, which is more important, provides it with a regular structure close to some extent to the structure of the class of extensions of minimal logic studied in the first part of the book. This is why we will work mainly with the logic $\mathbf{N4}^{\perp}$ and with $\mathbf{N4}^{\perp}$-lattices. However, all results in this chapter remains true if we replace "$\mathbf{N4}^{\perp}$-lattice" with "**N4**-lattice" and "Heyting algebra" with "implicative lattice".

Our investigation of $\mathbf{N4}^{\perp}$-lattices is based essentially on the representation of N-lattices obtained by A. Sendlewski [100]. N-lattices are exactly

[1]Parts of this chapter were originally published in [74] (Springer, Netherlands). Reprinted here by permission of the publisher.

$\mathbf{N4}^{\perp}$-lattices modelling **N3**. It is well known (see, [28, 110, 47]) that N-lattices can be represented as twist-structures over Heyting algebras satisfying the condition $a \wedge b = 0$ for any (a,b). We call such twist-structures *explosive*. D. Vakarelov [110] suggested the following intuitive interpretation of an explosive twist-structure: the underlying Heyting algebra $\mathcal{A}$ represents a set of assertions of an intuitionistic theory, i.e., it is a Lindenbaum algebra of this theory. Let a and b be assertions. We say that b is a counter-example of a if $a \wedge b$ is contradictory. For intuitionistic theory, the latter means exactly that $a \wedge b = 0$. Thus, an explosive twist-structure is a set of statements of an intuitionistic theory together with their counter-examples. The twist-operations determine how to construct a counter-example of a complex statement from counter-examples of components. For instance, a counter-example of a conjunction of two assertions is a disjunction of counter-examples of these assertions, $(a, b) \wedge (c, d) = (a \wedge b, c \vee d)$. Any statement can be considered as a counter-example of its counter-example, therefore, $\sim (a, b) = (b, a)$. Of course, any assertion has a counter-example, therefore, for any twist-structure $\mathcal{B}$ over $\mathcal{A}$, $\pi_1(\mathcal{B}) = \mathcal{A}$.

To represent arbitrary $\mathbf{N4}^{\perp}$-lattices we need arbitrary, not only explosive twist-structures over Heyting algebras. Therefore, the condition $a \wedge b = 0$ should be omitted and the Vakarelov interpretation should be modified. One of the possible solutions to this problem is as follows: we consider the notion of a counter-example as a primary notion, independent of the notion of contradiction in the underlying intuitionistic theory. As above, we consider a twist-structure $\mathcal{B}$ over $\mathcal{A}$ as some set of pairs (a, b) consisting of an assertion $a \in \mathcal{A}$ and its counter-example b. Of course, this set should be closed with respect to twist-operations. The contradictory assertion is now defined as an assertion of the form $a \wedge b$ for $(a, b) \in \mathcal{B}$, i.e., as a conjunction of an assertion and its counter-example.

A. Sendlewski [100] proved that an explosive twist-structure $\mathcal{B}$ over a Heyting algebra $\mathcal{A}$ is completely determined by a filter on $\mathcal{A}$ consisting of all elements of the form $a \vee b$, where $(a, b) \in \mathcal{B}$. On the other hand, for any filter ∇ over $\mathcal{A}$ containing the filter of dense elements, there exists an explosive twist-structure $\mathcal{B}$ over $\mathcal{A}$ such that $\nabla = \{a \vee b \mid (a, b) \in \mathcal{B}\}$. In what follows, we call ∇ a *filter of completions* of the twist-structure $\mathcal{B}$.

In this chapter, we generalize Sendlewski's characterization to arbitrary twist-structures over Heyting algebras. An explosive twist-structure is characterized by conditions on the join and meet of an assertion and its counter-example, i.e., on the join and meet of the first and second components of its elements. The join belongs to some fixed filter of the basic Heyting algebra and the meet is contradictory. The second condition means the same for any explosive twist-structure, namely, to be the least element of the basic Heyting algebra. But in case of arbitrary twist-structures, this condition may be varied. This leads to the hypothesis that a twist-structure $\mathcal{B}$ over a Heyting algebra $\mathcal{A}$ is determined by this algebra, by the set of possible joins of assertions and their counter-examples, which plausibly forms a filter on $\mathcal{A}$ as well as in Sendlewski's characterization, and by the set of contradictions, i.e., elements of $\mathcal{A}$ of the form $a \wedge b$ for $(a, b) \in \mathcal{B}$. It turns out that this set forms an ideal on the basic Heyting algebra, which we call an *ideal of contradictions*. One may assume that there is also some interrelation between the ideal of contradictions and the filter of completions but as we will see below, it is not the case.

We prove the main result of the chapter on the representation of twist-structures in Section 9.1. Moreover, we formulate and prove the embeddability criterium for $\mathbf{N4}^{\perp}$-lattices in terms of the above representation.

In Section 9.2, we deal with homomorphisms of $\mathbf{N4}^{\perp}$-lattices. In particular, we define special filters of the first kind (*sffk*) on $\mathbf{N4}^{\perp}$-lattices and prove that *sffk* are exactly kernels of homomorphisms of $\mathbf{N4}^{\perp}$-latices. Moreover, it is proved that *sffk* on an $\mathbf{N4}^{\perp}$-lattice $\mathcal{A}$ are in a one-to-one correspondence with filters on its basic Heyting algebra, which implies the isomorphism of congruence lattices $Con(\mathcal{A}) \cong Con(\mathcal{A}_{\bowtie})$. The latter fact allows one to obtain a characterization of subdirectly irreducible $\mathbf{N4}^{\perp}$-lattices. Namely, an $\mathbf{N4}^{\perp}$-lattice is subdirectly irreducible if and only if its basic Heyting algebra is subdirectly irreducible. A similar result on subdirectly irreducible N-lattices was proved in [99] and independently in [17], the proof was based on the Priestley duality.

The results of this and the previous chapters constitute the necessary semantic basis for the investigation of the lattice of logics $\mathcal{E}\mathbf{N4}^{\perp}$.

9.1 Structure of $\mathbf{N4}^{\perp}$-Lattices

In this section, we obtain a characterization of arbitrary twist-structures similar to that of A. Sendlewski [100].

First, we note that the condition $a \wedge b = 0$ can be replaced by the condition $a \wedge b \in \Delta$, where Δ is an arbitrary ideal of the basic Heyting algebra of a twist-structure.

Recall that an *ideal* on a Heyting algebra (implicative lattice) $\mathcal{A}$ is a non-empty subset $X \subseteq \mathcal{A}$ such that: 1) if $a, b \in X$, then $a \vee b \in X$; 2) if $a \in X$ and $b \leq a$, then $b \in X$. The set of all ideals on $\mathcal{A}$ is denoted by $\mathcal{I}(\mathcal{A})$.

Proposition 9.1.1 *Let $\mathcal{A}$ be a Heyting algebra, ∇ be a filter on $\mathcal{A}$ such that $\nabla \supseteq F_d(\mathcal{A})$, and let Δ be an ideal on $\mathcal{A}$. Then there exists a twist-structure $Tw(\mathcal{A}, \nabla, \Delta) \in S^{\bowtie}(\mathcal{A})$ with the universe*

$$|Tw(\mathcal{A}, \nabla, \Delta)| = \{(a, b) \mid a, b \in |\mathcal{A}|, a \vee b \in \nabla, a \wedge b \in \Delta\}.$$

Proof. Check that the set $T := |Tw(\mathcal{A}, \nabla, \Delta)|$ is closed under twist operations. Let $(a, b), (c, d) \in |\mathcal{A}| \times |\mathcal{A}|$ be such that $a \vee b, c \vee d \in \nabla$ and $a \wedge b, c \wedge d \in \Delta$.

For the disjunction $(a, b) \vee (c, d) = (a \vee c, b \wedge d)$, we have $(a \vee c) \vee (b \wedge d) = (a \vee c \vee b) \wedge (a \vee c \vee d)$. Both conjunction terms in the latter expression belong to ∇, and so $(a \vee c) \vee (b \wedge d) \in \nabla$. In a similar way, we have $(a \vee c) \wedge (b \wedge d) = (a \wedge b \wedge d) \vee (c \wedge b \wedge d) \in \Delta$. Thus, $(a, b) \vee (c, d) \in T$.

The case of conjunction can be considered similarly. Let us consider the implication $(a, b) \to (c, d) = (a \to c, a \wedge d)$. We have $(a \to c) \vee (a \wedge d) = ((a \to c) \vee a) \wedge ((a \to c) \vee d)$. The element $(a \to c) \vee a$ belongs to ∇, because it is dense. At the same time, $c \vee d \leq (a \to c) \vee d \in \nabla$. In this way, $(a \to c) \vee (a \wedge d) \in \nabla$. Further, $(a \to c) \wedge (a \wedge d) = a \wedge c \wedge d$. This element belongs to Δ by the condition $c \wedge d \in \Delta$. We have proved $(a, b) \to (c, d) \in T$.

It is obvious that $\sim (a, b) = (b, a) \in T$.

□

It turns out that any twist-structure can be represented in the form described in the last proposition.

Proposition 9.1.2 *Let $\mathcal{A}$ be a Heyting algebra and $\mathcal{B} \in S^{\bowtie}(\mathcal{A})$. We define*

$$I(\mathcal{B}) := \{a \vee \sim a \mid a \in \mathcal{B}\}, \ \nabla(\mathcal{B}) := \pi_1(I(\mathcal{B})), \ \Delta(\mathcal{B}) := \pi_2(I(\mathcal{B})).$$

Then $F_d(\mathcal{A}) \subseteq \nabla(\mathcal{B})$, $\nabla(\mathcal{B})$ is a filter on $\mathcal{A}$, and $\Delta(\mathcal{B})$ is an ideal on $\mathcal{A}$. Moreover,

$$\mathcal{B} = Tw(\mathcal{A}, \nabla(\mathcal{B}), \Delta(\mathcal{B})).$$

Proof. First, we obtain a more convenient definition for $I(\mathcal{B})$ and a representation of $\mathcal{B}$ via $I(\mathcal{B})$.

Lemma 9.1.3 *Let $\mathcal{A}$ be a Heyting algebra and $\mathcal{B} \in S^{\bowtie}(\mathcal{A})$. Then*

$$|\mathcal{B}| = \{(a, b) \mid a, b \in A, (a \vee b, a \wedge b) \in I(\mathcal{B})\},$$

$$I(\mathcal{B}) = \{(a \vee b, a \wedge b) \mid (a, b) \in \mathcal{B}\} = \{(a, b) \mid (a, b) \in \mathcal{B}, a \geq b\}.$$

Proof. Let us check the first equation. If $(a, b) \in \mathcal{B}$, then $(a \vee b, a \wedge b) = (a, b) \vee (b, a) = (a, b) \vee \sim (a, b) \in I(\mathcal{B})$.

To check the inverse inclusion take an element $(a, b) \in |\mathcal{A}| \times |\mathcal{A}|$ such that $(a \vee b, a \wedge b) \in I(\mathcal{B})$ and prove that $(a, b) \in \mathcal{B}$. By Definition 8.6.2 there is an element $c \in \mathcal{A}$ such that $(a, c) \in \mathcal{B}$. We have $I(\mathcal{B}) \subseteq \mathcal{B}$ by definition, therefore, $(a \wedge b, a \vee b) = \sim (a \vee b, a \wedge b) \in \mathcal{B}$. Further,

$$(a, c) \to (a \wedge b, a \vee b) = (a \to (a \wedge b), a \wedge (a \vee b)) = (a \to b, a) \in \mathcal{B}.$$

And finally, we have

$$(a \vee b, a \wedge b) \wedge (a \to b, a) = ((a \wedge (a \to b)) \vee (b \wedge (a \to b)), a) =$$

$$= ((a \wedge b) \vee b, a) = (b, a).$$

Thus, the elements (b, a) and $(a, b) = \sim (b, a)$ are in $\mathcal{B}$.

The representation for $I(\mathcal{B})$ immediately follows from the facts that for any $(a, b) \in \mathcal{B}$, $(a, b) \vee \sim (a, b) = (a \vee b, a \wedge b)$, and $(a, b) \vee \sim (a, b) = (a, b)$ whenever $a \geq b$.

□

This allows us to conclude that a twist-structure $\mathcal{B}$ is completely determined by the set $I(\mathcal{B})$.

Corollary 9.1.4 *Let $\mathcal{B}_1$ and $\mathcal{B}_2$ be twist-structures over a Heyting algebra $\mathcal{A}$. Assume $I(\mathcal{B}_1) = I(\mathcal{B}_2)$. Then $\mathcal{B}_1 = \mathcal{B}_2$.*

□

We now turn to the proof of the proposition and show that $\Delta(\mathcal{B})$ is an ideal of $\mathcal{A}$. We need one more lemma.

Lemma 9.1.5 $\Delta(\mathcal{B}) = \{a \in \mathcal{A} \mid (1, a) \in \mathcal{B}\} = \pi_2(D^{\mathcal{B}})$.

Proof. If $(1, a) \in \mathcal{B}$, then $(1, a) \vee \sim (1, a) = (1, a) \vee (a, 1) = (1, a)$, i.e., $(1, a) \in I(\mathcal{B})$ and $a \in \Delta(\mathcal{B})$. Conversely, let $a \in \Delta(\mathcal{B})$. This means that there is a $d = (b, c) \in \mathcal{B}$ such that $\pi_2(d \vee \sim d) = a$, i.e., $a = b \wedge c$. Further, $(b, c) \to (b, c) = (1, b \wedge c) = (1, a) \in \mathcal{B}$.

□

Let $a, b \in \Delta(\mathcal{B})$. In view of the previous lemma $(1, a), (1, b) \in \mathcal{B}$, whence $(1, a) \wedge (1, b) = (1, a \vee b) \in \mathcal{B}$, and so $a \vee b \in \Delta(\mathcal{B})$.

Now assume that $a \in \Delta(\mathcal{B})$ and $b \in \mathcal{A}$, $b \leq a$. Let $c \in \mathcal{A}$ be such that $(c, b) \in \mathcal{B}$. We have $(1, a) \vee (c, b) = (1, a \wedge b) = (1, b) \in \mathcal{B}$, which means that $b \in \Delta(\mathcal{B})$.

We have thus proved that $\Delta(\mathcal{B})$ is an ideal. Let us check that $\nabla(\mathcal{B})$ is a filter on $\mathcal{A}$ and $F_d(\mathcal{A}) \subseteq \nabla(\mathcal{B})$.

Let $a, b \in \nabla(\mathcal{B})$ and elements $c, d \in \Delta(\mathcal{B})$ are such that $(a, c), (b, d) \in I(\mathcal{B})$. In view of Lemma 9.1.3 we have $c \leq a$, $d \leq b$. The fact that $\Delta(\mathcal{B})$ is an ideal and Lemma 9.1.5 imply $(1, c \wedge d) \in I(\mathcal{B})$, whence the element $(a, c) \to (c \wedge d, 1) = (a \to (c \wedge d), a)$ is in $\mathcal{B}$. We have also $(a \to (c \wedge d), a) \wedge (a, c) = (a \wedge c \wedge d, a \vee c) = (c \wedge d, a) \in \mathcal{B}$. From this fact and Lemma 9.1.3 we have $(a, c \wedge d) \in I(\mathcal{B})$. Similarly, $(b, c \wedge d) \in I(\mathcal{B})$. Further, $(a, c \wedge d) \wedge (b, c \wedge d) = (a \wedge b, c \wedge d) \in I(\mathcal{B})$, from which we conclude $a \wedge b \in \nabla(\mathcal{B})$.

Assume $a \in \nabla(\mathcal{B})$, $b \in \mathcal{A}$, and $a \leq b$. In this case, there exist elements $c, d \in \mathcal{A}$ such that $a \geq c$, $(a, c) \in I(\mathcal{B})$, and $(b, d) \in \mathcal{B}$. We have then $(a \vee b, c \wedge d) = (b, c \wedge d) \in \mathcal{B}$. Moreover, $c \wedge d \leq a \leq b$, whence $(b, c \wedge d) \in I(\mathcal{B})$, i.e., $b \in \nabla(\mathcal{B})$.

Thus, $\nabla(\mathcal{B})$ is a filter. Let $a, b \in \mathcal{A}$ be such that $(a, b) \in \mathcal{B}$. Then

$$(a, b) \vee ((a, b) \to (0, 1)) = (a \vee (a \to 0), a \wedge b) \in B.$$

We have $a \wedge b \leq a \leq a \vee (a \to 0)$, consequently, $a \vee (a \to 0) \in \nabla(\mathcal{B})$ and we conclude that $F_d(\mathcal{A}) \subseteq \nabla(\mathcal{B})$.

It remains to prove the equality $\mathcal{B} = Tw(\mathcal{A}, \nabla(\mathcal{B}), \Delta(\mathcal{B}))$. As it follows directly from definitions $(a, b) \in Tw(\mathcal{A}, \nabla(\mathcal{B}), \Delta(\mathcal{B}))$ if and only if $(a \vee b, a \wedge b) \in I(\mathcal{B})$. In view of Lemma 9.1.3, the latter fact means that $(a, b) \in \mathcal{B}$. $\square$

This proposition can be generalized to arbitrary **N4**$^{\perp}$-lattices as follows.

Corollary 9.1.6 *Let $\mathcal{A} = \langle A, \vee, \wedge, \to, \sim, \perp \rangle$ be an* **N4**$^{\perp}$*-lattice. We define*

$$\nabla_l(\mathcal{A}) := e_{\bowtie}(\{a \vee \sim a \mid a \in A\}), \ \Delta_l(\mathcal{A}) := e_{\bowtie}(\{a \wedge \sim a \mid a \in A\}),$$

where $e_{\bowtie} : \langle A, \vee, \wedge, \to, \perp \rangle \to \mathcal{A}_{\bowtie}$ is a canonical epimorphism. Then $\nabla_l(\mathcal{A})$ is a filter on $\mathcal{A}_{\bowtie}$, $F_d(\mathcal{A}_{\bowtie}) \subseteq \nabla_l(\mathcal{A})$, and $\Delta_l(\mathcal{A})$ is an ideal on $\mathcal{A}_{\bowtie}$. Moreover, we have $i^{\bowtie}(\mathcal{A}) = Tw(\mathcal{A}_{\bowtie}, \nabla_l(\mathcal{A}), \Delta_l(\mathcal{A}))$, where $i^{\bowtie} : \mathcal{A} \hookrightarrow (\mathcal{A}_{\bowtie})^{\bowtie}$ is a canonical embedding, i.e.,

$$\mathcal{A} \cong Tw(\mathcal{A}_{\bowtie}, \nabla_l(\mathcal{A}), \Delta_l(\mathcal{A})).$$

Proof. Let us consider a twist-structure $\mathcal{A}' := i^{\bowtie}(\mathcal{A})$. Clearly, $\mathcal{A}' \in S^{\bowtie}(\mathcal{A}_{\bowtie})$ and it has the form $\mathcal{A}' = Tw(\mathcal{A}_{\bowtie}, \nabla(\mathcal{A}'), \Delta(\mathcal{A}'))$ by the previous proposition. In this way, the conclusion of the corollary will follow from the equations $\nabla_l(\mathcal{A}) = \nabla(\mathcal{A}')$ and $\Delta_l(\mathcal{A}) = \Delta(\mathcal{A}')$, which we prove now.

By Proposition 8.4.3 each element $b \in \mathcal{A}'$ is of the form $b = (e_{\bowtie}(a), e_{\bowtie}(\sim a))$ for some $a \in \mathcal{A}$. Thus,

$$b \vee \sim b = (e_{\bowtie}(a), e_{\bowtie}(\sim a)) \vee (e_{\bowtie}(\sim a), e_{\bowtie}(a)) = (e_{\bowtie}(a \vee \sim a), e_{\bowtie}(a \wedge \sim a)).$$

We have in this way the equalities:

$$\nabla(\mathcal{A}') = \pi_1(I(\mathcal{A}')) = \{e_{\bowtie}(a \vee \sim a) \mid a \in A\} = \nabla_l(\mathcal{A}),$$
$$\Delta(\mathcal{A}') = \pi_2(I(\mathcal{A}')) = \{e_{\bowtie}(a \wedge \sim a) \mid a \in A\} = \Delta_l(\mathcal{A}).$$

□

Remark. If $\mathcal{B} \in S^{\bowtie}(\mathcal{A})$ and $h : \mathcal{B}_{\bowtie} \to \mathcal{A}$ is an isomorphism given by the rule $h([(a, b)]_{\approx}) = a$ (see Proposition 8.4.2), then

$$h(\nabla_l(\mathcal{B})) = \nabla(\mathcal{B}) \text{ and } h(\Delta_l(\mathcal{B})) = \Delta(\mathcal{B}).$$

For this reason we omit later on the lower index "$-_l$" in the denotation $\nabla_l(\mathcal{A})$ and $\Delta_l(\mathcal{A})$.

As mentioned, all results in this chapter also hold true for **N4**-lattices. To formulate the above results for twist-structures over implicative lattices, we have to define a filter of dense elements for an arbitrary implicative lattice.

Let $\mathcal{A}$ be an implicative lattice. We define

$$F_d(\mathcal{A}) := \langle \{a \vee (a \to b) \mid a, b \in \mathcal{A}\} \rangle$$

and call $F_d(\mathcal{A})$ a *filter of dense elements* of $\mathcal{A}$. If $\mathcal{A}$ has the least element, the filter $F_d(\mathcal{A})$ will coincide with the filter of dense elements of $\mathcal{A}$ considered as a Heyting algebra. This explains why we use the same name and denotation.

It turns out that any element of $F_d(\mathcal{A})$ has the form $a \vee (a \to b)$.

Proposition 9.1.7 *For any implicative lattice $\mathcal{A}$,*

$$F_d(\mathcal{A}) = \{a \vee (a \to b) \mid a, b \in \mathcal{A}\}$$

Proof. Let $a \in F_d(\mathcal{A})$. Then one can find elements $a_1, \ldots, a_n, b_1, \ldots, b_n \in \mathcal{A}$ such that $a \geq \bigwedge_{i=1}^n a_i \vee (a_i \to b_i)$. For the element $d = \bigwedge_{i=1}^n a_i \wedge \bigwedge_{i=1}^n b_i$, consider the implicative lattice $\mathcal{A}^d$, a sublattice of $\mathcal{A}$ with the universe $\{a \in \mathcal{A} \mid a \geq d\}$. It is clear that $a \in F_d(\mathcal{A}^d) \subseteq F_d(\mathcal{A})$. On the other hand, $\mathcal{A}^d$ has the least element and the filter $F_d(\mathcal{A}^d)$ coincides with the filter of dense elements of the Heyting algebra $\mathcal{A}^d$. Therefore, for some $c \in \mathcal{A}^d$, $a = c \vee \neg c = c \vee (c \to d)$. We have thus obtained the desired representation for an arbitrary element of $F_d(\mathcal{A})$.

□

Analogs of Propositions 9.1.1 and 9.1.2 can be proved for twist-structures over implicative lattices in essentially the same way.

Proposition 9.1.8 *Let $\mathcal{A}$ be an implicative lattice, ∇ be a filter on $\mathcal{A}$ such that $\nabla \supseteq F_d(\mathcal{A})$, and let Δ be an ideal on $\mathcal{A}$. Then there exists a twist-structure $Tw(\mathcal{A}, \nabla, \Delta) \in S^{\bowtie}(\mathcal{A})$ with the universe*

$$|Tw(\mathcal{A}, \nabla, \Delta)| = \{(a, b) \mid a, b \in |\mathcal{A}|, a \vee b \in \nabla, a \wedge b \in \Delta\}.$$

□

Proposition 9.1.9 *Let $\mathcal{A}$ be an implicative lattice and $\mathcal{B} \in S^{\bowtie}(\mathcal{A})$. We define*

$$I(\mathcal{B}) := \{a \vee \sim a \mid a \in \mathcal{B}\}, \ \nabla(\mathcal{B}) := \pi_1(I(\mathcal{B})), \ \Delta(\mathcal{B}) := \pi_2(I(\mathcal{B})).$$

Then $F_d(\mathcal{A}) \subseteq \nabla(\mathcal{B})$, $\nabla(\mathcal{B})$ is a filter on $\mathcal{A}$, and $\Delta(\mathcal{B})$ is an ideal on $\mathcal{A}$. Moreover,

$$\mathcal{B} = Tw(\mathcal{A}, \nabla(\mathcal{B}), \Delta(\mathcal{B})).$$

An analog of Corollary 9.1.6 also holds. In the following we will formulate results only for twist-structures over Heyting algebras and for **N4**$^{\perp}$-lattices.

At the end of this section, we make some remarks on subalgebras of twist-structures.

Proposition 9.1.10

1. *Let $\mathcal{B} \in S^{\bowtie}(\mathcal{A})$. Then $\mathcal{B} = \mathcal{A}^{\bowtie}$ if and only if $\nabla(\mathcal{B}) = \Delta(\mathcal{B}) = \mathcal{A}$.*
2. *Let $\mathcal{A}_0$ and $\mathcal{A}_1$ be Heyting algebras, $\nabla_i \in \mathcal{F}(\mathcal{A}_i)$, $F_d(\mathcal{A}_i) \subseteq \nabla_i$, and $\Delta_i \in \mathcal{I}(\mathcal{A}_i)$, $i = 0, 1$. Moreover, let $\mathcal{A}_0 \leq \mathcal{A}_1, \nabla_0 \subseteq \nabla_1$, and $\Delta_0 \subseteq \Delta_1$. Then*
$$Tw(\mathcal{A}_0, \nabla_0, \Delta_0) \leq Tw(\mathcal{A}_1, \nabla_1, \Delta_1).$$
3. *Let $\mathcal{B} \in S^{\bowtie}(\mathcal{A})$ and $\mathcal{C} \leq \mathcal{B}$. Then $\mathcal{C} = Tw(\mathcal{A}_1, \nabla, \Delta)$, where $\mathcal{A}_1 \leq \mathcal{A}$, $\nabla \in \mathcal{F}(\mathcal{A}_1)$ and $\nabla \subseteq \nabla(\mathcal{B})$, $\Delta \in \mathcal{I}(\mathcal{A}_1)$ and $\Delta \subseteq \Delta(\mathcal{B})$.*
4. *Let $\mathcal{B} \in S^{\bowtie}(\mathcal{A})$ and $\mathcal{A}_1 \leq \mathcal{A}$. There exists a subalgebra $\mathcal{C}$ of $\mathcal{B}$ such that $\mathcal{C} \in S^{\bowtie}(\mathcal{A}_1)$.*
5. *Let $\mathcal{B} \in S^{\bowtie}(\mathcal{A})$ and $\mathcal{A}_1 \leq \mathcal{A}$. The greatest subalgebra $\mathcal{C}$ of $\mathcal{B}$ satisfying the condition $\mathcal{C} \in S^{\bowtie}(\mathcal{A}_1)$ has the following form*
$$\mathcal{C} = Tw(\mathcal{A}_1, \nabla(\mathcal{B}) \cap \mathcal{A}_1, \Delta(\mathcal{B}) \cap \mathcal{A}_1).$$

Proof. 1. Let $\mathcal{B} = \mathcal{A}^{\bowtie}$, i.e., $|\mathcal{B}| = |\mathcal{A}| \times |\mathcal{A}|$. Then

$$I(\mathcal{B}) = \{(a, b) \mid a \geq b, a, b \in \mathcal{A}\}.$$

This fact and the definition of $\nabla(\mathcal{B})$ and $\Delta(\mathcal{B})$ as projections of $I(\mathcal{B})$ onto the first and the second coordinates, respectively, imply $\nabla(\mathcal{B}) = \Delta(\mathcal{B}) = \mathcal{A}$.

If $\nabla(\mathcal{B}) = \Delta(\mathcal{B}) = \mathcal{A}$, then

$$|\mathcal{B}| = \{(a, b) \mid a \vee b \in \nabla(\mathcal{B}), a \wedge b \in \Delta(\mathcal{B})\} = \{(a, b) \mid a, b \in \mathcal{A}\}.$$

2. This item follows immediately from the definition of twist-structures $Tw(\mathcal{A}_i, \nabla_i, \Delta_i)$, $i = 0, 1$.

3. The operations $\vee, \wedge$, and $\to$ are componentwise with respect to the first coordinate, therefore, $\mathcal{A}_1 := \pi_1(\mathcal{C}) = \pi_2(\mathcal{C})$ is a subalgebra of $\mathcal{A}$. In this case $\mathcal{C} \in S^{\bowtie}(\mathcal{A}_1)$. Obviously, $I(\mathcal{C}) = \{a \vee \sim a \mid a \in \mathcal{C}\} \subseteq I(\mathcal{B})$, from which we have $\nabla(\mathcal{C}) = \pi_1(I(\mathcal{C})) \subseteq \pi_1(I(\mathcal{B})) = \nabla(\mathcal{B})$. Similarly, $\Delta(\mathcal{C}) \subseteq \Delta(\mathcal{B})$.

4. Note that $F_d(\mathcal{A}_1) \subseteq F_d(\mathcal{A})$, therefore, $F_d(\mathcal{A}_1) \subseteq \nabla(\mathcal{B}) \cap \mathcal{A}_1$. Moreover, $0 \in \Delta(\mathcal{B}) \cap \mathcal{A}_1 \neq \varnothing$. Since $\nabla(\mathcal{B}) \cap \mathcal{A}_1$ is obviously a filter and $\Delta(\mathcal{B}) \cap \mathcal{A}_1$ an ideal on $\mathcal{C}$, we can consider a twist-structure $\mathcal{C} = Tw(\mathcal{A}_1, \nabla(\mathcal{B}) \cap \mathcal{A}_1, \Delta(\mathcal{B}) \cap \mathcal{A}_1)$. That $\mathcal{C} \leq \mathcal{B}$ follows from Item 2.

5. As was noted above, $\mathcal{C}$ is a subalgebra of $\mathcal{B}$. By Item 2 it is the greatest subalgebra with the condition $\mathcal{C} \in S^{\bowtie}(\mathcal{A}_1)$.

□

9.2 Homomorphisms and Subdirectly Irreducible $\mathbf{N4}^{\perp}$-Lattices

In Definition 8.6.2 for any Heyting algebra $\mathcal{A}$, there was defined the $\mathbf{N4}^{\perp}$-lattice $\mathcal{A}^{\bowtie}$, the full twist-structure over $\mathcal{A}$. At the same time, for any $\mathbf{N4}^{\perp}$-lattice $\mathcal{A}$, there was defined the Heyting algebra $\mathcal{A}_{\bowtie}$. We extend these mappings to homomorphisms of Heyting algebras and of $\mathbf{N4}^{\perp}$-lattices. For any Heyting algebra homomorphism $h : \mathcal{A} \to \mathcal{B}$, the homomorphism of $\mathbf{N4}^{\perp}$-lattices $h^{\bowtie} : \mathcal{A}^{\bowtie} \to \mathcal{B}^{\bowtie}$ is defined by the rule: $h^{\bowtie}(a, b) := (h(a), h(b))$. That this is really a homomorphism follows immediately from the definition of twist-structures.

Now let $\mathcal{A}$ and $\mathcal{B}$ be $\mathbf{N4}^{\perp}$-lattices. For a homomorphism $h : \mathcal{A} \to \mathcal{B}$, we define a mapping $h_{\bowtie} : \mathcal{A}_{\bowtie} \to \mathcal{B}_{\bowtie}$ by the rule $h_{\bowtie}([a]_{\approx}) := [h(a)]_{\approx}$. The facts that h is a homomorphism and that the relation $\approx$ is definable via basic operations of an $\mathbf{N4}^{\perp}$-lattice guarantee that the above definition is correct. Due to the congruence properties of $\approx$, $h_{\bowtie}$ is a homomorphism of Heyting algebras. It turns out that we have defined a functor from the category $\mathcal{H}$ of Heyting algebras and their homomorphisms to the category $\mathcal{N}4^\perp$ of **N4**$^\perp$-lattices and their homomorphisms and the adjoint functor from $\mathcal{N}4^\perp$ to $\mathcal{H}$.

Theorem 9.2.1[2] *The mapping* $(\cdot)^{\bowtie}$ *is a covariant functor from the category* $\mathcal{H}$ *to the category* $\mathcal{N}4^\perp$. *The mapping* $(\cdot)_{\bowtie}$ *is a covariant functor from the category* $\mathcal{N}4^\perp$ *to the category* $\mathcal{H}$. *The functor* $(\cdot)_{\bowtie}$ *is left adjoint to* $(\cdot)^{\bowtie}$.

Proof. One can directly verify that $(\cdot)^{\bowtie}$ and $(\cdot)_{\bowtie}$ are covariant functors. We prove only that $(\cdot)_{\bowtie}$ is left adjoint to $(\cdot)^{\bowtie}$. To this end, it will be enough to show (see e.g. [12]) that the family of canonical embeddings $i^{\bowtie}_{\mathcal{B}}$, $\mathcal{B} \in \mathcal{N}4^\perp$, determines a natural transformation of an identical functor $1_{\mathcal{N}4^\perp}$ to the composition $((\cdot)_{\bowtie})^{\bowtie}$ such that for any $\mathcal{B} \in \mathcal{N}4^\perp$, the pair $(\mathcal{B}_{\bowtie}, i^{\bowtie}_{\mathcal{B}})$ is a reflection of $\mathcal{B}$ along $(\cdot)^{\bowtie}$. The latter means that for any Heyting algebra $\mathcal{A}$ and homomorphism $f : \mathcal{B} \to \mathcal{A}^{\bowtie}$ there exists a unique homomorphism $g : \mathcal{B}_{\bowtie} \to \mathcal{A}$ such that $f = g^{\bowtie} \circ i^{\bowtie}_{\mathcal{B}}$.

First, take a homomorphism $f : \mathcal{B} \to \mathcal{C}$ of **N4**$^\perp$-lattice and check the equality $i^{\bowtie}_{\mathcal{C}} \circ f = (f_{\bowtie})^{\bowtie} \circ i^{\bowtie}_{\mathcal{C}}$. For any $b \in \mathcal{B}$, we have

$$i^{\bowtie}_{\mathcal{C}} \circ f(b) = ([f(b)]_{\approx_\mathcal{C}}, [\sim f(b)]_{\approx_\mathcal{C}}) = ([f(b)]_{\approx_\mathcal{C}}, [f(\sim b)]_{\approx_\mathcal{C}}) =$$

$$= (f_{\bowtie}([b]_{\approx_\mathcal{B}}), f_{\bowtie}([\sim b]_{\approx_\mathcal{B}})) = (f_{\bowtie})^{\bowtie}([b]_{\approx_\mathcal{B}}, [\sim b]_{\approx_\mathcal{B}}) = (f_{\bowtie})^{\bowtie} \circ i^{\bowtie}_{\mathcal{C}}(b).$$

We have thus established that the family $i^{\bowtie}_{\mathcal{B}}$, $\mathcal{B} \in \mathcal{N}4^\perp$, determines the desired natural transformation.

Further, for a Heyting algebra $\mathcal{A}$ and a homomorphism $f : \mathcal{B} \to \mathcal{A}^{\bowtie}$ we define a mapping $g : \mathcal{B}_{\bowtie} \to \mathcal{A}$ by the rule $g([a]_{\approx_\mathcal{B}}) := \pi_1 f(a)$ for all $a \in \mathcal{B}$. This definition is correct due to the following observation. If $a, b \in \mathcal{B}$ and $a \approx_\mathcal{B} b$, then $f(a) \approx_{\mathcal{A}^{\bowtie}} f(b)$, from which we obtain $\pi_1 f(a) = \pi_1 f(b)$ by Proposition 8.4.2. That g is a Heyting algebra homomorphism follows from the congruence properties of $\approx_\mathcal{B}$ and the definition of positive operations on $\mathcal{A}^{\bowtie}$. For $b \in \mathcal{B}$, we calculate

$$g^{\bowtie} \circ i^{\bowtie}_{\mathcal{B}}(b) = (g([b]_{\approx_\mathcal{B}}), g([\sim b]_{\approx_\mathcal{B}})) = (\pi_1 f(b), \pi_1 f(\sim b)) =$$

$$= (\pi_1 f(b), \pi_1(\sim f(b))) = (\pi_1 f(b), \pi_2 f(b)) = f(b).$$

[2]This is the unique place where we use the special notions from the category theory. For this reason, we do not give all the necessary definitions and ask the reader to consult, e.g., [12]. Note that categorical properties of **N3**-lattices were studied by A. Sendlewski [100] and R. Cignoli [17].

On the other hand, for a homomorphism $h : \mathcal{B}_{\bowtie} \to \mathcal{A}$, the equality $f = h^{\bowtie} \circ i_{\mathcal{B}}^{\bowtie}$ immediately implies $h([a]_{\approx_{\mathcal{B}}}) := \pi_1 f(a)$ for $a \in \mathcal{B}$. Thus, g is a unique homomorphism with the desired property.

□

For Heyting algebras and N-lattices, interrelations of this kind were proved by V. Goranko [33] and A. Sendlewski [99, 100].

We note also that every homomorphism h of $\mathbf{N4}^{\perp}$-lattices is uniquely determined by the homomorphism $h_{\bowtie}$ of basic Heyting algebras.

Proposition 9.2.2 *Let $\mathcal{A}$ and $\mathcal{B}$ be $\mathbf{N4}^{\perp}$-lattices, and let $h^i : \mathcal{A} \to \mathcal{B}$, $i = 1, 2$, be homomorphisms such that $h^1_{\bowtie} = h^2_{\bowtie}$. Then $h^1 = h^2$.*

Proof. Let $g_i := (h^i_{\bowtie})^{\bowtie}$, $i = 1, 2$. Since $h^1_{\bowtie} = h^2_{\bowtie}$, we have $g_1 = g_2$. For any $a, b \in \mathcal{A}$,

$$g_i([a]_{\approx_{\mathcal{A}}}, [b]_{\approx_{\mathcal{A}}}) = (h^i_{\bowtie}([a]_{\approx_{\mathcal{A}}}), h^i_{\bowtie}([a]_{\approx_{\mathcal{A}}})) = ([h^i(a)]_{\approx_{\mathcal{B}}}, [h^i(b)]_{\approx_{\mathcal{B}}}).$$

Further, for $a \in \mathcal{A}$, we have

$$h^i(a) = (i_{\mathcal{B}}^{\bowtie})^{-1}([h^i(a)]_{\approx_{\mathcal{B}}}, [\sim h^i(a)]_{\approx_{\mathcal{B}}}) = (i_{\mathcal{B}}^{\bowtie})^{-1}([h^i(a)]_{\approx_{\mathcal{B}}}, [h^i(\sim a)]_{\approx_{\mathcal{B}}}).$$

In view of the above calculation, we obtain

$$h^i(a) = (i_{\mathcal{B}}^{\bowtie})^{-1} \circ g_i([a]_{\approx_{\mathcal{A}}}, [\sim a]_{\approx_{\mathcal{A}}}).$$

In this way, the equality $h^1 = h^2$ follows from $g_1 = g_2$.

□

We now introduce the notion of a special filter of the first kind on an $\mathbf{N4}^{\perp}$-lattice in exactly the same way as a special filter of the first kind on an N-lattice was defined by H. Rasiowa [92]. Let $\mathcal{A}$ be an $\mathbf{N4}^{\perp}$-lattice. A non-empty subset $\varnothing \neq \nabla \subseteq \mathcal{A}$ is called a *special filter of the first kind* (*sffk*) on $\mathcal{A}$ if: 1) $a \in \nabla$ and $b \in \nabla$ imply $a \wedge b \in \nabla$; 2) $a \in \nabla$ and $a \preceq b$ imply $b \in \nabla$. It is obvious that the set of all *sffk* on $\mathcal{A}$ forms a lattice, which we denote $\mathcal{F}^1(\mathcal{A})$. For a homomorphism $h : \mathcal{A} \to \mathcal{B}$ of $\mathbf{N4}^{\perp}$-lattices we define a *kernel* $Ker(h) := h^{-1}(D^{\mathcal{B}})$. As we can see from the following proposition, *sffk* are in one-to-one correspondence with congruences on an $\mathbf{N4}^{\perp}$-lattice and the least *sffk* on $\mathcal{A}$ coincides with $D^{\mathcal{A}}$.

Proposition 9.2.3 1. *For any $\mathbf{N4}^{\perp}$-lattice $\mathcal{A}$, $D^{\mathcal{A}}$ is an* sffk.

2. *Let $h : \mathcal{A} \to \mathcal{B}$ be an epimorphism of $\mathbf{N4}^{\perp}$-lattices. Then $Ker(h)$ is an* sffk. *For any $a, b \in \mathcal{A}$, $h(a) = h(b)$ if and only if $a \Leftrightarrow b \in Ker(h)$.*

3. *Let* $\mathcal{A}$ *be an* **N4**$^\perp$*-lattice and* ∇ *an* sffk *on* $\mathcal{A}$. *Then* $D^{\mathcal{A}} \subseteq \nabla$. *The relation* $\approx_\nabla$, $a \approx_\nabla b$ *iff* $a \Leftrightarrow b \in \nabla$, *is a congruence relation on* $\mathcal{A}$ *and* $\nabla = Ker(h)$, *where* $h : \mathcal{A} \to \mathcal{A}/{\approx_\nabla}$ *is a canonical epimorphism.*

Proof. 1. This item follows from the lemma below.

Lemma 9.2.4 *Let* $\mathcal{A}$ *be an* **N4**$^\perp$*-lattice. Then the following holds:*

1. *If* $a, b \in D^{\mathcal{A}}$, *then* $a \wedge b \in D^{\mathcal{A}}$.
2. *If* $a \in \mathcal{A}$ *and* $b \in D^{\mathcal{A}}$, *then* $a \preceq b$;
3. *For any* $a, b \in \mathcal{A}$, $a = b$ *if and only if* $a \Leftrightarrow b \in D^{\mathcal{A}}$.
4. *For any* $a, b \in \mathcal{A}$, $a \wedge (a \to b) \preceq b$.

Proof. Due to Proposition 8.4.3 we can identify $\mathcal{A}$ with a twist-structure over $\mathcal{A}_{\bowtie}$, in which case $D^{\mathcal{A}} = \{(1, a) \mid a \in \mathcal{A}_{\bowtie}\}$ and $(a, b) \preceq (c, d)$ is equivalent to $a \leq b$. The latter two facts easily imply all assertions of the lemma.

□

2. By the previous item $D^{\mathcal{B}}$ is an *sffk*, moreover, h preserves the relation $\preceq$ as definable via basic operations. These facts imply that $Ker(h)$ is an *sffk*. By Lemma 9.2.4 the equality $h(a) = h(b)$ is equivalent to $h(a \Leftrightarrow b) = h(a) \Leftrightarrow h(b) \in D^{\mathcal{B}}$, i.e., $a \Leftrightarrow b \in Ker(h)$.

3. The relation $D^{\mathcal{A}} \subseteq \nabla$ follows from Item 2 of Lemma 9.2.4. Recall that the strong equivalence $\Leftrightarrow$ has congruence properties in **N4**$^\perp$. This means, in particular, that the formula $(p \Leftrightarrow q) \to ((r \Leftrightarrow t) \to ((p \vee r) \Leftrightarrow (q \vee t)))$ is provable in **N4**$^\perp$. By Theorem 8.4.5 for any $a, b, c, d \in \mathcal{A}$,

$$(a \Leftrightarrow b) \to ((c \Leftrightarrow d) \to ((a \vee c) \Leftrightarrow (b \vee d))) \in D^{\mathcal{A}} \subseteq \nabla.$$

If we assume $a \Leftrightarrow b, c \Leftrightarrow d \in \nabla$, we may conclude by properties of *sffk* that $(a \vee c) \Leftrightarrow (b \vee d) \in \nabla$. Other congruence properties can be checked similarly.

Let $[a]$ be a coset of a with respect to $\approx_\nabla$. Assume that $a \in Ker(h)$, i.e., $[a] \in D^{\mathcal{A}/\approx_\nabla}$. This is equivalent to $[a \to a] = [a]$, i.e., $a \to a \approx_\nabla a$ or $(a \to a) \Leftrightarrow a \in \nabla$. Since $a \leq b$ implies $a \preceq b$, the relation $(a \to a) \Leftrightarrow a \in \nabla$ implies $(a \to a) \to a \in \nabla$. By $a \to a \in D^{\mathcal{A}} \subseteq \nabla$ and Item 4 of Lemma 9.2.4 we obtain $a \in \nabla$.

Conversely, let $a \in \nabla$. By identity 5^N (see Definition 8.5.1), we have $(a \to a) \to a \in \nabla$. It follows from $a \to a \in D^{\mathcal{A}} \subseteq \nabla$ and 5^N that $a \to (a \to a) \in \nabla$. At the same time,

$$\sim a \to \sim (a \to a) \approx \sim a \to (a \wedge \sim a) \approx \sim a \to a.$$

Since $a \in \nabla$, we have $\sim a \to a \in \nabla$ and so, $\sim a \to\sim (a \to a) \in \nabla$. Finally,

$$\sim (a \to a) \to\sim a \approx (\sim a \wedge a) \to\sim a \in D^{\mathcal{A}} \subseteq \nabla.$$

Whence $\sim (a \to a) \to\sim a \in \nabla$. In this way, $(a \to a) \Leftrightarrow a \in \nabla$ and $a \in Ker(h)$. We have thus proved the equality $\nabla = Ker(h)$.

□

Corollary 9.2.5 *Let $\mathcal{A}$ be an $\mathbf{N4}^{\perp}$-lattice and $\varnothing \neq \nabla \subseteq \mathcal{A}$. The set ∇ is an* sffk *on $\mathcal{A}$ if and only if: 1) $a, b \in \nabla$ implies $a \wedge b \in \nabla$; 2) $a, a \to b \in \nabla$ implies $b \in \nabla$.*

Proof. The direct implication follows from Item 4 of Lemma 9.2.4. The inverse implication follows from the fact that $D^{\mathcal{A}}$ is the least *sffk* on $\mathcal{A}$. Indeed, if $a \in \nabla$ and $a \preceq b$, then $a \to b \in D^{\mathcal{A}} \subseteq \nabla$.

□

In this way, *sffk* are exactly implicative filters on $\mathbf{N4}^{\perp}$-lattices.

Corollary 9.2.6 *For any $\mathbf{N4}^{\perp}$-lattice $\mathcal{A}$, $\mathcal{F}^1(\mathcal{A})$ is isomorphic to $Con(\mathcal{A})$ and mutually inverse isomorphisms are defined by the rules: $\nabla \mapsto \approx_{\nabla}$ for $\nabla \in \mathcal{F}^1(\mathcal{A})$ and $\theta \mapsto \{a \in \mathcal{A} \mid a\theta(a \to a)\}$ for $\theta \in Con(\mathcal{A})$.*

□

Corollary 9.2.7 *Let $h : \mathcal{A} \to \mathcal{B}$ be a homomorphism of $\mathbf{N4}^{\perp}$-lattices. Then h is a monomorphism if and only if $Ker(h) = D^{\mathcal{A}}$.*

Proof. Assume $h : \mathcal{A} \to \mathcal{B}$ is a monomorphism and $a \notin D^{\mathcal{A}}$, i.e., $a \to a \neq a$, then $h(a) \to h(a) \neq h(a)$. The equality $Ker(h) = D^{\mathcal{A}}$ now follows from the fact that $D^{\mathcal{A}}$ is the least *sffk* on $\mathcal{A}$.

If $Ker(h) = D^{\mathcal{A}}$ and $h(a) = h(b)$ for some $a, b \in \mathcal{A}$, then $a \Leftrightarrow b \in D^{\mathcal{A}}$ by Item 2 of Proposition 9.2.3 and we conclude that $a = b$ by Item 3 of Lemma 9.2.4.

□

As we can see from Proposition 9.2.2, homomorphisms of $\mathbf{N4}^{\perp}$-lattices are completely determined by homomorphisms of basic Heyting algebras. This leads to a suggestion that there is a close connection between *sffk* on an $\mathbf{N4}^{\perp}$-lattice and filters on the basic Heyting algebra. This is really so and we consider first the case of twist-structures.

Proposition 9.2.8 *Let* $\mathcal{B} \in S^{\bowtie}(\mathcal{A})$.

1. *For any* $\nabla \in \mathcal{F}(\mathcal{A})$, $\nabla^{\bowtie} := \pi_1^{-1}(\nabla) = \{(a,b) \in |\mathcal{A}| \times |\mathcal{A}| \mid a \in \nabla\}$ *is an* sffk *on* $\mathcal{B}$.

2. *For any* $\nabla \in \mathcal{F}^1(\mathcal{B})$, $\nabla_{\bowtie} := \pi_1(\nabla)$ *is a filter on* $\mathcal{A}$. *Moreover,* $\nabla = \{(a,b) \in |\mathcal{A}| \times |\mathcal{A}| \mid a \in \nabla_{\bowtie}\} = (\nabla_{\bowtie})^{\bowtie}$.

3. *The lattices* $\mathcal{F}(\mathcal{A})$ *and* $\mathcal{F}^1(\mathcal{B})$ *are isomorphic and the mappings* $\nabla \mapsto \nabla^{\bowtie}$, $\nabla \in \mathcal{F}(\mathcal{A})$ *and* $\nabla \mapsto \nabla_{\bowtie}$, $\nabla \in \mathcal{F}^1(\mathcal{B})$ *determine mutually inverse isomorphisms.*

Proof. Items 1 and 2 can be proved via direct verification, whereas Item 3 is an immediate consequence of them.

□

Recall that due to Proposition 8.4.3 and Definition 8.6.3 any **N4**$^\perp$-lattice $\mathcal{A}$ is isomorphic to a twist-structure over $\mathcal{A}_{\bowtie}$ and the canonical epimorphism $e_{\bowtie} : \mathcal{A} \to \mathcal{A}_{\bowtie}$ corresponds to the projection π_1 of the twist-structure onto the first coordinate. In this way, we generalize the previous statement to all **N4**$^\perp$-lattices.

Proposition 9.2.9 *Let* $\mathcal{A}$ *be an* **N4**$^\perp$*-lattice.*

1. *For any* $\nabla \in \mathcal{F}(\mathcal{A}_{\bowtie})$, $\nabla^{\bowtie} := (e_{\bowtie})^{-1}(\nabla)$ *is an* sffk *on* $\mathcal{A}$.

2. *For any* $\nabla \in \mathcal{F}^1(\mathcal{A})$, $\nabla_{\bowtie} := e_{\bowtie}(\nabla)$ *is a filter on* $\mathcal{A}_{\bowtie}$. *Moreover,* $\nabla = (\nabla_{\bowtie})^{\bowtie}$.

3. *The lattices* $\mathcal{F}(\mathcal{A}_{\bowtie})$ *and* $\mathcal{F}^1(\mathcal{A})$ *are isomorphic and the mappings* $\nabla \mapsto \nabla^{\bowtie}$, $\nabla \in \mathcal{F}(\mathcal{A}_{\bowtie})$ *and* $\nabla \mapsto \nabla_{\bowtie}$, $\nabla \in \mathcal{F}^1(\mathcal{A})$ *determine mutually inverse isomorphisms.*

□

Taking into account the dualities of congruences and filters on Heyting algebras and of congruences and *sffk* on **N4**$^\perp$-lattices, we obtain the following

Corollary 9.2.10 *For any* **N4**$^\perp$*-lattice* $\mathcal{A}$, $Con(\mathcal{A}) \cong Con(\mathcal{A}_{\bowtie})$.

□

The property of an algebra to be subdirectly irreducible is determined by the structure of its congruence lattice. Combining this fact with the previous corollary, we obtain

Corollary 9.2.11 *An* $\mathbf{N4}^{\perp}$*-lattice* $\mathcal{A}$ *is subdirectly irreducible if and only if the Heyting algebra* $\mathcal{A}_{\bowtie}$ *is subdirectly irreducible.*

□

The above description of the congruence lattice of an $\mathbf{N4}^{\perp}$-lattice allows one to prove that the lattice of $\mathbf{N4}^{\perp}$-extensions is distributive.

Proposition 9.2.12 $\mathcal{V}_{\mathbf{N4}^{\perp}}$ *is an arithmetic variety.*

Proof. The congruence distributivity of the variety of Heyting algebras and Corollary 9.2.10 allows one to conclude that the variety of $\mathbf{N4}^{\perp}$-lattices is congruence distributive. Prove that $\mathcal{V}_{\mathbf{N4}^{\perp}}$ is congruence permutable.

Let $\mathcal{A}$ be a Heyting algebra and $\mathcal{B} \in S^{\bowtie}(\mathcal{A})$. Consider some congruences $\theta_1, \theta_2 \in Con(\mathcal{B})$ determined by *sffk* $F_1^{\bowtie}$ and $F_2^{\bowtie}$, where $F_1, F_2 \in \mathcal{F}(\mathcal{A})$, i.e.,

$$((a,b),(c,d)) \in \theta_i \text{ iff } (a,b) \Leftrightarrow (c,d) \in F_i^{\bowtie}, \ \ i = 1,2.$$

The definition of filters $F_i^{\bowtie}$ and calculation with the help of twist-operations allows one to conclude that $(a,b) \Leftrightarrow (c,d) \in F_i^{\bowtie}$ is equivalent to $(a \leftrightarrow c) \wedge (b \leftrightarrow d) \in F_i$. In this way,

$$((a,b),(c,d)) \in \theta_i \text{ iff } (a,c),(b,d) \in \theta_{F_i}, \ \ i = 1,2,$$

where θ_{F_i} is a congruence on $\mathcal{A}$ determined by F_i. This equivalence easily implies that

$$((a,b),(c,d)) \in \theta_1 \circ \theta_2 \text{ iff } (a,c),(b,d) \in \theta_{F_1} \circ \theta_{F_2}.$$

The permutability of congruences on $\mathcal{A}$ imply that θ_1 and θ_2 are now permutable.

□

Proposition 9.2.13 *The lattice* $\mathcal{E}\mathbf{N4}^{\perp}$ *is distributive.*

Proof. This reasoning is similar to Corollary 2.2.11 but contains several peculiarities.

Consider a free ω-generated $\mathbf{N4}^{\perp}$-lattice $\mathcal{A}_\omega$ and its congruence lattice $Con(\mathcal{A}_\omega)$, which is distributive according to the previous proposition. Moreover, the congruences of $Con(\mathcal{A}_\omega)$ are permutable wrt composition. In this case, the elements of $\mathcal{A}_\omega$ can be identified with cosets of formulas wrt the strong equivalence in $\mathbf{N4}^{\perp}$,

$$|\mathcal{A}_\omega| = \{[\varphi]_{\Leftrightarrow} \mid \varphi \in \mathcal{F}\},$$

where $[\varphi]_{\Leftrightarrow} := \{\psi \mid \varphi \Leftrightarrow \psi \in \mathbf{N4}^{\perp}\}$. Each logic $L \in \mathcal{E}\mathbf{N4}^{\perp}$ corresponds to the congruence

$$\theta_L := \{([\varphi_0]_{\Leftrightarrow}, [\varphi_1]_{\Leftrightarrow}) \mid \varphi_0 \Leftrightarrow \varphi_1 \in L\}.$$

Clearly, the mapping $L \mapsto \theta_L$ is one-to-one and preserves the ordering. Check that this is a lattice embedding. First, we verify that for any $L_0, L_1 \in \mathcal{E}\mathbf{N4}^{\perp}$, the congruences $\theta_{L_0} \wedge \theta_{L_1}$ and $\theta_{L_0} \vee \theta_{L_1}$ are of the form θ_L for a suitable logic L. Note that θ_L is closed under substitution, i.e., $[\varphi_0]_{\Leftrightarrow}\theta_L[\varphi_1]_{\Leftrightarrow}$ implies $[\varphi_0(\psi_1, \ldots, \psi_n)]_{\Leftrightarrow}\theta_L[\varphi_1((\psi_1, \ldots, \psi_n))]_{\Leftrightarrow}$ for any formulas $\psi_1, \ldots, \psi_n$.

Let $\theta \in Con(\mathcal{A}_\omega)$ be closed under substitution. Put $L_\theta := \{\varphi \mid [\varphi]_{\Leftrightarrow}\theta[\varphi \to \varphi]_{\Leftrightarrow}\}$. Check that L_θ is a logic from $\mathcal{E}\mathbf{N4}^{\perp}$. Obviously, the set L_θ is closed under substitution. If $\varphi \in \mathbf{N4}^{\perp}$, it is not hard to prove that $\varphi \Leftrightarrow (\varphi \to \varphi) \in \mathbf{N4}^{\perp}$. Therefore, $\mathbf{N4}^{\perp} \subseteq L_\theta$. It remains to check that L_θ is closed under *modus ponens*. According to Corollary 9.2.6 $F_\theta = \{[\varphi]_{\Leftrightarrow} \mid [\varphi]_{\Leftrightarrow}\theta[\varphi \to \varphi]_{\Leftrightarrow}\}$ is an *sffk* on $\mathcal{A}_\omega$. If $\varphi, \varphi \to \psi \in L_\theta$, then $[\varphi]_{\Leftrightarrow}, [\varphi \to \psi]_{\Leftrightarrow} = [\varphi]_{\Leftrightarrow} \to [\psi]_{\Leftrightarrow} \in F_\theta$. From Corollary 9.2.5 we obtain $[\psi]_{\Leftrightarrow} \in F_\theta$, i.e., $\psi \in L_\theta$.

Check the equality $\theta = \theta_{L_\theta}$. It can easily be seen that this equality is equivalent to the following relation

$$([\varphi]_{\Leftrightarrow}, [\psi]_{\Leftrightarrow}) \in \theta \iff ([\varphi \Leftrightarrow \psi]_{\Leftrightarrow}, [(\varphi \Leftrightarrow \psi) \to (\varphi \Leftrightarrow \psi)]_{\Leftrightarrow}) \in \theta$$

for any φ and ψ, which is equivalent in turn to

$$[\varphi \Leftrightarrow \psi]_{\Leftrightarrow} \in F_\theta \iff [\varphi \Leftrightarrow \psi]_{\Leftrightarrow} \Leftrightarrow [(\varphi \Leftrightarrow \psi) \to (\varphi \Leftrightarrow \psi)]_{\Leftrightarrow} \in F_\theta.$$

The latter is a particular case of the relation

$$a \in F \iff a \Leftrightarrow (a \to a) \in F,$$

which was established in the proof of Item 3 of Proposition 9.2.3 for any *sffk* F on any $\mathbf{N4}^{\perp}$-lattice.

Thus, we have proved that the congruences of the form θ_L are exactly the congruences closed under substitution.

Check that $\theta_{L_0} \wedge \theta_{L_1}$ and $\theta_{L_0} \vee \theta_{L_1}$ are closed under substitution. Consider the non-trivial case of $\theta_{L_0} \vee \theta_{L_1}$. Since the lattice $\mathcal{A}_\omega$ is congruence permutable, $\theta_{L_0} \vee \theta_{L_1} = \theta_{L_0} \circ \theta_{L_1}$. Therefore $[\varphi_0]_{\Leftrightarrow}\theta_{L_0} \vee \theta_{L_1}[\varphi_1]_{\Leftrightarrow}$ iff there is a formula ψ such that $[\varphi_0]_{\Leftrightarrow}\theta_{L_0}[\psi]_{\Leftrightarrow}$ and $[\psi]_{\Leftrightarrow}\theta_{L_1}[\varphi_1]_{\Leftrightarrow}$. This equivalence immediately implies that $\theta_{L_0} \vee \theta_{L_1}$ is closed under substitution.

We have proved that the set of congruences θ_L forms a lattice. It is clear that the mapping $L \mapsto \theta_L$ is an order isomorphism of $\mathcal{E}\mathbf{N4}^{\perp}$ and the lattice of congruences of the form θ_L. If two lattices are isomorphic as orderings, they are isomorphic as lattices too.

□

We now prove the criterion when there exists a homomorphism from one $\mathbf{N4}^{\perp}$-lattice into the other, which generalizes the embeddability criterion from the previous section. But first, we point out some facts on interrelations of homomorphisms h and $h_{\bowtie}$.

Proposition 9.2.14 *Let $h : \mathcal{A} \to \mathcal{B}$ be a homomorphism of $\mathbf{N4}^{\perp}$-lattices.*

1. *$Ker(h_{\bowtie}) = (Ker(h))_{\bowtie}$ and $Ker(h) = (Ker(h_{\bowtie}))^{\bowtie}$.*
2. *h is a monomorphism if and only if $h_{\bowtie}$ is a monomorphism.*

Proof. 1. By definition $(Ker(h))_{\bowtie} = \{[a]_{\approx_{\mathcal{A}}} \mid h(a) \in D^{\mathcal{B}}\}$. The condition $h(a) \in D^{\mathcal{B}}$ is equivalent to $h_{\bowtie}([a]_{\approx_{\mathcal{A}}}) = [h(a)]_{\approx_{\mathcal{B}}} = 1_{\mathcal{B}_{\bowtie}}$. From this equivalence we immediately obtain $Ker(h_{\bowtie}) = (Ker(h))_{\bowtie}$. The second equality immediately follows from the first one and Proposition 9.2.9.

2. If h is a monomorphism, then $Ker(h) = D^{\mathcal{A}}$ by Corollary 9.2.7. Taking into account that $(D^{\mathcal{A}})_{\bowtie} = \{1_{\mathcal{A}_{\bowtie}}\}$ we infer from the previous item $Ker(h_{\bowtie}) = \{1_{\mathcal{A}_{\bowtie}}\}$, i.e., $h_{\bowtie}$ is a monomorphism. The inverse implication is similar.

□

Proposition 9.2.15 *Let $\mathcal{A}$ and $\mathcal{B}$ be $\mathbf{N4}^{\perp}$-lattices. There exists a homomorphism (monomorphism) $h : \mathcal{A} \to \mathcal{B}$ if and only if there exists a homomorphism (monomorphism) $g : \mathcal{A}_{\bowtie} \to \mathcal{B}_{\bowtie}$ such that $g(\nabla(\mathcal{A})) \subseteq \nabla(\mathcal{B})$ and $g(\Delta(\mathcal{A})) \subseteq \Delta(\mathcal{B})$.*

Proof. If $h : \mathcal{A} \to \mathcal{B}$ is a homomorphism, then $h_{\bowtie}$ can be chosen as g. Indeed, for $[a \vee \sim a]_{\approx_{\mathcal{A}}} \in \nabla(\mathcal{A})$, we have

$$h_{\bowtie}([a \vee \sim a]_{\approx_{\mathcal{A}}}) = [h(a \vee \sim a)]_{\approx_{\mathcal{B}}} = [h(a) \vee \sim h(a)]_{\approx_{\mathcal{B}}} \in \nabla(\mathcal{B}).$$

Analogously, $h_{\bowtie}(\Delta(\mathcal{A})) \subseteq \Delta(\mathcal{B})$.

If h is additionally a monomorphism, then $h_{\bowtie}$ is a monomorphism by Item 2 of the previous proposition.

Assume that there exists a homomorphism $g : \mathcal{A}_{\bowtie} \to \mathcal{B}_{\bowtie}$ such that $g(\nabla(\mathcal{A})) \subseteq \nabla(\mathcal{B})$ and $g(\Delta(\mathcal{A})) \subseteq \Delta(\mathcal{B})$. Let us consider $g^{\bowtie}$ and check $g^{\bowtie} \circ i^{\bowtie}_{\mathcal{A}}(\mathcal{A}) \subseteq i^{\bowtie}_{\mathcal{B}}(\mathcal{B})$.

Let $(a, b) \in i^{\bowtie}_{\mathcal{A}}(\mathcal{A})$. By Corollary 9.1.6 we have $a \vee b \in \nabla(\mathcal{A})$ and $a \wedge b \in \Delta(\mathcal{A})$. Applying to these facts our assumption we obtain

$g(a) \vee g(b) = g(a \vee b) \in \nabla(\mathcal{B})$ and $g(a) \wedge g(b) = g(a \wedge b) \in \Delta(\mathcal{B})$. Again, by Corollary 9.1.6 we have $g^{\bowtie}((a,b)) = (g(a), g(b)) \in i^{\bowtie}_{\mathcal{B}}(\mathcal{B})$.

Thus, the desired homomorphism from $\mathcal{A}$ to $\mathcal{B}$ can be defined as $h := (i^{\bowtie}_{\mathcal{B}})^{-1} \circ g^{\bowtie} \circ i^{\bowtie}_{\mathcal{A}}$.

If g is a monomorphism, then $g^{\bowtie}$ and h are also monomorphisms.

□

A homomorphic image of a twist-structure has the following presentation.

Proposition 9.2.16 *Let $\mathcal{B}$ be a Heyting algebra, F a filter on $\mathcal{B}$. Let $\mathcal{A} \in S^{\bowtie}(\mathcal{B})$ and $\mathcal{A} = Tw(\mathcal{B}, \nabla, \Delta)$. Then*

$$\mathcal{A}/F^{\bowtie} \cong Tw(\mathcal{B}/F, \nabla/F, \Delta/F).$$

Proof. Define a mapping $h : \mathcal{A}/F^{\bowtie} \to (\mathcal{B}/F)^{\bowtie}$ as follows. For any $(a,b) \in \mathcal{A}$, $h((a,b)/F^{\bowtie}) := (a/F, b/F)$. Clearly, h is a homomorphism. Check that this is a monomorphism. The equality $(a/F, b/F) = (c/F, d/F)$ is equivalent to $(a \leftrightarrow c) \wedge (b \leftrightarrow d) \in F$. At the same time, $(a,b)/F^{\bowtie} = (c,d)/F^{\bowtie}$ is equivalent by Proposition 9.2.3 to $(a,b) \Leftrightarrow (c,d) \in F^{\bowtie}$, which is equivalent in turn to $(a \leftrightarrow c) \wedge (b \leftrightarrow d) \in F$. Thus, h is a monomorphism and it remains to check that

$$h(\mathcal{A}/F^{\bowtie}) = Tw(\mathcal{B}/F, \nabla/F, \Delta/F),$$

Let $\mathcal{A}' := h(\mathcal{A}/F^{\bowtie}) = \{(a/F, b/F) \mid (a,b) \in \mathcal{A}\}$. Then

$$\nabla(\mathcal{A}') = \pi_1(I(\mathcal{A}')) = \pi_1(\{(a/F, b/F) \vee \sim (a/F, b/F) \mid (a,b) \in \mathcal{A}\}) =$$

$$\pi_1(\{(a \vee b/F, a \wedge b/F) \mid (a,b) \in \mathcal{A}\}) = \{a \vee b/F \mid (a,b) \in \mathcal{A}\} = \nabla/F.$$

In a similar way, $\Delta(\mathcal{A}') = \pi_2(I(\mathcal{A}')) = \Delta/F$.

□

Chapter 10

The Class of $\mathbf{N4}^{\perp}$-Extensions[1]

In this chapter, we study the structure of the lattice $\mathcal{E}\mathbf{N4}^{\perp}$ and discover a definite similarity to the structure of the lattice Jhn^{+} studied in the first part of the book. It should be noted that differences in the structure of these two classes of logics are also essential. Moreover, we give first applications of the developed theory: the class of extensions of the logic $\mathbf{N4}^{\perp}\mathbf{C}$ obtained by adding Dummett's linearity axiom $(p \to q) \vee (q \to p)$ to $\mathbf{N4}^{\perp}$ is completely described; two classical results by L.L. Maksimova, namely, the description of pretabular logics and the description of logics with Craig interpolation property, are transferred from the class of superintuitionistic logics to the class of $\mathbf{N4}^{\perp}$-extensions.

10.1 $\mathcal{E}\mathrm{N4}^{\perp}$ and Int^{+}

We start the investigation of the class $\mathcal{E}\mathbf{N4}^{\perp}$ with the question on interrelations between a logic in $\mathcal{E}\mathbf{N4}^{\perp}$ and its intuitionistic fragment. We define a mapping σ from $\mathcal{E}\mathbf{N4}^{\perp}$ into the class Int^{+} of extensions of intuitionistic logic **Li** so that $\sigma(L)$ is a $\langle \vee, \wedge, \to, \perp \rangle$-fragment of L. Later on, we call formulas of the language $\langle \vee, \wedge, \to, \perp \rangle$ *intuitionistic formulas* and denote the set of all intuitionistic formulas by $\mathcal{F}$. As in the first part of the book, Int denotes the lattice of intermediate logics, i.e., $\mathsf{Int}^{+} = \mathsf{Int} \cup \{\mathcal{F}\}$. The restriction of σ to the class $\mathcal{E}\mathbf{N3}$ was investigated by M. Kracht [47] and A. Sendlewski [101].

[1]Parts of this chapter were originally published in [77, 78, 79] (Springer, Netherlands). Reprinted here by permission of the publisher.

First, we point out that $\sigma(L)$ is determined by basic Heyting algebras of L-models. For a class $\mathcal{K}$ of $\mathbf{N4}^{\perp}$-lattices, put

$$\mathcal{K}_{\bowtie} := \{\mathcal{A}_{\bowtie} \mid \mathcal{A} \in \mathcal{K}\}.$$

Proposition 10.1.1 *For any $L \in \mathcal{E}\mathbf{N4}^{\perp}$ and a class $\mathcal{K}$ of $\mathbf{N4}^{\perp}$-lattices, if $L = L\mathcal{K}$, then $\sigma(L) = L\mathcal{K}_{\bowtie}$.*

Proof. Due to Proposition 8.4.3 it is enough to consider twist-structures isomorphic to elements of $\mathcal{K}$. Let $\mathcal{B} \in \mathsf{I}\mathcal{K}$ and $\mathcal{B} \in S^{\bowtie}(\mathcal{A})$ for some Heyting algebra $\mathcal{A}$. Assume that $\mathcal{A}(\cong \mathcal{B}_{\bowtie})$ is not a model for $\sigma(L)$. Let $\varphi \in \sigma(L)$ and $\mathcal{A}$-valuation v are such that $v(\varphi) \neq 1$. For any propositional variable p, there is an element $b_p \in |\mathcal{A}|$ with $(v(p), b_p) \in |\mathcal{B}|$. Let v' be a $\mathcal{B}$-valuation such that $v'(p) = (v(p), b_p)$ for any p. The twist-operations are componentwise wrt the first component, therefore, $\pi_1 v'(\varphi) = v(\varphi)$ for any φ. Thus, $\pi_1 v'(\varphi) \neq 1$ and $\mathcal{B} \not\models \varphi$, which conflicts with the assumption $\mathcal{B} \models \sigma(L) \subseteq L$.

We have thus proved the inclusion $\sigma(L) \subseteq L\mathcal{K}_{\bowtie}$. To check the inverse inclusion take an intuitionistic formula $\varphi \notin \sigma(L)$. Let $\mathcal{B} \in \mathsf{I}\mathcal{K}$, $\mathcal{B} \in S^{\bowtie}(\mathcal{A})$ and a $\mathcal{B}$-valuation v be such that $\pi_1 v(\varphi) \neq 1$. Then $\pi_1 v$ is an $\mathcal{A}$-valuation proving that $\mathcal{A} \not\models \varphi$.

□

The following fact was stated in the last proof and it makes sense to distinguish it as a separate lemma.

Lemma 10.1.2 *Let $\mathcal{A}$ be an $\mathbf{N4}^{\perp}$-lattice and φ an intuitionistic formula. Then*

$$\mathcal{A} \models \varphi \iff \mathcal{A}_{\bowtie} \models \varphi.$$

□

We now study the inverse image $\sigma^{-1}(L)$ for any $L \in \mathsf{Int}^{+}$, i.e., the class of all conservative extensions of L in the class $\mathcal{E}\mathbf{N4}^{\perp}$. We show that $\sigma^{-1}(L)$ forms an interval in the lattice $\mathcal{E}\mathbf{N4}^{\perp}$ and consider the mappings sending L to the end points of the interval $\sigma^{-1}(L)$. Let for any $L \in \mathsf{Int}^{+}$,

$$\eta(L) := \mathbf{N4}^{\perp} + L,$$

i.e., $\eta(L)$ is obtained by extending language and adding the strong negation axioms to L, and

$$\eta^{\circ}(L) = \eta(L) + \{\sim p \rightarrow (p \rightarrow q),\ \neg\neg(p \vee \sim p)\}.$$

Logics in $\mathcal{E}\mathbf{N4}^{\perp}$ having the form $\eta(L)$ we call *special* or *s*-logics, and logics of the form $\eta^{\circ}(L)$ are called *normal explosive* or *ne*-logics.

Prior to proving that $\eta(L)$ and $\eta^{\circ}(L)$ are the end points of the inverse image of L wrt σ, we describe models of $\eta(L)$ and $\eta^{\circ}(L)$ and classes of models generating such logics.

Proposition 10.1.3 *Let $L \in \mathsf{Int}^{+}$ and $\mathcal{A}$ be an $\mathbf{N4}^{\perp}$-lattice.*

1. $\mathcal{A} \models \eta(L)$ *iff* $\mathcal{A}_{\bowtie} \models L$
2. $\mathcal{A} \models \eta^{\circ}(L)$ *iff* $\mathcal{A}_{\bowtie} \models L$, $\nabla(\mathcal{A}) = F_d(\mathcal{A}_{\bowtie})$ *and* $\Delta(\mathcal{A}) = \{0\}$.

Proof. 1. The direct implication follows by Lemma 10.1.2. The inverse implication follows from the same lemma and the fact that the axioms of $\mathbf{N4}^{\perp}$ hold in any $\mathbf{N4}^{\perp}$-lattice.

2. If $\mathcal{A} \models \eta^{\circ}(L)$, then $\mathcal{A}$ is a model of $\mathbf{N3}$ and $\Delta(\mathcal{A}) = \{0\}$. The direct computation shows that for a twist-structure $\mathcal{A} \in S^{\bowtie}(\mathcal{B})$, the condition $\mathcal{A} \models \neg\neg(p \vee \sim p)$ is equivalent to $\neg\neg(a \vee b) = 1$ for any $(a, b) \in \mathcal{A}$, i.e., $\nabla(\mathcal{A}) \subseteq F_d(\mathcal{B})$. The inclusion $F_d(\mathcal{B}) \subseteq \nabla(\mathcal{A})$ holds for any twist-structure, which proves the direct implication.

We now assume that the right hand conditions of the equivalence are satisfied. Due to Lemma 10.1.2 the first condition implies $\mathcal{A} \models L$. We have $\mathcal{A} \models \sim p \to (p \to q)$ by $\Delta(\mathcal{A}) = \{0\}$. Finally, in view of the above considerations, $\nabla(\mathcal{A}) = F_d(\mathcal{A}_{\bowtie})$ implies $\mathcal{A} \models \neg\neg(p \vee \sim p)$.

□

For a Heyting algebra $\mathcal{A}$, define $\mathcal{A}^{\bowtie}_{\circ} := Tw(\mathcal{A}, F_d(\mathcal{A}), \{0\})$. This is the least twist-structure over $\mathcal{A}$. For any $\mathcal{B} \in S^{\bowtie}(\mathcal{A})$, $\mathcal{A}^{\bowtie}_{\circ} \leq \mathcal{B}$. Namely, such lattices are up to isomorphism models of logics of the form $\eta^{\circ}(L)$. If an $\mathbf{N4}^{\perp}$-lattice $\mathcal{A}$ is isomorphic to a lattice of the form $\mathcal{B}^{\bowtie}_{\circ}$, we call it a *normal* $\mathbf{N3}$-lattice. This is really an $\mathbf{N3}$-lattice, since in this case, $\Delta(\mathcal{A}) = \{0\}$. In the following, $\mathbf{N4}^{\perp}$-lattices isomorphic to full twist-structures will be called *special* $\mathbf{N4}^{\perp}$-lattices.

Further, we define translations of $\eta^{\circ}(L)$ and $\eta(L)$ into L. Extend the set $Prop$ of propositional variables adding a new propositional variable $\overline{p}$ for any $p \in Prop$. This extended set of variables will be denoted $\overline{Prop}$. To any formula φ in the language $\mathcal{L}^{\perp}$ we assign intuitionistic formulas $\varphi_{\bowtie}$ and $\varphi^{\circ}_{\bowtie}$ over $\overline{Prop}$ defined as follows. If φ is a *nnf*, then $\varphi_{\bowtie}$ is the result of replacement of every occurrence of $\sim p$ in φ by $\overline{p}$. If φ is not in *nnf*, then $\varphi_{\bowtie} := (\overline{\varphi})_{\bowtie}$. We define formulas $\widetilde{\varphi}$ and $\varphi^{\circ}_{\bowtie}$ as follows:

$$\widetilde{\varphi} := \bigwedge_{p \in var(\varphi)} \neg(p \wedge \overline{p}) \wedge \bigwedge_{p \in var(\varphi)} \neg\neg(p \vee \overline{p}),$$

$$\varphi^{\circ}_{\bowtie} := \widetilde{\varphi} \to \varphi_{\bowtie}.$$

Proposition 10.1.4 *Let $\mathcal{A}$ be a Heyting algebra and $\varphi \in For^{\perp}$. The following equivalences hold.*

1. *$\mathcal{A}^{\bowtie} \models \varphi$ if and only if $\mathcal{A} \models \varphi_{\bowtie}$.*
2. *$\mathcal{A}^{\bowtie}_{\circ} \models \varphi$ if and only if $\mathcal{A} \models \varphi^{\circ}_{\bowtie}$.*

Proof. In the course of the proof we assume that φ is a *nnf*.

1. For any $\mathcal{A}^{\bowtie}$-valuation v, we define $v_{\bowtie} : \overline{Prop} \to |\mathcal{A}|$ by the rule $v_{\bowtie}(p) := \pi_1 v(p)$ and $v_{\bowtie}(\overline{p}) := \pi_2 v(p)$. For any $v : \overline{Prop} \to |\mathcal{A}|$, we define $\mathcal{A}^{\bowtie}$-valuation $v^{\bowtie} : Prop \to |\mathcal{A}| \times |\mathcal{A}|$ by the rule $v^{\bowtie}(p) := (v(p), v(\overline{p}))$. Note that $(v_{\bowtie})^{\bowtie} = v$ for any $\mathcal{A}^{\bowtie}$-valuation v, and $(v^{\bowtie})_{\bowtie} = v$ for any $\mathcal{A}$-valuation v. We have thus defined a bijection between $\mathcal{A}^{\bowtie}$- and $\mathcal{A}$-valuations. Taking into account the fact that the action of intuitionistic operations on a twist-structure agrees with their action on the first component and the equality $\pi_1 v(\sim p) = \pi_2 v(p)$ we obtain

$$\pi_1 v(\varphi) = v_{\bowtie}(\varphi_{\bowtie}) \text{ and } \pi_1 v^{\bowtie}(\varphi) = v(\varphi_{\bowtie})$$

for any *nnf* φ. These relations and the above-stated bijection between $\mathcal{A}^{\bowtie}$- and $\mathcal{A}$-valuations imply the desired equivalence.

2. Let $\mathcal{A} \models \varphi^{\circ}_{\bowtie}$ and v be an $\mathcal{A}^{\bowtie}_{\circ}$-valuation. By definition of $\mathcal{A}^{\bowtie}_{\circ}$ we have $\pi_1 v(p) \wedge \pi_2 v(p) = 0$ and $\pi_1 v(p) \vee \pi_2 v(p) \in F_d(\mathcal{A})$ for any $p \in var(\varphi)$, which implies $v_{\bowtie}(\widetilde{\varphi}) = 1$, whence $v_{\bowtie}(\varphi_{\bowtie}) = 1$. By the equivalence from the previous item $v(\varphi) = 1$.

Assume now $\mathcal{A}^{\bowtie}_{\circ} \models \varphi$ and $v(\varphi^{\circ}_{\bowtie}) \neq 1$ for some $\mathcal{A}$-valuation v. Then $v(\widetilde{\varphi}) \not\leq v(\varphi_{\bowtie})$. Put $F = \langle v(\widetilde{\varphi}) \rangle$ and consider the quotient algebra $\mathcal{A}/F$ and the quotient valuation $v' := v/F$. We have $v'(\widetilde{\varphi}) = 1$ and $v'(\varphi_{\bowtie}) \neq 1$. The equality $v'(\widetilde{\varphi}) = 1$ implies that $\pi_1 (v')^{\bowtie}(p) \wedge \pi_2 (v')^{\bowtie}(p) = 0$ and $\pi_1 (v')^{\bowtie}(p) \vee \pi_2 (v')^{\bowtie}(p) \in F_d(\mathcal{A})$ for any $p \in var(\varphi)$, i.e., $v^{\bowtie}$ is an $(\mathcal{A}/F)^{\bowtie}_{\circ}$-valuation. Arguing as in the previous item from $v'(\varphi_{\bowtie}) \neq 1$, we infer $(\mathcal{A}/F)^{\bowtie}_{\circ} \not\models \varphi$. It remains to note that $(\mathcal{A}/F)^{\bowtie}_{\circ} \cong (\mathcal{A})^{\bowtie}_{\circ}/F^{\bowtie}$ by Proposition 9.2.16.

□

Corollary 10.1.5 *Let $L \in \mathsf{Int}^{+}$ and $\varphi \in For^{\perp}$. The following equivalences hold.*

1. *$\varphi \in \eta(L)$ if and only if $\varphi_{\bowtie} \in L$.*
2. *$\varphi \in \eta^{\circ}(L)$ if and only if $\varphi^{\circ}_{\bowtie} \in L$.*

Proof. 1. Let $\varphi \in \eta(L)$ and $\mathcal{A} \models L$. Then $\mathcal{A}^{\bowtie} \models \eta(L)$ by Proposition 10.1.3, and $\mathcal{A} \models \varphi_{\bowtie}$ by Item 1 of the last proposition. Conversely, let $\varphi_{\bowtie} \in L$ and $\mathcal{A} \models \eta(L)$. Then $\mathcal{A}_{\bowtie} \models L$ by Proposition 10.1.3, and $(\mathcal{A}_{\bowtie})^{\bowtie} \models \eta(L)$. Since $\mathcal{A} \hookrightarrow (\mathcal{A}_{\bowtie})^{\bowtie}$, we have $\mathcal{A} \models \varphi$.

2. This item follows similarly from Item 2 of Proposition 10.1.3.

□

Corollary 10.1.6 *For any $L \in \mathsf{Int}^{+}$, the logics $\eta(L)$ and $\eta^{\circ}(L)$ are conservative extensions of L.*

Proof. Intuitionistic formulas are $\sim$-free, therefore, $\varphi_{\bowtie} = \varphi$ for any such formula. In this way, the fact that $\eta(L)$ is a conservative extension of L follows from Item 1 of Corollary 10.1.5.

For any intuitionistic φ, $\widetilde{\varphi} \in \eta^{\circ}(L)$. Thus, the equivalence $\varphi \leftrightarrow \varphi^{\circ}_{\bowtie}$ is provable in $\eta^{\circ}(L)$ for any intuitionistic φ. Now, Item 2 of Corollary 10.1.5 implies that $\eta^{\circ}(L)$ is conservative over L.

□

Let $\mathcal{K}$ be a class of Heyting algebras. Put

$$\mathcal{K}^{\bowtie} := \{\mathcal{A}^{\bowtie} \mid \mathcal{A} \in \mathcal{K}\} \text{ and } \mathcal{K}^{\bowtie}_{\circ} := \{\mathcal{A}^{\bowtie}_{\circ} \mid \mathcal{A} \in \mathcal{K}\}.$$

Proposition 10.1.3 states, in fact, that

$$\eta(L) = L(Mod(L))^{\bowtie} \text{ and } \eta^{\circ}(L) = L(Mod(L))^{\bowtie}_{\circ}.$$

The proposition below generalizes this result.

Proposition 10.1.7 *Let $L \in \mathsf{Int}^{+}$ and $L = L\mathcal{K}$. Then*

$$\eta(L) = L\mathcal{K}^{\bowtie} \text{ and } \eta^{\circ}(L) = L\mathcal{K}^{\bowtie}_{\circ}.$$

Proof. The inclusions $\eta(L) \subseteq L\mathcal{K}^{\bowtie}$ and $\eta^{\circ}(L) \subseteq L\mathcal{K}^{\bowtie}_{\circ}$ follow from Proposition 10.1.3.

Let $\varphi \notin \eta(L)$. Then $\varphi_{\bowtie} \notin L$ and there is $\mathcal{A} \in \mathcal{K}$ such that $\mathcal{A} \not\models \varphi_{\bowtie}$. It follows by Item 1 of Proposition 10.1.4 that $\mathcal{A}^{\bowtie} \not\models \varphi$, i.e., $\varphi \notin L\mathcal{K}^{\bowtie}$.

Similarly, we use Item 2 of Proposition 10.1.4 to prove the second equality.

□

Taking into account Proposition 10.1.1, the last proposition can be reworded as follows.

Corollary 10.1.8 *Let $L \in \mathcal{E}\mathbf{N4}^{\perp}$.*

1. *L is a special logic if and only if L is determined by some family of special $\mathbf{N4}^{\perp}$-lattices.*

2. [33] *L is an* ne*-logic if and only if L is determined by some family of normal* **N3***-lattices.*

□

Proposition 10.1.9 *For any $L \in \mathsf{Int}^{+}$, $\sigma^{-1}(L) = [\eta(L), \eta^{\circ}(L)]$.*

Proof. If $L_1 \in \sigma^{-1}(L)$ and an $\mathbf{N4}^{\perp}$-lattice $\mathcal{A}$ is such that $\mathcal{A} \models L_1$, then $\mathcal{A}_{\bowtie} \models L$ by Lemma 10.1.2 and $\mathcal{A} \models \eta(L)$ by Item 1 of Proposition 10.1.3. Thus, $\eta(L) \subseteq L_1$.

For any $\mathcal{A} \in Mod(L_1)$, $(\mathcal{A}_{\bowtie})^{\bowtie}_{\circ}$ embeds into $\mathcal{A}$ and belongs to $Mod(L_1)$. Let $\mathcal{K} = \{\mathcal{A}_{\bowtie} \mid \mathcal{A} \in Mod(L_1)\}$. Then $\mathcal{K}^{\bowtie}_{\circ} \subseteq Mod(L_1)$. By Proposition 10.1.1, $L = L\mathcal{K}$, and $\eta^{\circ}(L) = L\mathcal{K}^{\bowtie}_{\circ}$ by Proposition 10.1.7. We have thus proved $L_1 \subseteq \eta^{\circ}(L)$.

That $\eta(L)$ and $\eta^{\circ}(L)$ belong to $\sigma^{-1}(L)$ was stated in Corollary 10.1.6.

□

Remark. The analog of the above proposition does not hold for the lattice of **N4**-extensions. If $\sigma^{p}(L)$ is the positive fragment of $L \in \mathcal{E}\mathbf{N4}$, one can still prove that $\mathbf{N4} + L$ is the least element of the inverse image $(\sigma^{p})^{-1}(L)$, where L is some extension of positive logic. But in the general case, the set $(\sigma^{p})^{-1}(L)$ has no greatest element. This follows from the fact that the family of intermediate logics with the same positive fragment does not necessarily have a greatest element.

Following M. Kracht [47] we obtain another useful characterization of special and normal explosive logics.

Proposition 10.1.10 *Let $L \in \mathcal{E}\mathbf{N4}^{\perp}$.*

1. *L is a special logic if and only if all rules of the form $\psi/\psi_{\bowtie}$ are admissible in L.*

2. *L is an* ne*-logic if and only if $L \in \mathcal{E}\mathbf{N3}$ and $\neg\neg(p \vee \sim p) \in L$.*

Proof. 1. Let L be a special logic. According to Corollary 10.1.8 L is determined by some family of special $\mathbf{N4}^{\perp}$-lattices. The admissibility of rules $\psi/\psi_{\bowtie}$ now follows from Proposition 10.1.4 and Lemma 10.1.2.

Conversely, assume that all rules of the form $\psi/\psi_{\bowtie}$ are admissible in L.

Lemma 10.1.11 *All rules of the form* $\psi_{\bowtie}/\psi$ *are admissible in any* $L \in \mathcal{E}\mathbf{N4}^{\perp}$.

Proof. Let $\mathcal{A}$ be an $\mathbf{N4}^{\perp}$-lattice. If $\mathcal{A} \models \psi_{\bowtie}$, then $\mathcal{A}_{\bowtie} \models \psi_{\bowtie}$ by Lemma 10.1.2, and $(\mathcal{A}_{\bowtie})^{\bowtie} \models \psi$ by Proposition 10.1.4. Since $\mathcal{A} \hookrightarrow (\mathcal{A}_{\bowtie})^{\bowtie}$, we conclude $\mathcal{A} \models \psi$.

□

Taking into account this lemma and Corollary 10.1.5, we obtain for any φ, $\varphi \in L$ if and only if $\varphi_{\bowtie} \in L$ if and only if $\varphi \in \eta(\sigma(L))$. Thus, $L = \eta(\sigma(L))$ and L is a special logic.

2. If L is an *ne*-logic, then $\sim p \to (p \to q)$ and $\neg\neg(p \vee \sim p)$ belong to L by definition.

Let $L \in \mathcal{E}\mathbf{N3}$, $\neg\neg(p\vee \sim p) \in L$, and $\mathcal{A} \models L$. In this case, $\mathcal{A}$ is an $\mathbf{N3}$-lattice, i.e., $\Delta(\mathcal{A}) = \{0\}$. The validity of $\neg\neg(p\vee \sim p)$ is equivalent to $\nabla(\mathcal{A}) = F_d(\mathcal{A})$. We have thus proved that any model of L is a normal $\mathbf{N3}$-lattice. By Corollary 10.1.8 L is an *ne*-logic.

□

Denote $\mathbf{N3}^{\circ} := \mathbf{N3} + \{\neg\neg(p\vee \sim p)\}$.

Proposition 10.1.12 *1. $\sigma : \mathcal{E}\mathbf{N4}^{\perp} \to \mathsf{Int}^{+}$ is a lattice epimorphism commuting with infinite meets and joins.*

2. $\eta : \mathsf{Int}^{+} \to \mathcal{E}\mathbf{N4}^{\perp}$ is a lattice monomorphism commuting with infinite meets and joins.

3. η° is a lattice isomorphism of Int^{+} and $\mathcal{E}\mathbf{N3}^{\circ}$.

Proof. 1. It follows immediately from definition that σ commutes with infinite meets. Prove that it commutes with infinite joins. The inclusion $\Sigma_{i\in I}\sigma(L_i) \subseteq \sigma(\Sigma_{i\in I}L_i)$, where $L_i \in \mathcal{E}\mathbf{N4}^{\perp}$, $i \in I$, is obvious. Check the inverse inclusion. Take a model $\mathcal{A}$ of $\Sigma_{i\in I}\sigma(L_i)$. By Proposition 10.1.3 we have $\mathcal{A}_{\circ}^{\bowtie} \models \eta^{\circ}\sigma(L_i)$ for $i \in I$. Since $L_i \subseteq \eta^{\circ}\sigma(L_i)$ by Proposition 10.1.9, we obtain $\mathcal{A}_{\circ}^{\bowtie} \models L_i$ for all $i \in I$, and so $\mathcal{A}_{\circ}^{\bowtie} \models \Sigma_{i\in I}L_i$. According to Lemma 10.1.2 $\mathcal{A} \models \sigma(\Sigma_{i\in I}L_i)$, which proves the inverse inclusion.

2. That η is one-to-one follows from Corollary 10.1.6. Check that η commutes with infinite meets. Let $L^* := \bigcap_{i\in I} L_i$. Since $\sigma(\bigcap_{i\in I}\eta(L_i))$ is obviously equal to L^*, the equality $\eta(L^*) = \bigcap_{i\in I}\eta(L_i)$ follows from the fact that $\bigcap_{i\in I}\eta(L_i)$ is a special logic. Each of $\eta(L_i)$ is closed under all rules of the form $\psi/\psi_{\bowtie}$ by Proposition 10.1.10. Obviously, the intersection $\bigcap_{i\in I}\eta(L_i)$ is also closed under such rules and it is also a special logic.

3. η° embeds Int into $\mathcal{E}\mathbf{N3}^{\circ}$ by Corollary 10.1.6. That η° is onto follows from Item 2 of Proposition 10.1.10.

□

We have thus presented the class $\mathcal{E}\mathbf{N4}^{\perp}$ as a union of disjoint intervals of the form $\sigma^{-1}(L)$:

$$\mathcal{E}\mathbf{N4}^{\perp} = \bigcup_{L\in\mathsf{Int}^{+}} [\eta(L), \eta^{\circ}(L)].$$

Now, we establish interrelations between these intervals. It turns out that if $L_1 \subseteq L_2$, then $\sigma^{-1}(L_2)$ is embedded into $\sigma^{-1}(L_1)$ as upper subinterval, at the same time, $\sigma^{-1}(L_2)$ is isomorphic to a homomorphic image of $\sigma^{-1}(L_1)$.

Let $L_1, L_2 \in \mathsf{Int}^{+}$ and $L_1 \subseteq L_2$. Mappings $r_{L_2,L_1} : \sigma^{-1}(L_2) \to \mathcal{E}\mathbf{N4}^{\perp}$ and $e_{L_1,L_2} : \sigma^{-1}(L_1) \to \mathcal{E}\mathbf{N4}^{\perp}$ are defined as follows:

$$r_{L_2,L_1}(L) = L \cap \eta^{\circ}(L_1) \text{ and } e_{L_1,L_2}(L) = L + \eta(L_2)$$

Proposition 10.1.13 *Let $L_1, L_2 \in \mathsf{Int}^{+}$ and $L_1 \subseteq L_2$. The following facts hold.*

1. *For any $L \in \sigma^{-1}(L_2)$, we have $e_{L_1,L_2} r_{L_2,L_1}(L) = L$.*
2. *For any $L \in \sigma^{-1}(L_1)$, we have*

$$r_{L_2,L_1} e_{L_1,L_2}(L) = L + r_{L_2,L_1}(\eta(L_2)).$$

3. *e_{L_1,L_2} is a lattice epimorphism from $\sigma^{-1}(L_1)$ onto $\sigma^{-1}(L_2)$.*
4. *r_{L_2,L_1} is a lattice monomorphism from $\sigma^{-1}(L_2)$ into $\sigma^{-1}(L_1)$ and*

$$r_{L_2,L_1}(\sigma^{-1}(L_2)) = [r_{L_2,L_1}(\eta(L_2)), \eta^{\circ}(L_1)].$$

5. *For any $L_3 \in \mathsf{Int}^{+}$ such that $L_2 \subseteq L_3$, we have*

$$e_{L_1,L_2} e_{L_2,L_3} = e_{L_1,L_3} \text{ and } r_{L_3,L_2} r_{L_2,L_1} = r_{L_3,L_1}.$$

Proof. 1. Let $L \in \sigma^{-1}(L_2)$. We calculate

$$e_{L_1,L_2} r_{L_2,L_1}(L) = (L + \eta(L_2)) \cap (\eta^{\circ}(L_1) + \eta(L_2)).$$

We have $L + \eta(L_2) = L$ since $\eta(L_2)$ is the least element of $\sigma^{-1}(L_2)$. Consider $L' := \eta^{\circ}(L_1) + \eta(L_2)$. Due to homomorphism properties of σ we have $\sigma(L') = L_2$. Moreover, $L' \in \mathcal{E}\mathbf{N3}^{\circ}$ since $\eta^{\circ}(L_1) \subseteq L'$. By Item 3 of Proposition 10.1.12

we conclude $L' = \eta^{\circ}(L_2)$. Finally, $L \cap \eta^{\circ}(L_2) = L$ since $\eta^{\circ}(L_2)$ is the greatest element of $\sigma^{-1}(L_2)$.

2. Again for $L \in \sigma^{-1}(L_1)$, we have

$$r_{L_2,L_1} e_{L_1,L_2}(L) = (L \cap \eta^{\circ}(L_1)) + (\eta(L_2) \cap \eta^{\circ}(L_1)),$$

where the first term is equal to L since $\eta^{\circ}(L_1)$ is the greatest element of $\sigma^{-1}(L_1)$ and the second term is exactly $r_{L_2,L_1}(\eta(L_2))$.

3. That e_{L_1,L_2} is a lattice homomorphism follows from the distributivity of $\mathcal{E}\mathbf{N4}^{\perp}$. Let $L \in \sigma^{-1}(L_1)$. Since σ is a homomorphism, we have

$$\sigma e_{L_1,L_2}(L) = \sigma(L) + \sigma\eta(L_2) = L_1 + L_2 = L_2.$$

Thus, the range of e_{L_1,L_2} is contained in $\sigma^{-1}(L_2)$. Item 1 implies that e_{L_1,L_2} is onto.

4. As above, we use the distributivity of $\mathcal{E}\mathbf{N4}^{\perp}$ and the homomorphism properties of σ to prove that r_{L_2,L_1} is a lattice homomorphism from $\sigma^{-1}(L_2)$ into $\sigma^{-1}(L_1)$. If $r_{L_2,L_1}(L) = r_{L_2,L_1}(L')$, then applying the formula of Item 1 we obtain $L = L'$. In this way, r_{L_2,L_1} is a monomorphism. The equality $r_{L_2,L_1}(\sigma^{-1}(L_2)) = [r_{L_2,L_1}(\eta(L_2)), \eta^{\circ}(L_1)]$ follows from Item 2.

5. This follows immediately from definitions.

□

10.2 The Lattice Structure of $\mathcal{E}\mathbf{N4}^{\perp}$

It was proved in the previous section that the lattice $\mathcal{E}\mathbf{N4}^{\perp}$ decomposes into a union of disjoint intervals $\sigma^{-1}(L)$, where $L \in \mathsf{Int}^{+}$ and that $\sigma^{-1}(L_1)$ is isomorphic to an upper subinterval of $\sigma^{-1}(L_2)$, whenever $L_2 \subseteq L_1$. Since $\mathbf{Lk}$ is the greatest point of Int, an isomorphic copy of $\sigma^{-1}(\mathbf{Lk})$ is an upper part of any interval of the form $\sigma^{-1}(L)$ for $L \neq \mathcal{F}$. For this reason, we start the study of $\mathcal{E}\mathbf{N4}^{\perp}$ with a description of the interval $\sigma^{-1}(\mathbf{Lk})$.

Consider subdirectly irreducible models of logics in $\sigma^{-1}(\mathbf{Lk})$. According to Corollary 9.2.11, any such model is isomorphic to an element of $S^{\bowtie}(\mathcal{A})$, where $\mathcal{A}$ is a subdirectly irreducible model of $\mathbf{Lk}$. Any subdirectly irreducible model of $\mathbf{Lk}$ is isomorphic to the two-element Boolean algebra $\mathbf{2}$. Therefore, we have exactly four subdirectly irreducible models of logics in $\sigma^{-1}(\mathbf{Lk})$:

$$\mathbf{2}^{\bowtie} = Tw(\mathbf{2}, \{0,1\}, \{0,1\}),\ \mathbf{2}^{\bowtie}_{\circ} = Tw(\mathbf{2}, \{1\}, \{0\}),$$
$$\mathbf{2}^{\bowtie}_{R} = Tw(\mathbf{2}, \{1\}, \{0,1\}),\ \mathbf{2}^{\bowtie}_{L} = Tw(\mathbf{2}, \{0,1\}, \{0\}).$$

These lattices define logics closely related to well-known finite valued logics. Consider matrices of the form $M(\mathcal{A}) = \langle \mathcal{A}, D^{\mathcal{A}} \rangle$ corresponding to the above $\mathbf{N4}^{\perp}$-lattices. The lattice $\mathbf{2}_{\circ}^{\bowtie}$ is two-element. Its elements $(0,1)$ and $(1,0)$ can be identified with the classical truth-values f and t respectively. The strong negation $\sim$ coincides with the classical one and $D^{\mathcal{A}} = \{t\}$ in this case. Thus, the matrix corresponding $\mathbf{2}_{\circ}^{\bowtie}$ defines, in fact, the classical logic with an additional negation $\sim$, such that $\sim p \leftrightarrow \neg p$. We denote $\widetilde{\mathbf{Lk}} := L\mathbf{2}_{\circ}^{\bowtie}$.

The matrix $M(\mathbf{2}^{\bowtie})$ has additional truth-values $\top := (1,1)$ and $\perp := (0,0)$. It has two distinguished values, $D^{\mathbf{2}^{\bowtie}} = \{t, \top\}$, and the following lattice structure.

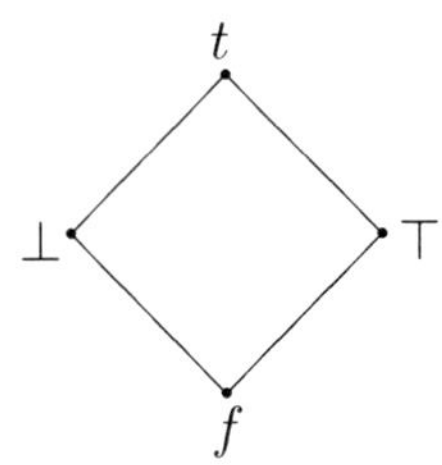

In fact, $M(2^{\bowtie})$ can be considered as a four-valued Belnap's matrix ([8, 9]) enriched with the weak implication $\rightarrow$ and the constant $\perp$. We denote $\mathbf{B}_4^{\rightarrow} := L\mathbf{2}^{\bowtie}$.

The logic of $\mathbf{2}_R^{\bowtie}$ differs from the maximal extension RM_3 of the relevant logic RM [3, 4] only by an additional constant $\perp$. Indeed, RM_3 can be defined via the matrix

$$M = \langle \{f, t, \top\}, \vee, \wedge, \supset, \sim, \{t, \top\} \rangle,$$

where $f \leq \top \leq t$, $\vee$ and $\wedge$ are the usual lattice operations, $\sim$ is an order reversing involution and the implication is defined by the rule:

$$a \supset b := \begin{cases} b, & \text{if } a \in D^M \\ t, & \text{if } a \notin D^M. \end{cases} \tag{10.1}$$

We have $|\mathbf{2}_R^{\bowtie}| = \{f, t, \top\}$ and $D^{\mathbf{2}_R^{\bowtie}} = \{t, \top\}$. The lattice structures of M and $\mathbf{2}_R^{\bowtie}$ are identical, the strong negation is also an order-reversing involution, finally, the direct computation shows that $\supset$ coincides with the implication of $\mathbf{2}_R^{\bowtie}$. Thus, $L\mathbf{2}_R^{\bowtie}$ is RM_3 enriched with the constant $\perp$. We denote $RM_3^{\perp} := L\mathbf{2}_R^{\bowtie}$.

It was pointed out in [110] that the logic of $\mathbf{2}^{\bowtie}_L$ is definition equivalent to the three-valued logic Ł3 of Łukasiewicz. We have $|\mathbf{2}^{\bowtie}_L| = \{f, \perp, t\}$. Let us set

$$a \supset b := (a \to b) \wedge ({\sim} b \to {\sim} a)$$

and calculate the truth tables for $\supset$ and $\sim$ on $\{f, \perp, t\}$:

$\to$	f	$\perp$	t
f	t	t	t
$\perp$	$\perp$	t	t
t	f	$\perp$	t

$\sim$	
f	t
$\perp$	$\perp$
t	f

Thus, $\langle \{f, \perp, t\}, \supset, \sim, \{t\} \rangle$ is the well-known matrix for Ł3. All operations of $\mathbf{2}^{\bowtie}_L$ can be defined through $\supset$ and $\sim$ as follows:

$$\neg a := a \supset {\sim} a, \;\; a \vee b := (a \supset b) \supset b,$$

$$a \wedge b := {\sim} ({\sim} a \vee {\sim} b), \;\; a \to b := a \supset (a \supset b).$$

For this reason, we denote $L\mathbf{2}^{\bowtie}_L$ as Ł3.

Note that $\mathbf{N4}^{\perp}$-lattices $\mathbf{2}^{\bowtie}$, $\mathbf{2}^{\bowtie}_{\circ}$, $\mathbf{2}^{\bowtie}_R$, and $\mathbf{2}^{\bowtie}_L$ have no non-trivial homomorphic images and are embedded one into another as follows

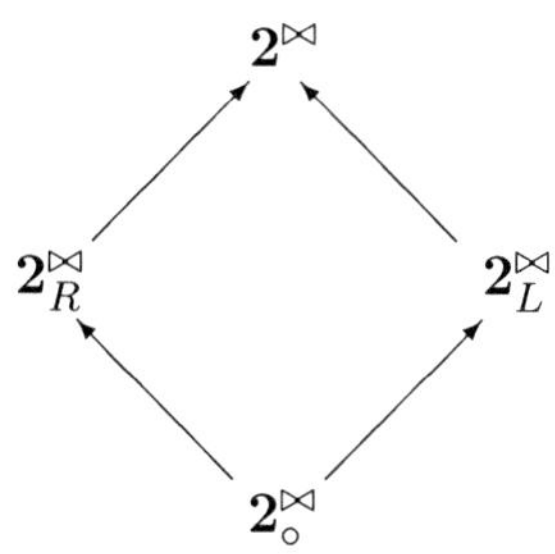

Thus, the interval $\sigma^{-1}(\mathbf{Lk})$ contains exactly five logics determined by the following sets of subdirectly irreducible $\mathbf{N4}^{\perp}$-lattices:

$$\{\mathbf{2}^{\bowtie}\}, \; \{\mathbf{2}^{\bowtie}_R, \mathbf{2}^{\bowtie}_L\}, \; \{\mathbf{2}^{\bowtie}_R\}, \; \{\mathbf{2}^{\bowtie}_L\}, \; \{\mathbf{2}^{\bowtie}_{\circ}\}.$$

We have thus proved the following

Proposition 10.2.1 *The interval $\sigma^{-1}(\mathbf{Lk})$ has the following structure:*

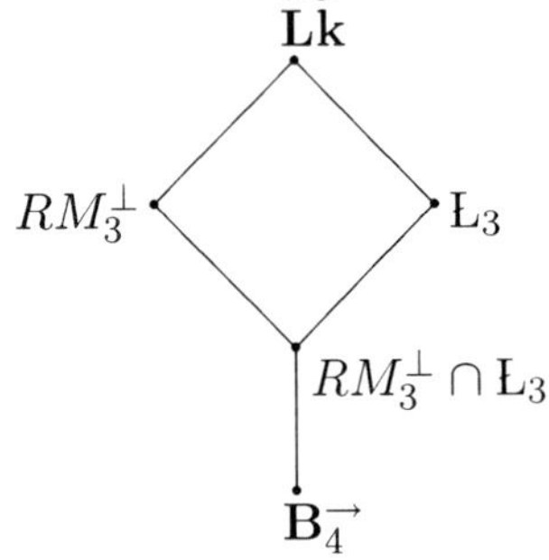

In particular, $\eta(\mathbf{Lk}) = \mathbf{B}_4^{\rightarrow}$ and $\eta^{\circ}(\mathbf{Lk}) = \widetilde{\mathbf{Lk}}$.

□

Taking into account that $L \subseteq \eta^{\circ}\sigma(L)$ for any $L \in \mathcal{E}\mathbf{N4}^{\perp}$ and that $\eta^{\circ}(L_1) \subseteq \eta^{\circ}(L_2)$ whenever $L_1 \subseteq L_2$, we conclude

Corollary 10.2.2 *1. Any non-trivial extension of $\mathbf{N4}^{\perp}$ is contained in the logic $\widetilde{\mathbf{Lk}}$.*

2. The logic $\mathbf{N4}^{\perp}$ has no contradictory non-trivial extension.

□

We have thus pointed out the first essential difference between the structure of $\mathcal{E}\mathbf{N4}^{\perp}$ and the structure of Jhn^{+}. Minimal logic has the subclass of contradictory extensions isomorphic to the class of extensions of positive logic, whereas in the case of $\mathbf{N4}^{\perp}$, no contradiction can be added to $\mathbf{N4}^{\perp}$ as a scheme. Further, unlike Jhn, the class $\mathcal{E}\mathbf{N4}^{\perp} \setminus \{For^{\perp}\}$ of non-trivial $\mathbf{N4}^{\perp}$-extensions forms a lattice with the unit element $\widetilde{\mathbf{Lk}}$, and our next step is to describe coatoms of the lattice $\mathcal{E}\mathbf{N4}^{\perp} \setminus \{For^{\perp}\}$.

We know two examples of coatoms. It follows from Proposition 10.2.1 that $RM_3^{\perp}$ and Ł3 are coatoms in $\mathcal{E}\mathbf{N4}^{\perp} \setminus \{For^{\perp}\}$. A further example of coatoms provides the twist-structure $\mathbf{3}_{\circ}^{\bowtie}$, where $\mathbf{3}$ is a three-element linearly ordered Heyting algebra, $|\mathbf{3}| = \{0, 1, 2\}$, $0 \leq 1 \leq 2$. Since $F_d(\mathbf{3}) = \{1, 2\}$, the lattice $\mathbf{3}_{\circ}^{\bowtie}$ has four elements $(0, 1)$, $(1, 0)$, $(0, 2)$, and $(2, 0)$. The lattice structure and the action of strong negation on $\mathbf{3}_{\circ}^{\bowtie}$ are presented in the diagram below.

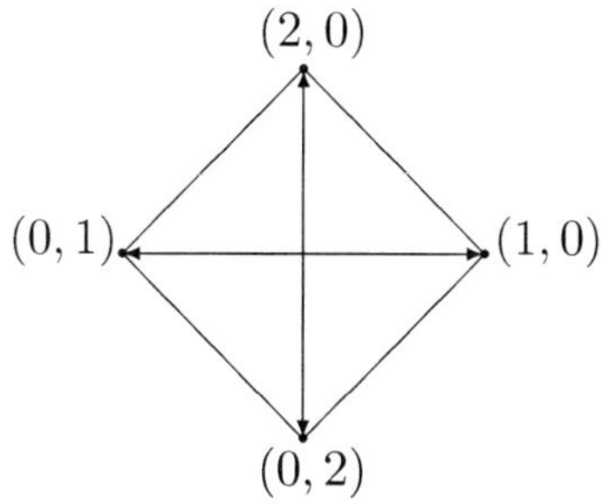

We can see that $\mathbf{3}_{\circ}^{\bowtie}$ is isomorphic to $\mathbf{2}^{\bowtie}$ as a lattice, but not as an implicative lattice, the negation also acts on $\mathbf{3}_{\circ}^{\bowtie}$ in another way.

Proposition 10.2.3 *The lattice $\mathcal{E}\mathbf{N4}^{\perp} \setminus \{For^{\perp}\}$ has exactly three coatoms: $RM_3^{\perp}$, Ł$_3$, and $L\mathbf{3}_{\circ}^{\bowtie}$. Each element of $\mathcal{E}\mathbf{N4}^{\perp} \setminus \{For^{\perp}\}$ different from $\widetilde{\mathbf{Lk}}$ is contained in one of them.*

Proof. Let $L \in \mathcal{E}\mathbf{N4}^{\perp}\setminus\{For^{\perp}\}$ and $L \neq \widetilde{\mathbf{Lk}}$. Assume L is contained neither in $RM_3^{\perp}$ nor in Ł$_3$. By Proposition 10.2.1 $\sigma(L) \neq \mathbf{Lk}$ in this case. It follows that L has a model $\mathcal{A} = Tw(\mathcal{B}, \nabla, \Delta)$, where $\mathcal{B}$ is not a Boolean algebra, in which case $\nabla \neq \{1\}$. Take an element a in $\mathcal{B}$ such that $a \neq 0, 1$ and $a \in \nabla$, and consider a twist-structure $\mathcal{A}_0 := Tw(\{0, a, 1\}, \{a, 1\}, \{0\})$. Clearly, $\mathcal{A}_0 \leq \mathcal{A}$ and $\mathcal{A}_0 \cong \mathbf{3}_{\circ}^{\bowtie}$. Thus, $\mathbf{3}_{\circ}^{\bowtie} \models L$ and $L \subseteq L\mathbf{3}_{\circ}^{\bowtie}$.

$L\mathbf{3}_{\circ}^{\bowtie}$ is incomparable with $RM_3^{\perp}$ and Ł$_3$. On the one hand, the intuitionistic fragment of $L\mathbf{3}_{\circ}^{\bowtie}$ is different from classical logic. On the other hand, $\sim p \to (p \to q) \in L\mathbf{3}_{\circ}^{\bowtie} \setminus RM_3^{\perp}$ and $\neg\neg(p\vee \sim p) \in L\mathbf{3}_{\circ}^{\bowtie} \setminus$ Ł$_3$.

□

Recall that an element x of a lattice $\mathcal{A}$ is called a *splitting element* if there exists an y such that for every element $z \in \mathcal{A}$, exactly one of the following conditions holds: $z \leq x$ or $y \leq z$. If x is a splitting element, the corresponding y is called the *splitting* of $\mathcal{A}$ by x and is denoted by $\mathcal{A}/x$. We write $\mathcal{A}/\{x, y\}$ for $\mathcal{A}/x \vee \mathcal{A}/y$.

Proposition 10.2.4 *The logics $RM_3^{\perp}$, Ł$_3$, and $L\mathbf{3}_{\circ}^{\bowtie}$ are splitting elements in the lattice $\mathcal{E}\mathbf{N4}^{\perp}$ and the following holds:*

$$\mathcal{E}\mathbf{N4}^{\perp}/RM_3^{\perp} = \mathbf{N3},\ \mathcal{E}\mathbf{N4}^{\perp}/\text{Ł}_3 = \mathbf{N4}^{\perp} + \neg\neg(p\vee \sim p),\ \mathcal{E}\mathbf{N4}^{\perp}/L\mathbf{3}_{\circ}^{\bowtie} = \mathbf{B}_4^{\rightarrow}$$

and

$$\mathcal{E}\mathbf{N4}^{\perp}/\{RM_3^{\perp}, \text{Ł}_3\} = \mathbf{N3}^{\circ}.$$

Proof. If $L \in \mathcal{E}\mathbf{N3}$, then $\Delta(\mathcal{A}) = \{0\}$ for any $\mathcal{A} \models L$. Therefore, $\mathbf{2}_R^{\bowtie}$ is not a model of L and $L \not\subseteq RM_3^{\perp}$. If $L \notin \mathcal{E}\mathbf{N3}$, there is $\mathcal{A} = Tw(\mathcal{B}, \nabla, \Delta)$ such that $\mathcal{A} \models L$ and $\Delta \neq \{0\}$. Choose an $a \in \Delta$ such that $a \neq 0$ and consider the quotient $\mathcal{A}_0 := \mathcal{A}/\langle a\rangle^{\bowtie}$. By Proposition 9.2.16

$$\mathcal{A}_0 \cong \mathcal{A}_1 := Tw(\mathcal{B}/\langle a\rangle, \nabla/\langle a\rangle, \Delta/\langle a\rangle).$$

Since $a \in \Delta$, $\Delta/\langle a\rangle = \mathcal{B}/\langle a\rangle$. Consequently, $\mathcal{A}_1$ contains a subalgebra $Tw(\{0/\langle a\rangle, 1/\langle a\rangle\}, \{1/\langle a\rangle\}, \{0/\langle a\rangle, 1/\langle a\rangle\})$, which is isomorphic to $\mathbf{2}_R^{\bowtie}$. Thus, $L \subseteq RM_3^{\perp}$ and we have proved the equality $\mathcal{E}\mathbf{N4}^{\perp}/RM_3^{\perp} = \mathbf{N3}$.

If $\neg\neg(p\vee \sim p) \in L$, then every model of L is up to isomorphism of the form $Tw(\mathcal{B}, F_d(\mathcal{B}), \Delta)$. Obviously, $\mathbf{2}_L^{\bowtie}$ does not satisfy this condition since $\nabla(\mathbf{2}_L^{\bowtie}) = \mathbf{2}$. Therefore, $L \not\models \mathbf{2}_L^{\bowtie}$, and L is not contained in Ł$_3$.

If $\neg\neg(p\vee \sim p) \notin L$, then L has a model $\mathcal{A} = Tw(\mathcal{B}, \nabla, \Delta)$ such that $\nabla \neq F_d(\mathcal{B})$. Passing to the quotient $\mathcal{A}/(F_d(\mathcal{B}))^{\bowtie}$ we obtain a model $\mathcal{A}_0$ of L such that $\mathcal{A}_0 = Tw(\mathcal{B}_0, \nabla_0, \Delta_0)$, $\mathcal{B}_0$ is a Boolean algebra and $\nabla_0 \neq \{1\}$. Let $a \in \nabla_0$ and $a \neq 1$. Let $\overline{a}$ be a Boolean complement of a in $\mathcal{B}_0$. By Proposition 9.2.16 we have

$$\mathcal{A}_0/\langle \overline{a}\rangle^{\bowtie} \cong \mathcal{A}_2 := Tw(\langle a\rangle, \langle a\rangle, \Delta_1).$$

Since $a \neq 1$, the twist-structure $\mathcal{A}_2$ contains a subalgebra $Tw(\{1, a\}, \{1, a\}, \{a\})$ that is isomorphic to $\mathbf{2}_L^{\bowtie}$. Therefore, $L \models \mathbf{2}_L^{\bowtie}$. We have thus proved $\mathcal{E}\mathbf{N4}^{\perp}/$Ł$_3 = \mathbf{N4}^{\perp} + \neg\neg(p\vee \sim p)$.

To prove $\mathcal{E}\mathbf{N4}^{\perp}/L\mathbf{3}_{\circ}^{\bowtie} = \mathbf{B}_4^{\rightarrow}$, we notice that $L\mathbf{3}_{\circ}^{\bowtie} = \eta^{\circ}(L\mathbf{3})$ according to Proposition 10.1.7. Thus, if $\sigma(L) \subseteq L\mathbf{3}$, then $L \subseteq \eta^{\circ}(L) \subseteq \eta^{\circ}(L\mathbf{3})$. It is well known that $L\mathbf{3}$ is the opremum of Int. Consequently, if $\sigma(L) \not\subseteq L\mathbf{3}$, then $\sigma(L) = \mathbf{Lk}$ and L extends $\eta(\mathbf{Lk})$. It remains to notice that $\eta(\mathbf{Lk}) = \mathbf{B}_4^{\rightarrow}$ by Proposition 10.2.1.

The last equality immediately follows from the first and the second.

□

We denote $\mathbf{N4}^{\mathsf{N}} := \mathbf{N4}^{\perp} + \{\neg\neg(p\vee \sim p)\}$ and distinguish in $\mathcal{E}\mathbf{N4}^{\perp}$ the following subclasses:

$$\mathsf{Exp} := \{L \in \mathcal{E}\mathbf{N4}^{\perp} \mid \sim p \rightarrow (p \rightarrow q) \in L\},$$

$$\mathsf{Nor} := \{L \in \mathcal{E}\mathbf{N4}^{\perp} \mid \neg\neg(p\vee \sim p) \in L\},$$

$$\mathsf{Gen} := \mathcal{E}\mathbf{N4}^{\perp} \setminus (\mathsf{Exp} \cup \mathsf{Nor}).$$

Let $L \in \mathcal{E}\mathbf{N4}^{\perp}$. We say that L is *explosive* if $L \in \mathsf{Exp}$, we call L *normal* if $L \in \mathsf{Nor}$. Finally, if $L \in \mathsf{Gen}$, we say that L is a logic of *general form*. In the proposition below, we collect a series of simple facts on the introduced classes.

Proposition 10.2.5 *1.* $\mathsf{Exp} \cap \mathsf{Nor} = \mathcal{E}\mathbf{N3}^{\circ}$ *and* $L3_{\circ}^{\bowtie}$ *is an opremum in* $(\mathsf{Exp} \cap \mathsf{Nor}) \setminus \{For^{\perp}\}$.

2. $\mathsf{Exp} = \mathcal{E}\mathbf{N3}$.

3. $\mathsf{Exp} \setminus \mathsf{Nor} = [\mathbf{N3}, \text{Ł}_3]$

4. $L \in \mathsf{Exp}$ *iff* $L \not\subseteq RM_3^{\perp}$ *iff for any* $\mathcal{A} \in Mod(L)$, $\Delta(\mathcal{A}) = \{0\}$.

5. $\mathsf{Nor} = \mathcal{E}\mathbf{N4}^{\mathsf{N}}$.

6. $\mathsf{Nor} \setminus \mathsf{Exp} = [\mathbf{N4}^{\mathsf{N}}, RM_3^{\perp}]$

7. $L \in \mathsf{Nor}$ *iff* $L \not\subseteq \text{Ł}_3$ *iff for any* $\mathcal{A} \in Mod(L)$, $\nabla(\mathcal{A}) = F_d(\mathcal{A}_{\bowtie})$.

8. $\mathsf{Gen} = [\mathbf{N4}^{\perp}, RM_3^{\perp} \cap \text{Ł}_3]$.

□

All statements in this proposition easily follow from Propositions 10.2.3 and 10.2.4. Figure 10.1 presents the structure of $\mathcal{E}\mathbf{N4}^{\perp}$.

Remark. Let us note the similarity of classes Exp, Nor, Gen and respectively, the classes Int, Neg, Par studied in the first part of the book. Both classes Exp and Int consist of explosive logics. The similarity of classes Nor and Neg is not so obvious. The class Neg consists of contradictory logics, which represent the structures of contradictions in all **Lj**-extensions in the following sense. If $L_1 \in \mathsf{Neg}$ is a negative counterpart of an **Lj**-extension L, then L_1 is embedded into L by the contradiction operator (see Section 4.2.1). According to Corollary 10.2.2 the logic $\mathbf{N4}^{\perp}$ has no contradictory extension and the structures of contradictions cannot be explicated in the same way as for **Lj**-extensions. However, they can be represented semantically as ideals of contradictions $\Delta(\mathcal{A})$. Every $\mathbf{N4}^{\perp}$-lattice is determined by its basic algebra $\mathcal{A}_{\bowtie}$, the filter of completions $\nabla(\mathcal{A})$ and the ideal of contradictions $\Delta(\mathcal{A})$. By Proposition 10.1.1 basic algebras of L-models for $L \in \mathcal{E}\mathbf{N4}^{\perp}$ determine the intuitionistic fragment of L. According to the last proposition, the ideal of contradictions is trivial in models of explosive logics and so, a logic from Exp is characterized only by its intuitionistic fragment and by filters of completions of its models. At the same time, the filter of completions has a fixed

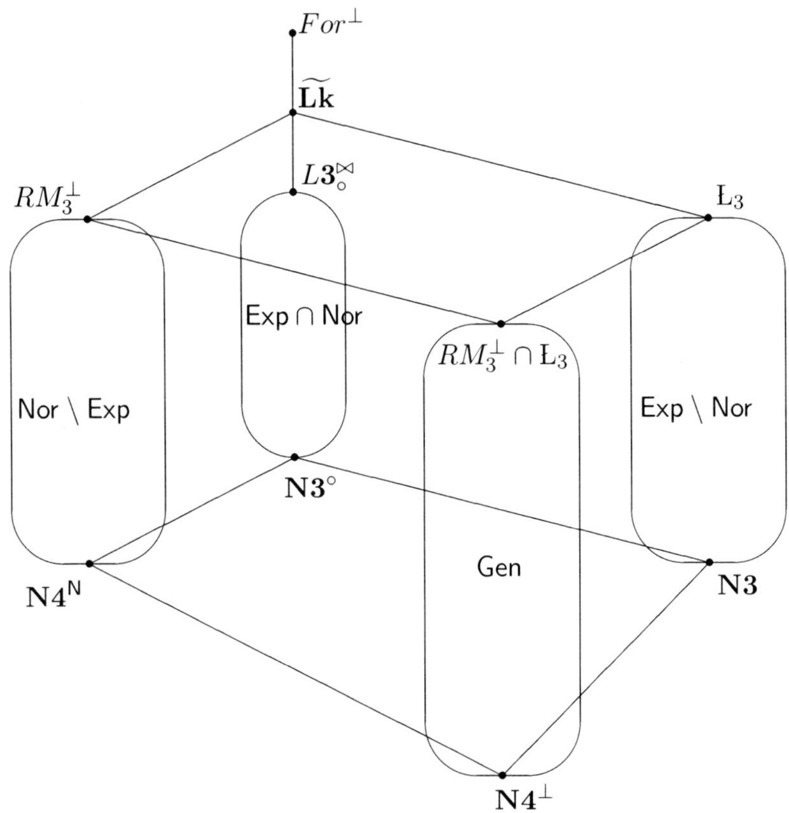

Figure 10.1

value, the least possible one, in models of normal logics. Thus, a normal logic is characterized by its intuitionistic fragment and by ideals of contradictions of its models. In this sense, the normal logics represent the structures of contradictions in all $\mathbf{N4}^{\perp}$-extensions and are similar to negative logics in the class of $\mathbf{Lj}$-extensions. It will be shown in the next section that the class Gen is connected to the classes Exp and Nor via the system of counterparts in a similar way to the connection of the class Par with the classes Int and Neg.

Our plans for the rest of this section are as follows. First, we consider the restrictions of the operator σ to classes Exp and Nor and point out the perfect analogy with the situation described in Section 10.1. For any $L \in \mathsf{Int}^{+}$, the inverse image of L with respect to the corresponding restriction of σ forms an interval in the class Exp (Nor), and the end points of such interval can be translated into L. Further, we study interrelations of the class Gen and classes Exp and Nor.

Denote

$$\sigma^3 := \sigma \restriction_{\mathsf{Exp}} \quad \text{and} \quad \sigma^n := \sigma \restriction_{\mathsf{Nor}} .$$

The mappings $\eta^3 : \mathsf{Int}^+ \to \mathsf{Exp}$ and $\eta^n : \mathsf{Int}^+ \to \mathsf{Nor}$ are defined as follows. For every $L \in \mathsf{Int}^+$,

$$\eta^3(L) := \eta(L) + \{\sim p \to (p \to q)\} \text{ and } \eta^n(L) := \eta(L) + \{\neg\neg(p \vee \sim p)\}.$$

Clearly, $\eta^\circ(L) = \eta^3(L) + \eta^n(L)$.

Logics in Exp having the form $\eta^3(L)$ we call *special explosive* or *se*-logics, and logics in Nor of the form $\eta^n(L)$ are called *special normal* or *sn*-logics. Models of *se*- and *sn*-logics are described in the following.

Proposition 10.2.6 *Let $L \in \mathsf{Int}^+$ and $\mathcal{A}$ be an $\mathbf{N4}^{\perp}$-lattice.*

1. *$\mathcal{A} \models \eta^3(L)$ iff $\mathcal{A}_{\bowtie} \models L$ and $\Delta(\mathcal{A}) = \{0\}$.*
2. *$\mathcal{A} \models \eta^n(L)$ iff $\mathcal{A}_{\bowtie} \models L$ and $\nabla(\mathcal{A}) = F_d(\mathcal{A}_{\bowtie})$.*

Proof. It can be proved similarly to Proposition 10.1.3.

□

In the following, we omit the proofs if they can be obtained similarly to the results of the previous section.

For a Heyting algebra $\mathcal{A}$, we define

$$\mathcal{A}_3^{\bowtie} := Tw(\mathcal{A}, \mathcal{A}, \{0\}) \text{ and } \mathcal{A}_n^{\bowtie} := Tw(\mathcal{A}, F_d(\mathcal{A}), \mathcal{A}).$$

If an $\mathbf{N3}$-lattice $\mathcal{A}$ is isomorphic to a lattice of the form $\mathcal{B}_3^{\bowtie}$, we call it a *special* $\mathbf{N3}$-lattice. If an $\mathbf{N4}^{\perp}$-lattice $\mathcal{A}$ is isomorphic to a lattice of the form $\mathcal{B}_n^{\bowtie}$, we call it a *special normal* $\mathbf{N4}^{\perp}$-lattice.

Translations of $\eta^3(L)$ and $\eta^n(L)$ into L can be defined in the following way. The extended set $\overline{Prop}$ of propositional variables and formula $\varphi_{\bowtie}$ were defined in the previous section. Put

$$\varphi_{\bowtie}^3 := \bigwedge_{p \in var(\varphi)} \neg(p \wedge \overline{p}) \to \varphi_{\bowtie},$$

$$\varphi_{\bowtie}^n := \bigwedge_{p \in var(\varphi)} \neg\neg(p \vee \overline{p}) \to \varphi_{\bowtie}.$$

Proposition 10.2.7 *Let $\mathcal{A}$ be a Heyting algebra and φ a formula in the language $\mathcal{L}^{\perp}$. The following equivalences hold.*

1. $\mathcal{A}_3^{\bowtie} \models \varphi$ *if and only if* $\mathcal{A} \models \varphi_{\bowtie}^3$.
2. $\mathcal{A}_n^{\bowtie} \models \varphi$ *if and only if* $\mathcal{A} \models \varphi_{\bowtie}^n$.

□

Corollary 10.2.8 *Let $L \in \mathsf{Int}^+$ and $\varphi \in For^{\perp}$. The following equivalences hold.*

1. $\varphi \in \eta^3(L)$ *if and only if* $\varphi_{\bowtie}^3 \in L$.
2. $\varphi \in \eta^n(L)$ *if and only if* $\varphi_{\bowtie}^n \in L$.

□

Let $\mathcal{K}$ be a class of Heyting algebras. Put

$$\mathcal{K}_3^{\bowtie} := \{\mathcal{A}_3^{\bowtie} \mid \mathcal{A} \in \mathcal{K}\} \text{ and } \mathcal{K}_n^{\bowtie} := \{\mathcal{A}_n^{\bowtie} \mid \mathcal{A} \in \mathcal{K}\}.$$

Proposition 10.2.9 *Let $L \in \mathsf{Int}^+$ and $L = L\mathcal{K}$. Then*

$$\eta^3(L) = L\mathcal{K}_3^{\bowtie} \text{ and } \eta^n(L) = L\mathcal{K}_n^{\bowtie}.$$

□

Corollary 10.2.10 *Let $L \in \mathcal{E}\mathbf{N4}^{\perp}$.*

1. *L is a special explosive logic if and only if L is determined by some family of special* **N3***-lattices.*
2. *L is a special normal logic if and only if L is determined by some family of special normal* $\mathbf{N4}^{\perp}$*-lattices.*

□

Proposition 10.2.11 *For any $L \in \mathsf{Int}^+$,*

$$(\sigma^3)^{-1}(L) = [\eta^3(L), \eta^{\circ}(L)] \text{ and } (\sigma^n)^{-1}(L) = [\eta^n(L), \eta^{\circ}(L)].$$

□

By analogy with Proposition 10.1.10, special explosive and special normal logics can be characterized via the admissibility of inference rules.

Proposition 10.2.12 *Let $L \in \mathcal{E}\mathbf{N4}^{\perp}$.*

1. *L is a special explosive logic if and only if $L \in \mathsf{Exp}$ and all rules of the form $\psi/\psi^3_{\bowtie}$ are admissible in L.*

2. *L is a special normal logic if and only if $L \in \mathsf{Nor}$ and all rules of the form $\psi/\psi^n_{\bowtie}$ are admissible in L.*

□

This characterization allows one to prove an analog of Proposition 10.1.12. The statements on σ^3 and η^3 were established earlier in [47].

Proposition 10.2.13 *1. $\sigma^3 : \mathsf{Exp} \rightarrow \mathsf{Int}^+$ and $\sigma^n : \mathsf{Nor} \rightarrow \mathsf{Int}^+$ are lattice epimorphisms commuting with infinite meets and joins.*

2. $\eta^3 : \mathsf{Int}^+ \rightarrow \mathsf{Exp}$ and $\eta^n : \mathsf{Int}^+ \rightarrow \mathsf{Nor}$ are lattice monomorphisms commuting with infinite meets and joins.

□

10.3 Explosive and Normal Counterparts

The decomposition of $\mathcal{E}\mathbf{N4}^{\perp}$ into classes Exp, Nor, and Gen is very similar to the decomposition of the class Jhn of non-trivial extensions of minimal logic into subclasses of intermediate, negative and properly paraconsistent logics. Our next step is to define explosive and normal counterparts for logics in $\mathcal{E}\mathbf{N4}^{\perp}$, in exactly the same way as we defined intuitionistic and negative counterparts for extensions of minimal logic.

The mappings $(-)_{exp} : \mathcal{E}\mathbf{N4}^{\perp} \rightarrow \mathsf{Exp}$, $(-)_{nor} : \mathcal{E}\mathbf{N4}^{\perp} \rightarrow \mathsf{Nor}$, and $(-)_{ne} : \mathcal{E}\mathbf{N4}^{\perp} \rightarrow \mathsf{Exp} \cap \mathsf{Nor}$ are defined by the rules:

$$L_{exp} := L + \mathbf{N3}, \quad L_{nor} := L + \mathbf{N4}^{\mathsf{N}}, \quad L_{ne} := L + \mathbf{N3}^{\circ},$$

where $L \in \mathcal{E}\mathbf{N4}^{\perp}$. Obviously, $L_{ne} := L_{exp} + L_{nor}$. Call the logic L_{exp} an *explosive counterpart* of L, L_{nor} a *normal counterpart* of L, and L_{ne} a *normal explosive counterpart* of L. Thus, by definition the explosive (normal) counterpart of a logic $L \in \mathcal{E}\mathbf{N4}^{\perp}$ is the least explosive (normal) logic containing L.

Notice that $\mathbf{N4}^{\perp}$ has the following counterparts:

$$\mathbf{N4}^{\perp}_{exp} = \mathbf{N3}, \quad \mathbf{N4}^{\perp}_{nor} = \mathbf{N4}^{\mathsf{N}}, \quad \mathbf{N4}^{\perp}_{ne} = \mathbf{N3}^{\circ}.$$

Some simple properties of counterparts are collected in the proposition below.

Proposition 10.3.1 *1. $(-)_{exp}$, $(-)_{not}$, and $(-)_{ne}$ are lattice epimorphisms.*

2. $L \in \mathsf{Exp}$ iff $L = L_{exp}$ iff $L_{nor} = L_{ne}$.

3. $L \in \mathsf{Nor}$ iff $L = L_{nor}$ iff $L_{exp} = L_{ne}$.

4. $L \in \mathcal{E}\mathbf{N3}^{\circ}$ iff $L_{exp} = L_{nor}$.

5. $\sigma(L) = \sigma(L_{exp}) = \sigma(L_{nor}) = \sigma(L_{ne})$ for every $L \in \mathcal{E}\mathbf{N4}^{\perp}$.

6. $L_{ne} = \eta^{\circ}\sigma(L)$ for every $L \in \mathcal{E}\mathbf{N4}^{\perp}$.

Proof. Item 1 follows from the distributivity of the lattice $\mathcal{E}\mathbf{N4}^{\perp}$. Items 2 and 3 hold by the definition of counterparts. Item 4 follows from the relation $\mathbf{N3}^{\circ} = \mathbf{N3} + \mathbf{N4}^{\mathsf{N}}$. Homomorphism properties of σ imply Item 5.

We prove the last item. By Item 5 $\sigma(L_{ne}) = \sigma(L)$ and $L_{ne} \in \mathcal{E}\mathbf{N3}^{\circ}$ by definition. Thus, the equality $L_{ne} = \eta^{\circ}\sigma(L)$ follows from the fact that η° is a lattice isomorphism of Int^{+} and $\mathcal{E}\mathbf{N3}^{\circ}$ (see Proposition 10.1.12).

□

The semantic characterization of logics $\mathbf{N3}$, $\mathbf{N4}^{\mathsf{N}}$, and $\mathbf{N3}^{\circ}$ allows one to characterize models of counterparts as follows.

Proposition 10.3.2 *Let $L \in \mathcal{E}\mathbf{N4}^{\perp}$ and $\mathcal{A}$ be an $\mathbf{N4}^{\perp}$-lattice.*

1. $\mathcal{A} \models L_{exp}$ if and only if $\mathcal{A} \models L$ and $\Delta(\mathcal{A}) = \{0\}$.

2. $\mathcal{A} \models L_{nor}$ if and only if $\mathcal{A} \models L$ and $\nabla(\mathcal{A}) = F_d(\mathcal{A})$.

3. $\mathcal{A} \models L_{ne}$ if and only if $\mathcal{A} \models L$, $\Delta(\mathcal{A}) = \{0\}$, and $\nabla(\mathcal{A}) = F_d(\mathcal{A})$.

□

We now study how the counterparts of L can be defined in L. For any formula φ, we put

$$\varphi^{e} := \bigwedge_{p \in var(\varphi)} \neg(p \wedge \sim p), \quad \varphi^{n} := \bigwedge_{p \in var(\varphi)} \neg\neg(p \vee \sim p),$$

and

$$\varphi^{exp} := \varphi^{e} \to \varphi, \quad \varphi^{nor} := \varphi^{n} \to \varphi, \quad \varphi^{ne} := (\varphi^{e} \wedge \varphi^{n}) \to \varphi.$$

Let $\mathcal{A}$ be a twist-structure, $\mathcal{A} = Tw(\mathcal{B}, \nabla, \Delta)$. We associate with $\mathcal{A}$ the following substructures:

$$\mathcal{A}_{exp} = Tw(\mathcal{B}, \nabla, \{0\}), \quad \mathcal{A}_{nor} = Tw(\mathcal{B}, F_d(\mathcal{B}), \Delta),$$

$$\mathcal{A}_{ne} = Tw(\mathcal{B}, F_d(\mathcal{B}), \{0\}).$$

According to Proposition 10.3.2 if $\mathcal{A}$ is a model of L, then $\mathcal{A}_{exp}$, $\mathcal{A}_{nor}$, and $\mathcal{A}_{ne}$ are models of explosive, normal and normal explosive counterparts of L, respectively. The validity of formulas on $\mathcal{A}_{exp}$, $\mathcal{A}_{nor}$, and $\mathcal{A}_{ne}$ can be simulated in $\mathcal{A}$ as follows.

Proposition 10.3.3 *Let $\mathcal{A} \in S^{\bowtie}(\mathcal{B})$ and φ be a formula.*

1. *$\mathcal{A}_{exp} \models \varphi$ if and only if $\mathcal{A} \models \varphi_{exp}$.*
2. *$\mathcal{A}_{nor} \models \varphi$ if and only if $\mathcal{A} \models \varphi_{nor}$.*
3. *$\mathcal{A}_{ne} \models \varphi$ if and only if $\mathcal{A} \models \varphi_{ne}$.*

Proof. 1. Let $\mathcal{A} \models \varphi_{exp}$ and v be an $\mathcal{A}_{exp}$-valuation. Note that v is also an $\mathcal{A}$-valuation, since $\mathcal{A}_{exp} \leq \mathcal{A}$. For any $(a,b) \in \mathcal{A}_{exp}$, $a \wedge b = 0$. Consequently, $\pi_1 v(\bigwedge_{p \in var(\varphi)} \neg(p \wedge \sim p)) = 1$. And since $\mathcal{A} \models \varphi_{exp}$, we have $\pi_1 v(\varphi) = 1$. We proved $\mathcal{A}_{exp} \models \varphi$.

Assume now that $\mathcal{A}_{exp} \models \varphi$ but $\mathcal{A} \not\models \varphi_{exp}$. Let v be an $\mathcal{A}$-valuation such that $\pi_1 v(\varphi_{exp}) \neq 1$. Put $a := \pi_1 v(\bigwedge_{p \in var(\varphi)} \neg(p \wedge \sim p))$ and $F := \langle a \rangle^{\bowtie}$. Then $a \nleq \pi_1 v(\varphi)$. Consider the quotient structure $\mathcal{A}/F$ and the quotient valuation v/F. By Proposition 9.2.16

$$\mathcal{A}/F \cong \mathcal{A}_1 := Tw(\mathcal{B}/\langle a \rangle, \nabla/\langle a \rangle, \Delta/\langle a \rangle).$$

Applying the same proposition we obtain $(\mathcal{A}_1)_{exp} \cong \mathcal{A}_{exp}/F$. Clearly, $\pi_1 v/F(\bigwedge_{p \in var(\varphi)} \neg(p \wedge \sim p)) = 1$, whence v/F can be considered as a valuation in $(\mathcal{A}_1)_{exp}$. Since $a \nleq \pi_1 v(\varphi)$, we have $\pi_1 v/F(\varphi) \neq 1$. Thus, $\mathcal{A}_{exp}/F \not\models \varphi$, which conflicts with the assumption $\mathcal{A}_{exp} \models \varphi$.

The proofs of Items 2 and 3 are similar and are based on the following facts. For any $\mathcal{A}$-valuation v, v is an $\mathcal{A}_{nor}$-valuation iff $\pi_1 v(\neg\neg(p \vee \sim p)) = 1$ for all p, and v is an $\mathcal{A}_{ne}$-valuation iff for all p,

$$\pi_1 v(\neg(p \wedge \sim p)) = 1 \text{ and } \pi_1 v(\neg\neg(p \vee \sim p)) = 1.$$

□

Proposition 10.3.4 *Let $L \in \mathcal{E}\mathbf{N4}^{\perp}$ and φ be a formula.*

1. *$\varphi \in L_{exp}$ if and only if $\varphi_{exp} \in L$.*
2. *$\varphi \in L_{nor}$ if and only if $\varphi_{nor} \in L$.*
3. *$\varphi \in L_{ne}$ if and only if $\varphi_{ne} \in L$.*

Proof. We prove only the first item. The proofs of others are similar and are based on the respective items of the last proposition.

Assume $\varphi \in L_{exp}$. Let $\mathcal{A}$ be a twist-structure such that $L \models \mathcal{A}$. Then $\mathcal{A}_{exp} \models L_{exp}$. Indeed, $\mathcal{A}_{exp} \models L$ as a substructure of $\mathcal{A}$ and $\Delta(\mathcal{A}_{exp}) = 0$. By Item 1 of the previous proposition we have $\mathcal{A} \models \varphi_{exp}$. Finally, since every model of L can be represented as a twist-structure, we conclude $\varphi_{exp} \in L$.

Conversely, let $\varphi_{exp} \in L$. Consider a twist-structure $\mathcal{A} \models L_{exp}$. By Proposition 10.3.2 $\Delta(\mathcal{A}) = 0$, i.e., $\mathcal{A} = \mathcal{A}_{exp}$. By Item 1 of the last proposition $\mathcal{A} \models \varphi$ if and only if $\mathcal{A} \models \varphi_{exp}$, from which we conclude $\varphi \in L_{exp}$.

□

Let $\mathcal{K}$ be a class of twist-structures and $\tau \in \{exp, nor, ne\}$. We put

$$\mathcal{K}_\tau := \{\mathcal{A}_\tau \mid \mathcal{A} \in \mathcal{K}\}.$$

Proposition 10.3.5 *Let $\mathcal{K}$ be a class of twist-structures, $L = L\mathcal{K}$, and $\tau \in \{exp, nor, ne\}$. Then $L_\tau = L\mathcal{K}_\tau$.*

Proof. Consider the case of explosive counterparts. Obviously, $L_{exp} \subseteq L\mathcal{K}_{exp}$. If $\varphi \notin L_{exp}$, then $\varphi_{exp} \notin L$ by the previous proposition. Consequently, there is $\mathcal{A} \in \mathcal{K}$ such that $\mathcal{A} \not\models \varphi_{exp}$. By Proposition 10.3.3 $\mathcal{A}_{exp} \not\models \varphi$.

□

At the end of this section, we consider classes of logics having given explosive and normal logics as explosive and normal counterparts. According to Proposition 10.3.1 explosive and normal counterparts of a logic have the same intuitionistic fragment. Therefore, for given $L_1 \in \mathsf{Exp}$ and $L_2 \in \mathsf{Nor}$, there is a logic L with $L_{exp} = L_1$ and $L_{nor} = L_2$ if and only if $\sigma(L_1) = \sigma(L_2)$.

For $L_1 \in \mathsf{Exp}$ and $L_2 \in \mathsf{Nor}$ with $\sigma(L_1) = \sigma(L_2)$, we define the family of logics

$$Spec(L_1, L_2) := \{L \in \mathcal{E}\mathbf{N4}^{\perp} \mid L_{exp} = L_1 \text{ and } L_{nor} = L_2\}$$

and the logic

$$L_1 * L_2 := \mathbf{N4}^{\perp} + \{\varphi_{exp} \mid \varphi \in L_1\} \cup \{\varphi_{nor} \mid \varphi \in L_2\}.$$

First of all, we note that all logics of $Spec(L_1, L_2)$ have the same intuitionistic fragment and that if one of the logics L_1 or L_2 belongs to the intersection $\mathsf{Exp} \cap \mathsf{Nor}$, the class $Spec(L_1, L_2)$ is one-element.

Proposition 10.3.6 *Let* $L_1 \in \mathsf{Exp}$ *and* $L_2 \in \mathsf{Nor}$ *be such that* $L = \sigma(L_1) = \sigma(L_2)$.

1. $Spec(L_1, L_2) \subseteq \sigma^{-1}(L)$.
2. *If* $L_2 \in \mathsf{Exp}$, *then* $Spec(L_1, L_2) = \{L_1\}$.
3. *If* $L_1 \in \mathsf{Nor}$, *then* $Spec(L_1, L_2) = \{L_2\}$.

Proof. 1. If $L' \in Spec(L_1, L_2)$, then $L'_{exp} = L_1$ and

$$L = \sigma(L'_{exp}) = \sigma(L') + \sigma(\mathbf{N3}) = \sigma(L') + \mathbf{Li} = \sigma(L').$$

2. If $L_2 \in \mathsf{Exp} \cap \mathsf{Nor} = \mathcal{E}\mathbf{N3}^\circ$, then $L_2 = \eta^\circ(L)$. Since $\eta^\circ(L)$ is the greatest element of $\sigma^{-1}(L)$, we obtain $L_1 \subseteq L_2$.

Let $L' \in Spec(L_1, L_2)$. We then have $L' + \mathbf{N4}^{\mathsf{N}} = L_2 \supseteq L_1 \supseteq \mathbf{N3}$. We claim that $\mathbf{N3} \subseteq L'$ in this case. Assume that L' is not explosive. There then is a twist-structure $\mathcal{A} \models L'$ with $\Delta(\mathcal{A}) \neq \{0\}$. By Proposition 10.3.2 $\mathcal{A}_{nor} \models L_2 = L'_{nor}$ and $\Delta(\mathcal{A}_{nor}) = \Delta(\mathcal{A}) \neq \{0\}$. The latter is impossible, since L_2 is explosive. From $\mathbf{N3} \subseteq L'$ and $L_1 = L' + \mathbf{N3}$ we obtain $L' = L_1$.

3. From $L_1 \in \mathsf{Nor}$ we conclude $L_2 \subseteq L_1$ and finish the proof similar to the previous item.

□

In the case when neither of L_1 or L_2 is normal explosive, the class $Spec(L_1, L_2)$ forms an interval in the lattice $\mathcal{E}\mathbf{N4}^\perp$ containing at least two points.

Proposition 10.3.7 *Let* $L_1 \in \mathsf{Exp}$ *and* $L_2 \in \mathsf{Nor}$ *be such that* $\sigma(L_1) = \sigma(L_2)$. *Then* $Spec(L_1, L_2) = [L_1 * L_2, L_1 \cap L_2]$. *If* $L_1 \notin \mathsf{Nor}$ *and* $L_2 \notin \mathsf{Exp}$, *then* $L_1 * L_2 \neq L_1 \cap L_2$.

Proof. It follows by definition that $L \subseteq L_{exp} \cap L_{nor}$ for all $L \in \mathcal{E}\mathbf{N4}^\perp$, i.e., $L \subseteq L_1 \cap L_2$ for all $L \in Spec(L_1, L_2)$. Let us check that $L_1 \cap L_2 \in Spec(L_1, L_2)$. We calculate

$$(L_1 \cap L_2)_{exp} = (L_1 + \mathbf{N3}) \cap (L_2 + \mathbf{N3}) = L_1 \cap \eta^\circ\sigma(L_1) = L_1.$$

That $(L_1 \cap L_2)_{nor} = L_2$, can be checked similarly.

We have thus proved that $L_1 \cap L_2$ is the greatest point of $Spec(L_1, L_2)$. Let us consider the logic $L_1 * L_2$. If $L \in Spec(L_1, L_2)$, then

$$\{\varphi_{exp} \mid \varphi \in L_1\} \cup \{\varphi_{nor} \mid \varphi \in L_2\} \subseteq L$$

by Proposition 10.3.4, i.e., $L_1 * L_2 \subseteq L$. It remains to verify that $L_1 * L_2$ belongs to $Spec(L_1, L_2)$.

Let $L^* := L_1 * L_2$. The inclusions $L_1 \subseteq L^*_{exp}$ and $L_2 \subseteq L^*_{nor}$ follow from the definition of $L_1 * L_2$ and Proposition 10.3.4. Let us prove the inverse inclusions.

By definition

$$L^*_{exp} = \mathbf{N3} + \{\varphi_{exp} \mid \varphi \in L_1\} \cup \{\varphi_{nor} \mid \varphi \in L_2\} = L_1 + \{\varphi_{ne} \mid \varphi \in L_2\}.$$

The last equality is due to the fact for any φ,

$$\mathbf{N3} \vdash \varphi \leftrightarrow \varphi_{exp} \text{ and } \mathbf{N3} \vdash \varphi_{nor} \leftrightarrow \varphi_{ne},$$

which follows from $\neg(p\wedge \sim p) \in \mathbf{N3}$. By Proposition 10.3.4 a formula of the form φ_{ne} belongs to L^*_{exp} if and only if $\varphi \in (L^*_{exp})_{ne}$. The equality $\sigma(L_1) = \sigma(L_2)$ implies $L_2 \subseteq (L_2)_{ne} = (L_1)_{ne}$, whence, $\varphi \in L_2$ implies $\varphi_{ne} \in L_1$. Thus, $L^*_{exp} = L_1$. That $L^*_{nor} = L_2$, can be proved similarly.

Assume $L_1 \notin \mathsf{Nor}$ and $L_2 \notin \mathsf{Exp}$. Prove that in this case

$$(\sim p \to (p \to q)) \vee \neg\neg(r\vee \sim r) \in (L_1 \cap L_2) \setminus (L_1 * L_2).$$

Since $\sim p \to (p \to q) \in L_1$ and $\neg\neg(r\vee \sim r) \in L_2$, the disjunction of these formulas belongs to the intersection $L_1 \cap L_2$. To prove $(\sim p \to (p \to q)) \vee \neg\neg(r\vee \sim r) \notin L_1 * L_2$ note that on the one hand, $L_1 * L_2 \subseteq \mathbf{B}_4^{\to} = \text{Ł}_3 * RM_3^{\perp}$, and on the other hand,

$$(\sim p \to (p \to q)) \vee \neg\neg(r\vee \sim r) \notin \mathbf{B}_4^{\to}.$$

□

As a consequence, we obtain a semantic characterization of logics of the form $L_1 * L_2$.

Corollary 10.3.8 *Let $L_1 \in \mathsf{Exp}$ and $L_2 \in \mathsf{Nor}$ be such that $\sigma(L_1) = \sigma(L_2)$. For every $\mathbf{N4}^{\perp}$-lattice $\mathcal{A}$ holds the equivalence*

$$\mathcal{A} \models L_1 * L_2 \;\Leftrightarrow\; \mathcal{A}_{exp} \models L_1 \textit{ and } \mathcal{A}_{nor} \models L_2.$$

Proof. The direct implication follows from the fact that $L_1 * L_2 \in Spec(L_1, L_2)$ stated above. The inverse implication is by the definition of $L_1 * L_2$.

□

10.4 The Structure of $\mathcal{E}$N4C and $\mathcal{E}$N4$^{\perp}$C

In this section, we completely describe the extensions of the logics **N4**$^{\perp}$**C** := **N4**$^{\perp}$ + {**C**} and **N4C** := **N4** + {**C**}, which are obtained by adding to **N4**$^{\perp}$ and respectively to **N4** Dummett's linearity axiom $\mathbf{C} = (p \to q) \vee (q \to p)$. This result is interesting for the following reasons.

First, comparison of the structures of $\mathcal{E}$**N4**$^{\perp}$**C** and $\mathcal{E}$**N4C** vividly demonstrates how the regular structure of the lattice **N4**$^{\perp}$ is collapsed if we delete the intuitionistic negation from the language. In particular, one cannot define normal logics in the class $\mathcal{E}$**N4**.

Second, it is worth comparing the structures of the lattices $\mathcal{E}$**N4**$^{\perp}$**C**, $\mathcal{E}$**N3C** and $\mathcal{E}$**LC**, where **N3C** := **N3** + {**C**} and **LC** is Dummett's logic obtained by adding the linearity axiom to intuitionistic logic, i.e. **LC** = **Lil** in the denotation of the first part of the book. Dummett's logic is the first example of superintuitionistic pretabular logic, whose structure of extensions was completely described [27]. Recall that a logic is pretabular if any of its extensions is tabular, i.e., is determined by a finite algebra, but the logic is not tabular itself. In fact, the article [27] by M.J. Dunn and R. Meyer was published earlier then the notion of pretabular logic was introduced and generated an interest in pretabular logics as logics whose class of extensions admits an exhaustive description. M. Kracht [47] described the lattices of **N3C**-extensions and proved that although this logic is not pretabular, it preserves the most essential properties of pretabular logics. Namely, all extensions of **N3C** are finitely axiomatizable and decidable. Moreover, given a formula φ, one can determine which of the **N3C**-extensions is axiomatized by φ. It turns out that the class of **N4**$^{\perp}$**C**-extensions also satisfies all these properties. Moreover, comparing lattices $\mathcal{E}$**N4**$^{\perp}$**C** and $\mathcal{E}$**N3C** demonstrates the complexity of the lattice of **N4**$^{\perp}$-extensions.

Finally, notice that the description of the lattice **N3C** by M. Kracht [47] was motivated by M. Gelfond's question on the structure of the upper part of the lattice $\mathcal{E}$**N3**. This is connected with thc so-called answer set semantics for logic programs. The logic of "here-and-there" with strong negation lying in the upper part of $\mathcal{E}$**N3** plays an important role in the study of the answer set semantics. For this reason, it is important to have same information on the surroundings of this logic in the lattice $\mathcal{E}$**N3**. In the study of paraconsistent answer sets [82] there arises a paraconsistent version of the logic "here-and-there" with strong negation, which belongs to the upper part of $\mathcal{E}$**N4**$^{\perp}$. This naturally leads to the question on the description of the lattice $\mathcal{E}$**N4**$^{\perp}$**C**.

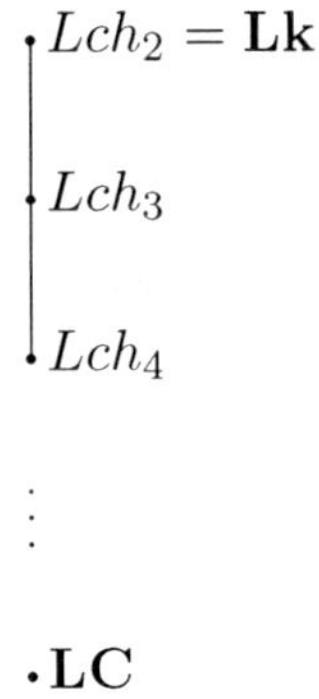

Figure 10.2

Recall that the class $\mathcal{E}\mathbf{LC}$ of extensions of Dummett's logic has the following structure. Let ch_n be a linearly ordered n-element Heyting algebra. Algebras ch_n are all up to isomorphism finitely generated subdirectly irreducible models of $\mathbf{LC}$. Obviously, for any $n \in \omega$, $ch_n \hookrightarrow ch_{n+1}$. Therefore, every proper $\mathbf{LC}$-extension has the form Lch_n for some $n \in \omega$, and $\mathcal{E}\mathbf{LC}$ is isomorphic to a linear order of type $(\omega + 1)^*$ (see Figure 10.2).

Given a logic L, consider the class $Mod_{fsi}(L)$ of its finitely generated subdirectly irreducible models. Define on $Mod_{fsi}(L)$ the following preordering:

$$\mathcal{A} \sqsubseteq \mathcal{B} \Leftrightarrow L\mathcal{A} \subseteq L\mathcal{B}.$$

The following theorem by Jonsson is well known. In a congruence distributive variety, all subdirectly irreducible elements of the subvariety generated by the class $\mathcal{K}$ of algebras belong to $\mathsf{HSUp}(\mathcal{K})$. If an algebra $\mathcal{A}$ is finite, then $\mathsf{HSUp}(\{\mathcal{A}\}) = \mathsf{HS}(\{\mathcal{A}\})$. The varieties of $\mathbf{N4}$-lattices and $\mathbf{N4}^{\perp}$-lattices are congruence distributive, therefore, for any finite $\mathbf{N4}$-($\mathbf{N4}^{\perp}$-)lattices $\mathcal{A}$ and $\mathcal{B}$, holds the equivalence

$$L\mathcal{A} \subseteq L\mathcal{B} \Leftrightarrow \mathcal{B} \in \mathsf{HS}(\{\mathcal{A}\}).$$

As was noted above, all finitely generated subdirectly irreducible models of Dummett's logic $\mathbf{LC}$ are finite. According to Corollary 9.2.11, an $\mathbf{N4}$($\mathbf{N4}^{\perp}$)-lattice $\mathcal{A}$ is subdirectly irreducible if and only if $\mathcal{A}_{\bowtie}$ is subdirectly irreducible. The fact that if $\mathcal{A}$ is finitely generated, then $\mathcal{A}_{\bowtie}$ is finitely generated, follows from the lemma below.

Lemma 10.4.1 *If an* $\mathbf{N4}$($\mathbf{N4}^{\perp}$)*-lattice* $\mathcal{A}$ *is finitely generated, then* $\mathcal{A}_{\bowtie}$ *is finitely generated.*

Proof. If $\mathcal{A}$ is generated by the set $\{a_1, \dots, a_n\}$, then its reduct $\langle|\mathcal{A}|, \vee, \wedge, \to\rangle$ ($\langle|\mathcal{A}|, \vee, \wedge, \to, \perp\rangle$) is generated by the set $\{a_1, \dots, a_n, \sim a_1, \dots, \sim a_n\}$. This follows from the fact that every formula is equivalent in **N4**(**N4**$^\perp$) to an *nnf*. By definition $\mathcal{A}_{\bowtie} = \langle|\mathcal{A}|, \vee, \wedge, \to\rangle / \approx$ ($\langle|\mathcal{A}|, \vee, \wedge, \to, \perp\rangle / \approx$). Consequently, $\mathcal{A}_{\bowtie}$ is generated by the set

$$\{[a_1]_{\approx}, \dots, [a_n]_{\approx}, [\sim a_1]_{\approx}, \dots, [\sim a_n]_{\approx}\}.$$

□

Thus, every finitely generated subdirectly irreducible model of **N4C** (**N4**$^\perp$**C**) is isomorphic to a twist-structure over ch_n for suitable $n \in \omega$, therefore, all finitely generated subdirectly irreducible models of **N4C** and **N4**$^\perp$**C** are finite. In this section, we consider only extensions of the logics **N4C** and **N4**$^\perp$**C**, therefore, in the following for $\mathcal{A}, \mathcal{B} \in Mod_{fsi}(L)$, holds the equivalence:

$$\mathcal{A} \sqsubseteq \mathcal{B} \;\Leftrightarrow\; \mathcal{B} \in \mathsf{HS}(\{\mathcal{A}\}).$$

Every extension of a logic L is determined by the class of its finitely generated subdirectly irreducible models. This class forms a cone in the preordering $(Mod_{fsi}(L), \sqsubseteq)$. Of course, not every cone of $(Mod_{fsi}(L), \sqsubseteq)$ can be represented in the form $Mod_{fsi}(L')$ for a suitable L-extension L' in the general case. However, we prove that for every cone U of $(Mod_{fsi}(\mathbf{N4}^\perp\mathbf{C}), \sqsubseteq)$, there exists $L \in \mathcal{E}\mathbf{N4}^\perp\mathbf{C}$ such that $Mod_{fsi}(L) = U$. Cones of the preordering $(Mod_{fsi}(\mathbf{N4}^\perp\mathbf{C}), \sqsubseteq)$ are closed under isomorphism. Note that the isomorphism relation, coincides with the equivalence determined by the preordering $\sqsubseteq$. Factoring out the preordering $(Mod_{fsi}(\mathbf{N4}^\perp\mathbf{C}), \sqsubseteq)$ wrt the isomorphism relation, we obtain the partial ordering of isomorphism types of finitely generated subdirectly irreducible **N4**$^\perp$**C**-models, which we denote $(Mod^*_{fsi}(\mathbf{N4}^\perp\mathbf{C}), \sqsubseteq^*)$. In this way, the description of the lattice $\mathcal{E}\mathbf{N4}^\perp\mathbf{C}$ is reduced to the description of the ordering $(Mod^*_{fsi}(\mathbf{N4}^\perp\mathbf{C}), \sqsubseteq^*)$.

Let $|ch_n| = \{0, 1, \dots, n-1\}$, $0 \leq 1 \leq \dots \leq n-1$. Ideals of ch_n are exactly non-empty initial segments of ch_n. The filter $F_d(ch_n)$ equals $\{1, \dots, n-1\}$. Therefore, if $\mathcal{A} = Tw(ch_n, \nabla, \Delta)$, then $\nabla = ch_n$ or $\nabla = F_d(ch_n)$, and $\Delta = \{0, \dots, m\}$ for some $m \leq n-1$. We have thus proved $|S^{\bowtie}(ch_n)| = 2n$. Denote

$$\mathsf{ch}_n(k, +) := Tw(ch_n, ch_n, \{0, \dots, k-1\}),$$

$$\mathsf{ch}_n(k, -) := Tw(ch_n, F_d(ch_n), \{0, \dots, k-1\}),$$

where $1 \leq k \leq n$.

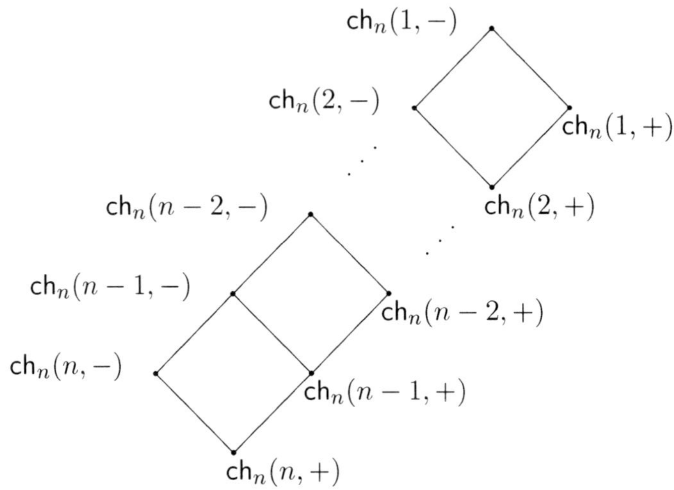

Figure 10.3

Let $\mathcal{A}, \mathcal{B} \in S^{\bowtie}(ch_n)$. As it follows from Propositions 9.2.8 and 9.2.16 every proper homomorphic image of $\mathcal{B}$ is isomorphic to a twist-structure over a proper homomorphic image of ch_n, i.e., over ch_m, $m < n$. In this way, $\mathcal{B} \sqsubseteq \mathcal{A}$ if and only if $\mathcal{A} \hookrightarrow \mathcal{B}$. Thus, according to Proposition 9.1.10 $\mathcal{B} \sqsubseteq \mathcal{A}$ if and only if $|\nabla(\mathcal{A})| \leq |\nabla(\mathcal{B})|$ and $|\Delta(\mathcal{A})| \leq |\Delta(\mathcal{B})|$. The structure of $S^{\bowtie}(ch_n)$ ordered by $\sqsubseteq$ is depicted in Figure 10.3.

Proposition 9.2.2 states that a homomorphic image $h(\mathcal{A})$ of an $\mathbf{N4}^{\perp}$-lattice $\mathcal{A}$ is uniquely determined by the homomorphic image $h_{\bowtie}(\mathcal{A}_{\bowtie})$ of the basic Heyting algebra. Since every filter on ch_n has the form $F_m := \{m, \ldots, n-1\}$ and $ch_n/F_m \cong ch_{m+1}$, Proposition 9.2.16 allows one to conclude that quotients of twist-structures over ch_n have the following form:

$$\mathsf{ch}_n(k, \epsilon)/(F_m)^{\bowtie} \cong \mathsf{ch}_{m+1}(min\{k, m+1\}, \epsilon), \;\; m \leq n-1, \;\; \epsilon \in \{+, -\}.$$

Consider the question of embedding twist-structures of the form $\mathsf{ch}_n(k, \epsilon)$. If $h : \mathsf{ch}_{n_1}(k_1, \epsilon_1) \hookrightarrow \mathsf{ch}_{n_2}(k_2, \epsilon_2)$, then according to Proposition 9.2.15 $h_{\bowtie}$ embeds ch_{n_1} into ch_{n_2} and the inclusions

$$h_{\bowtie}(\nabla(\mathsf{ch}_{n_1}(k_1, \epsilon_1))) \subseteq \nabla(\mathsf{ch}_{n_2}(k_2, \epsilon_2)), \; h_{\bowtie}(\Delta(\mathsf{ch}_{n_1}(k_1, \epsilon_1))) \subseteq \Delta(\mathsf{ch}_{n_2}(k_2, \epsilon_2))$$

hold. Consequently, $n_1 \leq n_2$ and $k_1 \leq k_2$. From $h_{\bowtie}(0) = 0$ and first of the above inclusions we conclude that $\epsilon_1 = +$ implies $\epsilon_2 = +$.

We set $\epsilon_1 \leq \epsilon_2$ if $\epsilon_1 = \epsilon_2$, or $\epsilon_1 = -$ and $\epsilon_2 = +$.

It is easy to see that the conditions $\epsilon_1 \leq \epsilon_2$, $n_1 \leq n_2$, and $k_1 \leq k_2$ guarantee that there exists an embedding $g : ch_{n_1} \hookrightarrow ch_{n_2}$ such that

$g(\{0,\dots,k_1-1\}) \subseteq \{0,\dots,k_2-1\}$. Again, by Proposition 9.2.15 we conclude that there exists an embedding

$$h : \mathsf{ch}_{n_1}(k_1,\epsilon_1) \hookrightarrow \mathsf{ch}_{n_2}(k_2,\epsilon_2)$$

such that $h_{\bowtie} = g$.

We have thus proved the following statement.

Lemma 10.4.2 *Let $n_1, n_2, k_1, k_2 \in \omega$, $k_1 \leq n_1$, $k_2 \leq n_2$, and $\epsilon_1, \epsilon_2 \in \{+,-\}$.*

1. $\mathsf{ch}_{n_1}(k_1,\epsilon_1) \hookrightarrow \mathsf{ch}_{n_2}(k_2,\epsilon_2)$ *if and only if* $n_1 \leq n_2$, $k_1 \leq k_2$, *and* $\epsilon_1 \leq \epsilon_2$.
2. *If* $\mathsf{ch}_{n_1}(k_1,\epsilon_1) \in \mathsf{H}(\mathsf{ch}_{n_2}(k_2,\epsilon_2))$, *then* $\mathsf{ch}_{n_1}(k_1,\epsilon_1) \hookrightarrow \mathsf{ch}_{n_2}(k_2,\epsilon_2)$.

□

Let $\mathbf{T} := \bigcup_{n\in\omega} S^{\bowtie}(ch_n)$. Combining Lemma 10.4.2 with the structure of $(S^{\bowtie}(ch_n), \sqsubseteq)$ presented in Figure 10.3, we obtain that $\mathbf{T}$ is ordered by $\sqsubseteq$ as depicted in Figure 10.4.

Since every finitely generated subdirectly irreducible model of $\mathbf{N4}^\perp\mathbf{C}$ is isomorphic to some element of $\mathbf{T}$, we arrive at the following statement.

Proposition 10.4.3 *The preordering $(Mod^*_{fsi}(\mathbf{N4}^\perp\mathbf{C}), \sqsubseteq^*)$ has the structure depicted in Figure 10.4.*

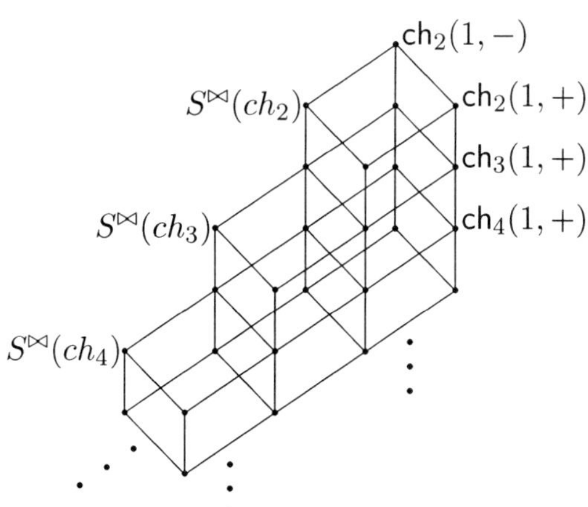

Figure 10.4

Describe now the cones of $(\mathbf{T}, \sqsubseteq)$.

Let $\mathcal{A} \in \mathbf{T}$. If $\mathcal{A} = \mathsf{ch}_n(k, \epsilon)$, we say that $\mathcal{A}$ is of type (n, k, ϵ) and write $tp(\mathcal{A}) = (n, k, \epsilon)$. By $tp_i(\mathcal{A})$, $i = 1, 2, 3$, we denote the i-th coordinate of $tp(\mathcal{A})$.

For a cone U of $(\mathbf{T}, \sqsubseteq)$ and $i = 1, 2$, we put

$$\begin{array}{rcl} tp_i^-(U) & := & max\{tp_i(\mathcal{A}) \mid \mathcal{A} \in U \text{ and } tp_3(\mathcal{A}) = -\}, \\ tp_i^+(U) & := & max\{tp_i(\mathcal{A}) \mid \mathcal{A} \in U \text{ and } tp_3(\mathcal{A}) = +\}. \end{array}$$

Clearly, $tp_i^+(U) \leq tp_i^-(U)$ for $i = 1, 2$.

Further, we distinguish the following cones:

$$\begin{array}{rcl} \mathbf{T}^- & := & \{\mathcal{A} \in \mathbf{T} \mid tp_3(\mathcal{A}) = -\}, \\ \mathbf{T}_k & := & \{\mathcal{A} \in \mathbf{T} \mid tp_2(\mathcal{A}) \leq k\}, \\ \mathbf{T}_k^- & := & \{\mathcal{A} \in \mathbf{T} \mid tp_3(\mathcal{A}) = - \text{ and } tp_2(\mathcal{A}) \leq k\}. \end{array}$$

Proposition 10.4.4 *Every cone of $(\mathbf{T}, \sqsubseteq)$ can be represented as $U \cup V$, where U is a finite cone and V is one of the following cones:*

$$\varnothing,\ \mathbf{T},\ \mathbf{T}^-,\ \mathbf{T}_k,\ \mathbf{T}_k^-,\ \mathbf{T}^- \cup \mathbf{T}_k,\ \mathbf{T}_k \cup \mathbf{T}_{k'}^-,$$

where $k < k'$.

Proof. Let U be a cone in $(\mathbf{T}, \sqsubseteq)$. If $tp_1^-(U) < \omega$ and $tp_2^-(U) < \omega$, it can easily be seen from Figure 10.4 that U is finite or empty.

If $tp_1^+(U) = \omega$ and $tp_2^+(U) = \omega$, then $U = \mathbf{T}$.

Let $tp_1^-(U) = \omega$ and $tp_2^-(U) = \omega$. Then $\mathbf{T}^- \subseteq U$. If additionally $tp_1^+(U) < \omega$ and $tp_2^+(U) < \omega$, then $U = \mathbf{T}^- \cup U'$, where U' is finite.

If $tp_1^+(U) = \omega$ and $tp_2^+(U) < \omega$, there exists $m \leq tp_2^+(U)$ such that $\mathsf{ch}_n(m, +) \in U$ for all n. Let

$$k := max\{m \mid \mathsf{ch}_n(m, +) \in U \text{ for all } n\}.$$

Then $U = \mathbf{T}^- \cup \mathbf{T}_k \cup U'$, where U' is finite.

Let $tp_1^-(U) = \omega$ and $tp_2^-(U) < \omega$. Put

$$k := max\{m \mid \mathsf{ch}_n(m, -) \in U \text{ for all } n\}.$$

Then $\mathbf{T}_k^- \subseteq U$ and arguing as above, we obtain that either $U = \mathbf{T}_k^- \cup U'$, or $U = \mathbf{T}_k^- \cup \mathbf{T}_{k'} \cup U'$, $k' < k$, or $U = \mathbf{T}_k \cup U'$, where U' is some finite cone. □

Theorem 10.4.5 *The lattice $\mathcal{E}\mathbf{N4}^{\perp}\mathbf{C}$ is isomorphic to the lattice of cones of the partial ordering presented in Figure 10.4. All elements of $\mathcal{E}\mathbf{N4}^{\perp}\mathbf{C}$ are finitely axiomatized and decidable.*

Proof. In view of Proposition 10.4.3, the lattice $\mathcal{E}\mathbf{N4}^{\perp}\mathbf{C}$ is embedded into the lattice of cones of the partial order presented in Figure 10.4. We have to prove that this embedding is onto, i.e., that for any cone of $(\mathbf{T}, \sqsubseteq)$, there is $L \in \mathcal{E}\mathbf{N4}^{\perp}\mathbf{C}$ such that $\mathbf{T}(L) := Mod_{fsi}(L) \cap \mathbf{T} = U$. Since $\mathbf{T}$ contains all up to isomorphism finitely generated subdirectly irreducible models of $\mathbf{N4}^{\perp}\mathbf{C}$, the equality $\mathbf{T}(L) = U$ implies $LU = L$.

First, we note that the empty cone corresponds to the trivial logic $For^{\perp}$. According to Corollary 10.2.2 every non-trivial extension of $\mathbf{N4}^{\perp}$ is consistent, therefore, every contradiction axiomatizes $For^{\perp}$ over $\mathbf{N4}^{\perp}\mathbf{C}$. Consider $p_0 \wedge \sim p_0$ as a "standard" axiom of the trivial logic.

Prove the equality $\mathbf{T}^{-} = \mathbf{T}(\mathbf{N4}^{\perp}\mathbf{C} + \{\neg\neg(p \vee \sim p)\})$. By definition the cone $\mathbf{T}^{-}$ contains exactly algebras from $\mathbf{T}$ satisfying $\nabla(\mathcal{A}) = F_d(\mathcal{A})$. This condition is equivalent to the validity of $\neg\neg(p \vee \sim p)$ on $\mathcal{A}$, which implies the desired equality.

Consider the formulas

$$D_n := \bigvee_{1 \leq k < m \leq n+1} (p_k \wedge \sim p_k) \leftrightarrow (p_m \wedge \sim p_m),$$

$$E_n := \bigvee_{1 \leq k < m \leq n+1} p_k \leftrightarrow p_m.$$

Let $\mathcal{A} \in \mathbf{T}$. Recall that by definition $\Delta(\mathcal{A}) = \pi_2\{a \vee \sim a \mid a \in \mathcal{A}\}$. We have $\pi_2(a \vee \sim a) = \pi_1(\sim (a \vee \sim a)) = \pi_1(a \wedge \sim a)$. Thus, $\Delta(\mathcal{A}) = \pi_1(\{a \wedge \sim a \mid a \in \mathcal{A}\})$. In this way, $\mathcal{A} \models D_n$ is equivalent to the fact that $v(E_n) = 1$ for any $\mathcal{A}_{\bowtie}$-valuation such that $v(p_k) \in \Delta(\mathcal{A})$ for $1 \leq k \leq n+1$. Since $\mathcal{A}_{\bowtie}$ is linearly ordered, we immediately obtain that $\mathcal{A} \models D_n$ if and only if $\Delta(\mathcal{A})$ contains at most n elements. In other words, $\mathcal{A} \models D_n$ if and only if $tp_2(\mathcal{A}) \leq n$. We have thus proved $\mathbf{T}_n = \mathbf{T}(\mathbf{N4}^{\perp}\mathbf{C} + \{D_n\})$.

Since $\mathbf{T}_n^{-} = \mathbf{T}^{-} \cap \mathbf{T}_n$, this cone is distinguished in $\mathbf{T}$ by the set of axioms $\{\neg\neg(p \vee \sim p),\ D_n\}$.

For $\mathcal{A} \in \mathbf{T}$, the condition $tp_1(\mathcal{A}) \leq n$ is equivalent to $\mathcal{A}_{\bowtie} = ch_m$, $m \leq n$. The class of models with this property is distinguished in $Mod_{fsi}(\mathbf{LC})$ by the formula E_n.

By Lemma 10.1.2 the validity of an intuitionistic formula φ on a twist-structure over a Heyting algebra $\mathcal{B}$ is equivalent to $\mathcal{B} \models \varphi$. In particular, $\mathcal{A} \models E_n$ if and only if $\mathcal{A}_{\bowtie} \models E_n$. The cone of algebras satisfying the condition $tp_1(\mathcal{A}) \leq n$ corresponds to the logic $\mathbf{N4}^{\perp}\mathbf{C} + \{E_n\}$. Note that by Lemma 10.4.2 this cone is generated by $\mathsf{ch}_n(n, +)$, i.e., by the full twist-structure over ch_n.

The cone generated by $\mathsf{ch}_n(k,+)$ consists of algebras $\mathsf{ch}_m(s,\epsilon)$ satisfying the conditions $m \le n$ and $s \le k$. Therefore, it is distinguished in $\mathbf{T}$ by the formulas $\{E_n, D_k\}$. Algebras over $\mathsf{ch}_n(k,-)$ satisfy additionally the condition $\epsilon = -$. Therefore, the cone generated by $\mathsf{ch}_n(k,-)$ is distinguished in $\mathbf{T}$ by the formulas $\{E_n, D_k, \neg\neg(p\vee \sim p)\}$.

Note that all logics corresponding to the cones up to this moment are finitely axiomatizable.

In view of Proposition 10.4.4 every cone of $(\mathbf{T}, \sqsubseteq)$ can be represented as a finite union of cones of the forms considered above. The logic of a union of cones is the intersection of logics of cones. If $L_1 = L + \{\varphi_1, \dots, \varphi_n\}$ and $L_2 = L + \{\psi_1, \dots, \psi_n\}$, moreover, the formulas $\varphi_1, \dots, \varphi_n$ do not contain propositional variables occurring $\psi_1, \dots, \psi_n$ and *vice versa*, then

$$L_1 \cap L_2 := L + \{\varphi_i \vee \psi_j \mid 1 \le i \le n,\ 1 \le j \le m\}.$$

In the case when L, L_1 and L_2 are extensions of positive or minimal logic, this result was stated in Proposition 2.1.5. Its proof is based on the facts that *modus ponens* is the only inference rule and that L, L_1 and L_2 contain axioms of positive logics. Therefore, the result remains true for extensions of $\mathbf{N4}^{\perp}$. In this way, Proposition 10.4.4 implies that every extension of $\mathbf{N4}^{\perp}\mathbf{C}$ is finitely axiomatizable.

We know axiomatics for any logic in $\mathcal{E}\mathbf{N4}^{\perp}\mathbf{C}$. Given a logic $L \in \mathcal{E}\mathbf{N4}^{\perp}\mathbf{C}$ with a finite list of axioms A and some formula φ, we enumerate all formulas deducible from A and from $A \cup \{\varphi\}$. At a finite step we deduce φ from A or we deduce from $A \cup \{\varphi\}$ an axiom of some other logic not contained in L. Thus, L is decidable.

□

Note that every logic $L \in \mathcal{E}\mathbf{N4}^{\perp}\mathbf{C}$ such that $\mathbf{T}^- \subseteq Mod_{fsi}(L)$ is not tabular. Consequently, $\mathbf{N4}^{\perp}\mathbf{C}$ is not pretabular. Moreover, it has infinitely many non-tabular extensions.

It follows from the proof of Theorem 10.4.5 that we can construct such an enumeration $L_0, L_1, \dots$ of all logics from $\mathcal{E}\mathbf{N4}^{\perp}\mathbf{C}$ that given n one can effectively construct the list of axioms of L_n. Fix some enumeration of this kind.

Proposition 10.4.6 *There is an algorithm finding for any formula φ a logic $L_n \in \mathcal{E}\mathbf{N4}^{\perp}\mathbf{C}$ such that $L_n = \mathbf{N4}^{\perp}\mathbf{C} + \{\varphi\}$.*

Proof. Let $\psi_0, \psi_1, \dots$ be an effective enumeration of all formulas inferable from $\mathbf{N4}^{\perp}\mathbf{C} + \{\varphi\}$. The desired algorithm works as follows.

Step n. For all $m \leq n$, check whether all axioms of L_m are contained in the list of formulas $\psi_0, \dots, \psi_n$. If we have a positive answer for some $m_0 \leq n$, i.e., we have established the inclusion $L_{m_0} \subseteq \mathbf{N4}^{\perp}\mathbf{C} + \{\varphi\}$, we then check whether $\varphi \in L_{m_0}$. In case of the second positive answer, we have proved $L_{m_0} = \mathbf{N4}^{\perp}\mathbf{C} + \{\varphi\}$ and the algorithm terminates. Otherwise, we pass on to the next step.

Since the list $L_0, L_1, \dots$ contains all logics from $\mathcal{E}\mathbf{N4}^{\perp}\mathbf{C}$, and the list $\psi_0, \psi_1, \dots$ all formulas from $\mathbf{N4}^{\perp}\mathbf{C} + \{\varphi\}$, at some finite step we meet a number n such that $L_n = \mathbf{N4}^{\perp}\mathbf{C} + \{\varphi\}$.

□

We have thus proved that given a formula, one can determine which of the extensions of $\mathbf{N4}^{\perp}\mathbf{C}$ is axiomatized by this formula.

In conclusion to this section, we consider briefly how the situation changes if we pass on to the logic **N4C** in the restricted language $\mathcal{L} = \langle \vee, \wedge, \rightarrow, \sim \rangle$ and to the explosive logic **N3C**.

Consider the logic $\mathbf{LpC} := \mathbf{Lp} + \{(p \rightarrow q) \vee (q \rightarrow p)\}$. Finitely generated subdirectly irreducible models of **LpC** are finite linearly ordered implicative lattices, the reducts of Heyting algebras ch_n to the language $\langle \vee, \wedge, \rightarrow \rangle$. As above, we have $F_d(ch_n) = \{1, \dots, n-1\}$ (recall that for implicative lattices $F_d(\mathcal{A}) = \langle \{a \vee (a \rightarrow b) \mid a, b \in |\mathcal{A}|\} \rangle$). The lattice $\mathcal{E}\mathbf{LpC}$ as well as $\mathcal{E}\mathbf{LC}$ is a linear order of type $(\omega + 1)^*$. We can identify twist-structures from $Mod_{fsi}(\mathbf{N4C})$ with elements of $\mathbf{T}$. As above $\mathsf{ch}_n(k, \epsilon)$ denotes a twist-structure $\mathcal{A}$ over n-element linearly ordered implicative lattice with k-element ideal $\Delta(\mathcal{A})$ and $\nabla(\mathcal{A}) = \mathcal{A}$ if $\epsilon = +$, $\nabla(\mathcal{A}) = F_d(\mathcal{A})$ if $\epsilon = -$.

However, the ordering $\sqsubseteq$ differs in this case (see Figure 10.5).

The structures $\mathsf{ch}_n(k, \epsilon)$ have the same homomorphic images but the embedding relation changes. In the language $\langle \vee, \wedge, \rightarrow, \sim \rangle$, the condition $h(0) = 0$ should not be satisfied for a homomorphism $h : \mathcal{A} \rightarrow \mathcal{B}$ of **N4**-lattices. Therefore, the additional embeddings $\mathsf{ch}_n(k, +) \hookrightarrow \mathsf{ch}_{n+1}(k+1, \epsilon)$ are possible. For this reason, the ordering $(\mathbf{T}, \sqsubseteq)$ has the structure depicted in Figure 10.5. This ordering has fewer cones than the ordering presented in Figure 10.4. For example, in denotation of Proposition 10.4.4 we have $\mathbf{T}_k \subseteq \mathbf{T}^-_{k+1}$. This shows that **N4**-lattices $\mathcal{A}$ satisfying the condition $\nabla(\mathcal{A}) = F_d(\mathcal{A})$ do not form a variety and we cannot define analogs of normal logics as was done in the class of $\mathbf{N4}^{\perp}$-extensions. Similarly to Theorem 10.4.5, we can prove the following statement.

Theorem 10.4.7 *The lattice* $\mathcal{E}\mathbf{N4C}$ *is isomorphic to the lattice of cones of the partial order presented in Figure 10.5. All elements of* $\mathcal{E}\mathbf{N4C}$ *are finitely axiomatizable and decidable.*

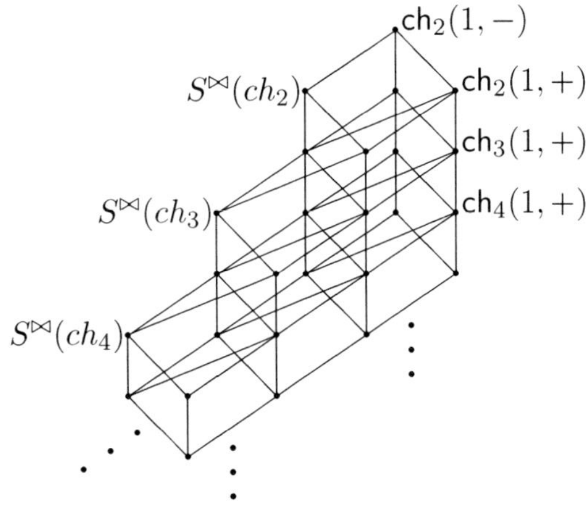

Figure 10.5

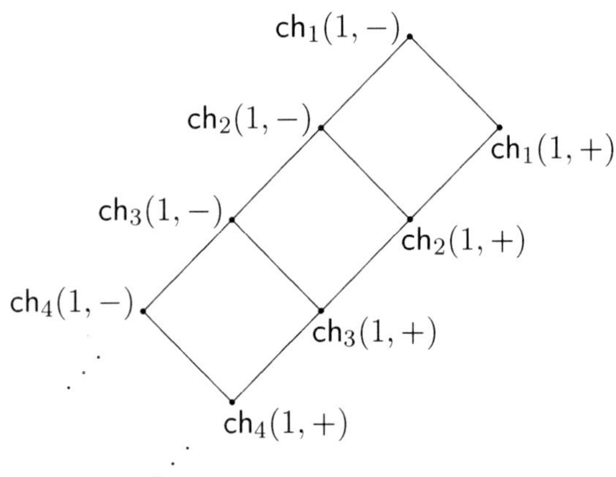

Figure 10.6

Let us consider the logic **N3C**. Only twist-structures of the form $\mathsf{ch}_n(1,\epsilon)$ are models of **N3C**. Therefore, the isomorphism types of algebras from $Mod_{fsi}(\mathbf{N3C})$ are ordered wrt $\sqsubseteq$ as shown in Figure 10.6.

According to Theorem 10.4.5, the lattice $\mathcal{E}\mathbf{N3C}$ is isomorphic to the lattice of cones of the above ordering. It is easy to see that $\mathcal{E}\mathbf{N3C}$ has the structure depicted in Figure 10.7. This diagram was obtained for the first time by M. Kracht [47].

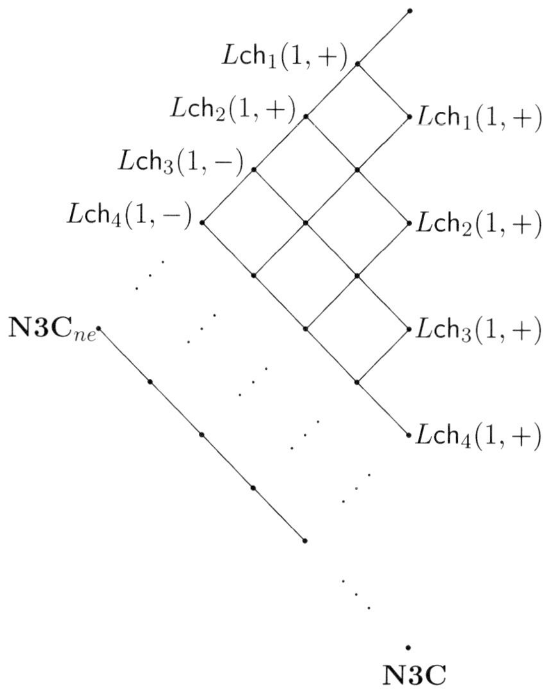

Figure 10.7

10.5 Some Transfer Theorems for the Class of N4$^{\perp}$-Extensions

In this section we transfer several results well known for the class of intermediate logics to the class of **N4**$^{\perp}$-extensions. We obtain the characterization of tabular and pretabular logics and logics with the Craig interpolation property in the class $\mathcal{E}$**N4**$^{\perp}$. The corresponding results for Int we mention without proofs.

First, we describe tabular and pretabular logics in $\mathcal{E}$**N4**$^{\perp}$, which generalizes the result by A. Sendlewski [99] for $\mathcal{E}$**N3**.

Let $\mathcal{E}$ be a lattice of logics and $L \in \mathcal{E}$. We say that L is *of finite codimension* [93] if there is no infinite descending chain of logics containing L. It is known that

Theorem 10.5.1 [93] *Let* $L \in \mathsf{Int}$. *L is tabular if and only if L is of finite codimension in* Int.

□

It turns out that an **N4**$^{\perp}$-extension is tabular if and only if its intuitionistic fragment is tabular.

Proposition 10.5.2 *Let $L \in \mathcal{E}\mathbf{N4}^{\perp}$. L is tabular if and only if $\sigma(L)$ is tabular.*

Proof. Let $L = L\mathcal{A}$ for some finite $\mathbf{N4}^{\perp}$-lattice $\mathcal{A}$. Then $\sigma(L) = L\mathcal{A}_{\bowtie}$ by Proposition 10.1.1. Thus, if L is tabular, $\sigma(L)$ is tabular too.

Assume $\sigma(L) = L\mathcal{B}$ for some finite Heyting algebra $\mathcal{B}$. Since Heyting algebras are congruence distributive, all finitely generated subdirectly irreducible $\sigma(L)$-models belong to $\mathsf{HS}(\mathcal{B})$. Consequently, $\sigma(L)$ has only a finite number of finitely generated subdirectly irreducible models $\{\mathcal{B}_1, \ldots, \mathcal{B}_n\}$ and all these models are finite. For every finitely generated subdirectly irreducible model $\mathcal{A}$ of L, $\mathcal{A}_{\bowtie}$ is a subdirectly irreducible model of $\sigma(L)$ by Corollary 9.2.11, and $\mathcal{A}_{\bowtie}$ is finitely generated by Lemma 10.4.1. Consequently, every element of $Mod_{fsi}(L)$ is isomorphic to some $\mathcal{A} \in S^{\bowtie}(\mathcal{B}_i)$, $i = 1, \ldots, n$. Thus, $Mod_{fsi}(L)$ contains only a finite number of algebras up to isomorphism, which means that L is tabular.

□

This fact and Theorem 10.5.1 yield the description of tabular logics in $\mathcal{E}\mathbf{N4}^{\perp}$.

Theorem 10.5.3 *Let $L \in \mathcal{E}\mathbf{N4}^{\perp}$. L is tabular if and only if L is of finite codimension in $\mathcal{E}\mathbf{N4}^{\perp}$.*

Proof. If L is tabular, up to isomorphism it has only a finite number of finitely generated subdirectly irreducible models $\{\mathcal{A}_1, \ldots, \mathcal{A}_n\}$. This follows from the fact that the variety of $\mathbf{N4}^{\perp}$-lattices is congruence distributive and the above-mentioned result by Jonsson. Any extension of L is determined by some subset of $\{\mathcal{A}_1, \ldots, \mathcal{A}_n\}$. Therefore, L has finitely many extensions, in particular, it has a finite codimension.

Let L have a finite codimension in $\mathcal{E}\mathbf{N4}^{\perp}$. By Proposition 10.1.12 the mapping η° embeds the lattice $\mathcal{E}\sigma(L)$ into $\mathcal{E}L$. Therefore, $\sigma(L)$ also has a finite codimension. Now the tabularity of $\sigma(L)$ follows from Theorem 10.5.1 and the tabularity of L from Proposition 10.5.2.

□

In [55], L.L. Maksimova proved that there are exactly three pretabular intermediate logics.

Theorem 10.5.4 [55] *The class* Int *contains exactly three pretabular logics:*

$$\begin{array}{lcl}
\mathbf{LC} & := & \mathbf{Li} + \{(p \to q) \vee (q \to p)\}, \\
\mathbf{LJ} & := & \mathbf{Li} + \{p \vee (p \to (q \vee \neg q))\}, \\
\mathbf{LH} & := & \mathbf{Li} + \{\neg p \vee \neg\neg p,\ p \vee (p \to (q \vee (q \to (r \vee \neg r))))\}.
\end{array}$$

□

Proposition 10.5.5 *Let $L \in \mathcal{E}\mathbf{N4}^{\perp}$. L is pretabular if and only if $L = L_{ne}$ and $\sigma(L)$ is pretabular.*

Proof. Let L be pretabular. Then $\sigma(L)$ is pretabular by Proposition 10.5.2. Indeed, every proper extension of $\sigma(L)$ is of the form $\sigma(L')$ for some proper extension L' of L. By assumption L' is tabular, whence $\sigma(L')$ is tabular. At the same time, $\sigma(L)$ is not tabular since L is not tabular.

Assume $L \neq L_{ne}$. We have $\sigma(L_{ne}) = \sigma(L)$ and $L \subseteq L_{ne}$, consequently, L_{ne} is a non-tabular proper extension of L, which conflicts with the assumption that L is pretabular.

Assume $L = L_{ne}$ and $\sigma(L)$ is pretabular. Then $L \in \mathcal{E}\mathbf{N3}^{\circ}$. By Proposition 10.1.12, every proper L-extension is of the form $\eta^{\circ}(L')$ for some proper $\sigma(L)$-extension L'. By assumption L' is tabular, consequently, $\eta^{\circ}(L')$ is tabular by Proposition 10.5.2. Thus, L is pretabular.

□

By Item 6 of Proposition 10.3.1 if $L = L_{ne}$, then $L = \eta^{\circ}\sigma(L)$. This fact and the last two assertions imply.

Theorem 10.5.6 *The class $\mathcal{E}\mathbf{N4}^{\perp}$ contains exactly three pretabular logics: $\eta^{\circ}(\mathbf{LC})$, $\eta^{\circ}(\mathbf{LJ})$, and $\eta^{\circ}(\mathbf{LH})$.*

□

We say that a logic L possesses the *Craig interpolation property* (*CIP*) if $\varphi \to \psi \in L$ implies that there exists a formula χ such that $\varphi \to \chi \in L$ and $\chi \to \psi \in L$, and χ has occurrences of common variables of φ and ψ only.

The algebraic counterpart of this property is as follows. Let $\mathcal{K}$ be a class of algebras. We say that $\mathcal{K}$ has an *amalgamation property* if for any algebras $\mathcal{A}_0, \mathcal{A}_1, \mathcal{A}_2 \in \mathcal{K}$ and monomorphisms $i_1 : \mathcal{A}_0 \hookrightarrow \mathcal{A}_1$ and $i_2 : \mathcal{A}_0 \hookrightarrow \mathcal{A}_2$, there exist an algebra $\mathcal{A} \in \mathcal{K}$ and monomorphisms $\varepsilon_1 : \mathcal{A}_1 \hookrightarrow \mathcal{A}$ and $\varepsilon_2 : \mathcal{A}_2 \hookrightarrow \mathcal{A}$ such that $\varepsilon_1 i_1 = \varepsilon_2 i_2$, i.e., the diagram

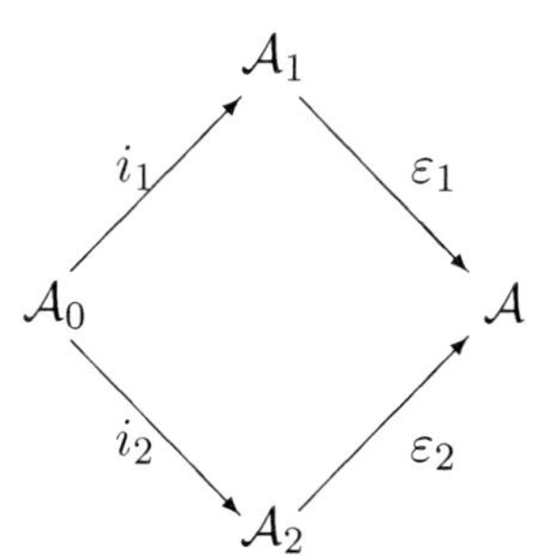

commutes. The triple $(\mathcal{A}, \varepsilon_1, \varepsilon_2)$ is called an *amalgam* of $\mathcal{A}_1$ and $\mathcal{A}_2$ over $\mathcal{A}_0$.

It was proved in [56] that an intermediate logic possesses CIP if and only if the corresponding variety of Heyting algebras has the amalgamation property. The next statement can be obtained by a natural modification of the proof from [56].

Proposition 10.5.7 *Let $L \in \mathcal{E}\mathbf{N4}^{\perp}$. L possesses CIP if and only if the variety $Var(L)$ has the amalgamation property.*

Proof. Assume that L possesses CIP and prove the amalgamation property for $Var(L)$. Let $\mathcal{A}_0, \mathcal{A}_1, \mathcal{A}_2 \in Var(L)$ be such that $\mathcal{A}_0$ is a common subalgebra of $\mathcal{A}_1$ and $\mathcal{A}_2$. Assign to each element $a \in \mathcal{A}_i$, $i = 0, 1, 2$, a propositional variable p_a^i. Assume additionally that $p_a^0 = p_a^1 = p_a^2$ for $a \in \mathcal{A}_0$ and all other variables are different. Denote by $\mathcal{F}_i$ the set of formulas with variables p_a^i. Put $\mathcal{F} := \mathcal{F}_0 \cup \mathcal{F}_1 \cup \mathcal{F}_2$. Let $v : \mathcal{F} \to \mathcal{A}_1 \cup \mathcal{A}_2$ be such that $v(p_a^i) = a$. For $i = 1, 2$, denote

$$T_i := \{\varphi \in \mathcal{F}_i \mid v(\varphi) \in D^{\mathcal{A}_i}\}.$$

It is easy to see that $\mathcal{F}_i \cap L \subseteq T_i$ and that the sets T_i are closed under *modus ponens*. Put

$$T := \{\varphi \in \mathcal{F} \mid T_1 \cup T_2 \vdash_L \varphi\}.$$

Lemma 10.5.8 *Let $\{i, j\} = \{1, 2\}$, $\varphi \in \mathcal{F}_i$, and $\psi \in \mathcal{F}_j$. Then*

$$\varphi \to \psi \in T \iff \exists \chi \in \mathcal{F}_0 (v(\varphi) \preceq v(\chi) \text{ and } v(\chi) \preceq v(\psi)).$$

Proof. If $v(\varphi) \preceq v(\chi)$ and $v(\chi) \preceq v(\psi)$, then $v(\varphi \to \chi) \in D^{\mathcal{A}_i}$ and $v(\chi \to \psi) \in D^{\mathcal{A}_j}$. Therefore, $\varphi \to \chi, \chi \to \psi \in T$, whence $\varphi \to \psi \in T$.

Conversely, let $\varphi \to \psi \in T$. There are finite subsets $\Gamma_i \subseteq T_i$ and $\Gamma_j \subseteq T_j$ such that $\Gamma_i, \Gamma_j \vdash_L \varphi \to \psi$. By Deduction Theorem

$$\vdash_L \bigwedge \Gamma_i \to (\bigwedge \Gamma_j \to (\varphi \to \psi)),$$

which is equivalent to

$$\vdash_L \bigwedge \Gamma_i \wedge \varphi \to (\bigwedge \Gamma_j \to \psi).$$

According to CIP, there is a $\chi \in \mathcal{F}_0$ such that $\vdash_L \bigwedge \Gamma_i \wedge \varphi \to \chi$ and $\vdash_L \chi \to (\bigwedge \Gamma_j \to \psi)$. We obtain $\Gamma_i \vdash_L \varphi \to \chi$ and $\Gamma_j \vdash_L \chi \to \psi$. Since $\Gamma_i \subseteq T_i$, we have $\varphi \to \chi \in T_i$. Consequently, $v(\varphi \to \chi) \in D^{\mathcal{A}_i}$, i.e., $v(\varphi) \preceq v(\chi)$. Similarly, $v(\chi) \preceq v(\psi)$.

□

Let $\psi \in \mathcal{F}_j$ be such that $T \vdash_L \psi$. Take some $\varphi \in L \cap \mathcal{F}_i$, then $T \vdash_L \varphi \to \psi$. By the previous lemma there is a formula $\chi \in \mathcal{F}_0$ such that $v(\varphi) \preceq v(\chi)$ and $v(\chi) \preceq v(\psi)$. It follows from $\varphi \in L \cap \mathcal{F}_i$ that $v(\varphi) \in D^{\mathcal{A}_i}$. Taking into account the first inequality we obtain $v(\chi) \in D^{\mathcal{A}_i}$, or, equivalently, $v(\chi \to \chi) = v(\chi)$. Consequently, we have also $v(\chi) \in D^{\mathcal{A}_j}$. Hence, $v(\psi) \in D^{\mathcal{A}_j}$, i.e., $\psi \in T_j$. In this way, we have proved $T \cap \mathcal{F}_j = T_j$, $j = 1, 2$.

Define on $\mathcal{F}$ the relation $\cong := \{(\varphi, \psi) \mid \varphi \Leftrightarrow \psi \in T\}$. Since the provable strong equivalence $\Leftrightarrow$ has in $\mathbf{N4}^{\perp}$ the congruence properties, $\cong$ is a congruence on $\mathcal{F}$.

Let $\varphi, \psi \in \mathcal{F}_i$. The equality $T \cap \mathcal{F}_i = T_i$ implies

$$\varphi \cong \psi \text{ if and only if } v(\varphi) = v(\psi) \qquad (*)$$

Put $\mathcal{A} := \mathcal{F}/\cong$. Clearly, $\mathcal{A}$ is an $\mathbf{N4}^{\perp}$-lattice. Let mappings $\varepsilon_i : \mathcal{A}_i \to \mathcal{A}$, $i = 1, 2$, be given by the rules

$$\varepsilon_i(a) := [p_a^i]_{\cong}, \; a \in \mathcal{A}_i.$$

The mapping ε_i is a homomorphism since $\cong$ is a congruence, and is one-to-one in view of $(*)$. Since $p_a^0 = p_a^1 = p_a^2$, we have $\varepsilon_1(a) = \varepsilon_2(a)$ for $a \in \mathcal{A}_0$. Thus, $(\mathcal{A}, \varepsilon_1, \varepsilon_2)$ is an amalgam of $\mathcal{A}_1$ and $\mathcal{A}_2$ over $\mathcal{A}_0$.

Prove the inverse implication. Assume that $Var(L)$ has the amalgamation property. We need one more lemma from [56].

Lemma 10.5.9 *Let $\mathcal{A}_0$ is a common subalgebra in Heyting algebras $\mathcal{A}_1$ and $\mathcal{A}_2$. If $a \in \mathcal{A}_1$ and $b \in \mathcal{A}_2$ are such that there is no $c \in \mathcal{A}_0$ with $a \leq_{\mathcal{A}_1} c$ and $c \leq_{\mathcal{A}_2} b$, then there are filters F_1 on $\mathcal{A}_1$ and F_2 on $\mathcal{A}_2$ such that $a \in F_1$, $b \notin F_2$, and $\mathcal{A}_0 \cap F_1 = \mathcal{A}_0 \cap F_2$.*

Proof. We write $\leq_i$ for $\leq_{\mathcal{A}_i}$, $i = 1, 2$. Consider the sets

$$\nabla := \{z \in \mathcal{A}_0 \mid a \leq_1 z\} \text{ and } \Delta := \{z \in \mathcal{A}_0 \mid z \leq_2 b\}.$$

By assumption $\nabla \cap \Delta = \varnothing$. Consider an ideal on $\mathcal{A}_2$ generated by b, i.e., the set $I = \{x \in \mathcal{A}_2 \mid x \leq_2 b\}$. Then $F_2 := \mathcal{A}_2 \setminus I$ is a filter on $\mathcal{A}_2$ such that $b \notin F_2$. Put

$$F_0 := F_2 \cap \mathcal{A}_0 \text{ and } I_0 := I \cap \mathcal{A}_0.$$

It is clear that $F_0 = \mathcal{A}_0 \setminus I_0$, $\nabla \subseteq F_0$, and $\Delta \subseteq I_0$. Further, consider a filter on $\mathcal{A}_1$ generated by $F_0 \cup \{a\}$, i.e.,

$$F_1 = \{x \in \mathcal{A}_1 \mid a \wedge z \leq_1 x \text{ for some } z \in F_0\}.$$

We claim that $F_1 \cap I_0 = \varnothing$. Indeed, if $x \in F_1 \cap I_0$, then $a \wedge z \leq_1 x$ for $z \in F_0$. The latter means that $a \leq_1 z \to x$. Since $x \in I_0 \subseteq \mathcal{A}_0$ and $z \in \mathcal{A}_0$, we have $z \to x \in \nabla \subseteq F_0$. Taking into account $z \in F_0$ we obtain $x \in F_0$, which conflicts with $F_0 \cap I_0 = \varnothing$.

From $F_0 \subseteq F_1$ and $F_1 \cap I_0 = \varnothing$ we immediately infer that $F_1 \cap \mathcal{A}_0 = F_0$, i.e., $\mathcal{A}_0 \cap F_1 = \mathcal{A}_0 \cap F_2$.

□

We adopt this statement for $\mathbf{N4}^{\perp}$-lattices.

Lemma 10.5.10 *Let an $\mathbf{N4}^{\perp}$-lattice $\mathcal{A}_0$ is a common subalgebra in $\mathbf{N4}^{\perp}$-lattices $\mathcal{A}_1$ and $\mathcal{A}_2$. If $a \in \mathcal{A}_1$ and $b \in \mathcal{A}_2$ are such that there is no $c \in \mathcal{A}_0$ with $a \preceq_{\mathcal{A}_1} c$ and $c \preceq_{\mathcal{A}_2} b$, then there are* sffk *$\Phi_1$ on $\mathcal{A}_1$ and Φ_2 on $\mathcal{A}_2$ such that $a \in \Phi_1$, $b \notin \Phi_2$, and $\mathcal{A}_0 \cap \Phi_1 = \mathcal{A}_0 \cap \Phi_2$.*

Proof. Without loss of generality we may assume that $\mathcal{A}_i \in S^{\bowtie}(\mathcal{B}_i)$, $i = 0, 1, 2$, $a = (a_1, a_2)$ and $b = (b_1, b_2)$. As we can see from Proposition 8.4.2 the algebras $\mathcal{B}_0$, $\mathcal{B}_1$, $\mathcal{B}_2$ and the elements $a_1 \in \mathcal{B}_1$ and $b_1 \in \mathcal{B}_2$ satisfy the conditions of the previous lemma. Let the conclusion of this lemma hold for filters F_1 on $\mathcal{B}_1$ and F_2 on $\mathcal{B}_2$. Then *sffk* $\Phi_1 := F_1^{\bowtie}$ and $\Phi_2 := F_2^{\bowtie}$ satisfy all the desired conditions. We check only the equality $\mathcal{A}_0 \cap \Phi_1 = \mathcal{A}_0 \cap \Phi_2$. If $(a, b) \in \mathcal{A}_0 \cap \Phi_1$, then $a \in \mathcal{B}_0 \cap F_1 = \mathcal{B}_0 \cap F_2$, whence $(a, b) \in F_2^{\bowtie}$. The inverse inclusion can be checked similarly.

□

Let x, y, z be disjoint tuples of variables and let for $\varphi(x, y)$ and $\psi(x, z)$ there is no $\chi(x)$ such that $\varphi(x, y) \to \chi(x) \in L$ and $\chi(x) \to \psi(x, z) \in L$. The relation $\varphi \to \psi \in L$ is equivalent to $\varphi \preceq \psi \in Eq(Var(L))$. Therefore, there is no $\chi(x)$ such that $Var(L) \models \varphi(x, y) \preceq \chi(x) \preceq \psi(x, z)$. Let $\mathcal{A}_x$ be a free algebra in $Var(L)$ with free generators x and $\mathcal{A}_{x,y,z}$ a free algebra with generators x, y, z. Consider $\mathcal{A}_x$ as a subalgebra of $\mathcal{A}_{x,y,z}$.

According to Lemma 10.5.10 there are *sffk* Φ_1 and Φ_2 on $\mathcal{A}_{x,y,z}$ such that $[\varphi(x, y)] \in \Phi_1$, $[\psi(x, z)] \notin \Phi_2$, and $\mathcal{A}_x \cap \Phi_1 = \mathcal{A}_x \cap \Phi_2$. Here $[\phi]$ denotes the element of the free algebra corresponding to the formula ϕ.

Put $\mathcal{A}_1 := \mathcal{A}_{x,y,z}/\Phi_1$ and $\mathcal{A}_2 := \mathcal{A}_{x,y,z}/\Phi_2$. Then $[\varphi(x, y)]/\Phi_1 \in D^{\mathcal{A}_1}$ and $[\psi(x, z)]/\Phi_2 \notin D^{\mathcal{A}_2}$. Recall that for any $u, v \in \mathcal{A}_x$, the equality $u/\Phi_1 = v/\Phi_1$ is equivalent to $u \Leftrightarrow v \in \Phi_1$, which is equivalent in turn to $u \Leftrightarrow v \in \Phi_2$ in view of $\mathcal{A}_x \cap \Phi_1 = \mathcal{A}_x \cap \Phi_2$. Thus,

$$u/\Phi_1 = v/\Phi_1 \text{ if and only if } u/\Phi_2 = v/\Phi_2.$$

Therefore, there is a natural embedding $i_2(u/\Phi_1) := u/\Phi_2$ of $\mathbf{N4}^{\perp}$-lattice $\mathcal{A}_0 := \{u/\Phi_1 \mid u \in \mathcal{A}_x\} \subseteq \mathcal{A}_1$ into $\mathbf{N4}^{\perp}$-lattice $\mathcal{A}_2$.

By the amalgamation property of $Var(L)$ there are $\mathcal{A} \in Var(L)$ and monomorphisms $\varepsilon_i : \mathcal{A}_i \hookrightarrow \mathcal{A}$, $i = 1, 2$, such that $\varepsilon_1 \restriction \mathcal{A}_0 = \varepsilon_2 i_2$. Put

$$h([x]) := \varepsilon_1([x]/\Phi_1),\ h([y]) := \varepsilon_1([y]/\Phi_1),\ h([z]) := \varepsilon_2([z]/\Phi_2),$$

where $h([x]) := h([p_1]), \dots, h([p_n])$ for $x = p_1, \dots, p_n$, etc. Extend this mapping to a homomorphism $h : \mathcal{A}_{x,y,z} \to \mathcal{A}$. We obtain

$$h([\varphi(x,y)]) = \varepsilon_1([\varphi(x,y)]/\Phi_1) \in D^{\mathcal{A}},\ h([\psi(x,z)]) = \varepsilon_2([\psi(x,z)]/\Phi_2) \notin D^{\mathcal{A}}.$$

Therefore, $Var(L) \not\models \varphi(x,y) \preceq \psi(x,z)$, i.e., $\varphi(x,y) \to \psi(x,z) \notin L$.

□

The description of intermediate logics with CIP looks as follows:

Theorem 10.5.11 [56] *In the class* Int, *there exists exactly 7 logics with* CIP*:*

$$\begin{array}{lcl} L_1 & := & \mathbf{Li}, \\ L_2 & := & \mathbf{Li} + \{\neg p \vee \neg\neg p\}, \\ L_3 & := & \mathbf{Li} + \{p \vee (p \to (q \vee \neg q))\}, \\ L_4 & := & L_3 + \{(p \to q) \vee (q \to p) \vee (p \leftrightarrow \neg q)\}, \\ L_5 & := & L_3 + \{\neg p \vee \neg\neg p\}, \\ L_6 & := & \mathbf{Li} + \{(p \to q) \vee (q \to p)\}, \\ L_7 & := & \mathbf{Lk}. \end{array}$$

□

We use the facts presented above to find all logics in $\mathcal{E}\mathbf{N4}^{\perp}$ possessing CIP. First of all, we notice that the intuitionistic fragment of a logic in $\mathcal{E}\mathbf{N4}^{\perp}$ inherits CIP.

Proposition 10.5.12 *If $L \in \mathcal{E}\mathbf{N4}^{\perp}$ possesses CIP, then $\sigma(L)$ possesses CIP.*

Proof. According to Proposition 10.5.7 the variety $Var(L)$ has the amalgamation property. We show that the class of $\sigma(L)$-models has the amalgamation property too. Let Heyting algebras $\mathcal{A}_0$, $\mathcal{A}_1$, and $\mathcal{A}_2$ model $\sigma(L)$, and $i_1 : \mathcal{A}_0 \hookrightarrow \mathcal{A}_1$ and $i_2 : \mathcal{A}_0 \hookrightarrow \mathcal{A}_2$ are embeddings.

Consider twist-structures $(\mathcal{A}_0)^{\bowtie}_{\circ}$, $(\mathcal{A}_1)^{\bowtie}_{\circ}$, and $(\mathcal{A}_2)^{\bowtie}_{\circ}$. According to Proposition 10.1.3 these twist-structures are models of $\eta^{\circ}\sigma(L)$. Logics L and $\eta^{\circ}\sigma(L)$ have the same intuitionistic fragment, therefore, $L \subseteq \eta^{\circ}\sigma(L)$ by Proposition 10.1.9. Thus, $(\mathcal{A}_0)^{\bowtie}_{\circ}$, $(\mathcal{A}_1)^{\bowtie}_{\circ}$, and $(\mathcal{A}_2)^{\bowtie}_{\circ}$ model L. Moreover,

$i_1^{\bowtie} : (\mathcal{A}_0)_{\circ}^{\bowtie} \hookrightarrow (\mathcal{A}_1)_{\circ}^{\bowtie}$ and $i_2^{\bowtie} : (\mathcal{A}_0)_{\circ}^{\bowtie} \hookrightarrow (\mathcal{A}_2)_{\circ}^{\bowtie}$ are embeddings according to Proposition 9.2.14. By assumption one can find an amalgam $(\mathcal{A}, \varepsilon_1, \varepsilon_2)$ of $(\mathcal{A}_1)_{\circ}^{\bowtie}$ and $(\mathcal{A}_2)_{\circ}^{\bowtie}$ over $(\mathcal{A}_0)_{\circ}^{\bowtie}$. In particular, $\varepsilon_1 i_1^{\bowtie} = \varepsilon_2 i_2^{\bowtie}$. Using functor properties of $(-)_{\bowtie}$ we obtain

$$(\varepsilon_1)_{\bowtie} i_1 = (\varepsilon_1 i_1^{\bowtie})_{\bowtie} = (\varepsilon_2 i_2^{\bowtie})_{\bowtie} = (\varepsilon_2)_{\bowtie} i_2.$$

Again by Proposition 9.2.14 $(\varepsilon_1)_{\bowtie} : \mathcal{A}_1 \hookrightarrow \mathcal{A}_{\bowtie}$ and $(\varepsilon_2)_{\bowtie} : \mathcal{A}_2 \hookrightarrow \mathcal{A}_{\bowtie}$ are embeddings. Thus, $(\mathcal{A}_{\bowtie}, (\varepsilon_1)_{\bowtie}, (\varepsilon_2)_{\bowtie})$ is an amalgam of $\mathcal{A}_1$ and $\mathcal{A}_2$ over $\mathcal{A}_0$. □

The implication in the above proposition can be reversed for some special logics from the interval $\sigma^{-1}(L)$.

Proposition 10.5.13 *Let $L \in$ Int. If L possesses CIP, then the logics $\eta(L)$, $\eta^3(L)$, $\eta^n(L)$, and $\eta^{\circ}(L)$ possess CIP.*

Proof. Recall that for any logic $L' \in \sigma^{-1}(L)$ for any $\mathbf{N4}^{\perp}$-lattice $\mathcal{B}$ modelling L', $\mathcal{B}$ is isomorphic to a twist-structure over a Heyting algebra $\mathcal{A}$ such that $\mathcal{A} \models L$.

Let $\mathcal{A}_0, \mathcal{A}_1, \mathcal{A}_2 \models L$. Consider twist-structures $\mathcal{B}_i := Tw(\mathcal{A}_i, \nabla_i, \Delta_i)$, $i = 0, 1, 2$, and their embeddings $i_1 : \mathcal{B}_0 \hookrightarrow \mathcal{B}_1$ and $i_2 : \mathcal{B}_0 \hookrightarrow \mathcal{B}_2$. Then $(i_1)_{\bowtie}$ embeds $\mathcal{A}_0$ into $\mathcal{A}_1$, and $(i_2)_{\bowtie}$ embeds $\mathcal{A}_0$ into $\mathcal{A}_2$. Since L possesses CIP, there is an amalgam $(\mathcal{A}, \varepsilon_1, \varepsilon_2)$ of $\mathcal{A}_1$ and $\mathcal{A}_2$ over $\mathcal{A}_0$. Proposition 9.2.15 allows us to lift up monomorphisms ε_1 and ε_2 so that we obtain the following diagram.

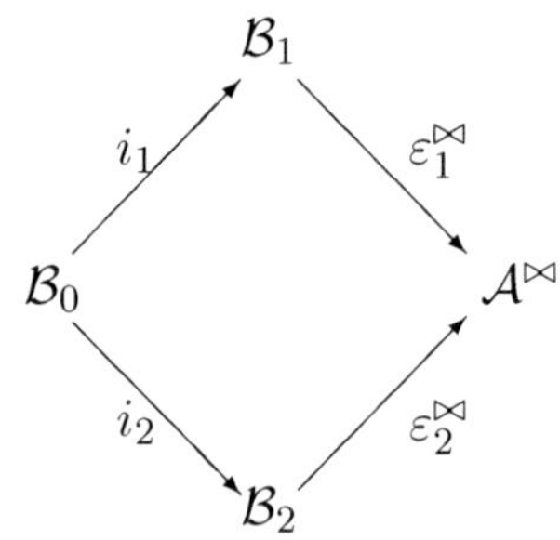

The functor properties of $(-)^{\bowtie}$ guarantee that this diagram commutes. The full twist-structure $\mathcal{A}^{\bowtie}$ is a model of $\eta(L)$ by Proposition 10.1.3 and we have thus proved that $\eta(L)$ possesses CIP.

If $\mathcal{B}_i \models \eta^3(L)$, then $\Delta_i = \{0\}$, $i = 0, 1, 2$, by Proposition 10.2.6. In this case, $\varepsilon_i^{\bowtie}$ embeds $\mathcal{B}_i$ into $\mathcal{A}_{exp}^{\bowtie}$. Again by Proposition 10.2.6 $\mathcal{A}_{exp}^{\bowtie} \models \eta^3(L)$, which proves that $\eta^3(L)$ possesses CIP.

In a similar way, applying Propositions 10.2.6 and 10.1.3, we prove that $\eta^n(L)$ and $\eta^{\circ}(L)$ possess CIP.

□

It turns out that all logic from $\sigma^{-1}(L)$ that can possess CIP were listed in the last proposition.

Proposition 10.5.14 *If $L \in \mathcal{E}\mathbf{N4}^{\perp}$ and L possesses CIP, then L is of the form $\eta(L')$, $\eta^3(L')$, $\eta^n(L')$, or $\eta^{\circ}(L')$ for some $L' \in \mathsf{Int}$.*

Proof. Let L possesses CIP. If $L = L_{ne}$, then $L = \eta^{\circ}\sigma(L)$. Let $L \neq L_{ne}$. Then one of the formulas $\neg(p \wedge \sim p)$ or $\neg\neg(p \vee \sim p)$ does not belong to L.

Assume $\neg(p \wedge \sim p) \in L$ and $\neg\neg(p \vee \sim p) \notin L$ and prove that $L = \eta^3\sigma(L)$ in this case.

Consider the Lindenbaum twist-structure of L, $\mathcal{L}_L^{\bowtie}$. Recall that

$$|\mathcal{L}_L^{\bowtie}| = \{([\varphi]_L, [\sim \psi]_L) \mid \psi \in [\varphi]_L\},$$

where $[\varphi]_L$ is a coset of the equivalence

$$\varphi \equiv_L \psi \;\Leftrightarrow\; \varphi \leftrightarrow \psi \in L.$$

Thus, $\mathcal{L}_L^{\bowtie}$ is a twist-structure over the Heyting algebra $\mathcal{L}_L$ of all cosets $[\varphi]_L$. Note that $\neg(p \wedge \sim p) \in L$ implies the equality $\Delta(\mathcal{L}_L^{\bowtie}) = \{0\}$. Put

$$F := \{[\varphi]_L \mid \neg(p_0 \vee \sim p_0) \to \varphi \in L\}.$$

Obviously, F is a filter on $\mathcal{L}_L$ and $F^{\bowtie}$ is an *sffk* on $\mathcal{L}_L^{\bowtie}$. Consider a quotient algebra $\mathcal{L}_0 := \mathcal{L}_L^{\bowtie}/F^{\bowtie}$.

Let $([\varphi_0]_L, [\psi_0]_L), ([\varphi_1]_L, [\psi_1]_L) \in \mathcal{L}_L^{\bowtie}$. By Proposition 9.2.3 the equality $([\varphi_0]_L, [\psi_0]_L)/F^{\bowtie} = ([\varphi_1]_L, [\psi_1]_L)/F^{\bowtie}$ is equivalent to

$$([\varphi_0]_L, [\psi_0]_L) \Leftrightarrow ([\varphi_1]_L, [\psi_1]_L) \in F^{\bowtie}.$$

By definition of $F^{\bowtie}$ the latter is equivalent to

$$[\varphi_0 \leftrightarrow \psi_0]_L, [\varphi_1 \leftrightarrow \psi_1]_L \in F.$$

Thus, an element of $\mathcal{L}_0$ can be identified with a pair of elements of the quotient $\mathcal{L}_L/F = \mathcal{L}_F$,

$$[([\varphi]_L, [\psi]_L)]_{F^{\bowtie}} = ([\varphi]_F, [\psi]_F).$$

Consider the element $a := ([p_0]_F, [\sim p_0]_F) \in \mathcal{L}_0$. Then

$$a \vee \sim a = ([p_0 \vee \sim p_0]_F, [p_0 \wedge \sim p_0]_F),$$

which means that $[p_0 \vee \sim p_0]_F \in \Delta(\mathcal{L}_0)$. By definition $[\neg(p_0 \vee \sim p_0)]_L \in F$, i.e., $[\neg(p_0 \vee \sim p_0)]_F = \neg[p_0 \vee \sim p_0]_F = 1$. Consequently, $[p_0 \vee \sim p_0]_F = 0$. We have thus proved that $\nabla(\mathcal{L}_0) = \mathcal{L}_F$, i.e., $\mathcal{L}_0$ is a special $\mathbf{N3}$-lattice.

Let us check the equality $L = L\mathcal{L}_0$. We have $L \subseteq L\mathcal{L}_0$, since $\mathcal{L}_0$ is a quotient of $\mathcal{L}_L^{\bowtie}$. Let $\mathcal{L}_0 \models \varphi(q_0, \ldots, q_n)$ and p_0 does not occur in φ. Then $[\varphi(q_0, \ldots, q_n)]_L \in F$, i.e.,

$$\neg(p_0 \vee \sim p_0) \to \varphi(q_0, \ldots, q_n) \in L.$$

By CIP we have $\neg(p_0 \vee \sim p_0) \to \phi \in L$ and $\phi \to \varphi(q_0, \ldots, q_n) \in L$, where ϕ is some formula without propositional variables, i.e., constructed from $\perp$ and logical connectives. Denote $\top := \sim \perp$. Then $\top \in \mathbf{N4}^{\perp}$. By an easy induction on the structure of formulas one can prove that $\phi \leftrightarrow \perp \in \mathbf{N4}^{\perp}$ or $\phi \leftrightarrow \top \in \mathbf{N4}^{\perp}$. In this way, we obtain $\neg(p_0 \vee \sim p_0) \to \perp \in L$ or $\top \to \varphi(q_0, \ldots, q_n) \in L$. In view of our assumption $\neg(p_0 \vee \sim p_0) \to \perp \notin L$, consequently, $\varphi(q_0, \ldots, q_n) \in L$.

We have thus proved that L is determined by the special $\mathbf{N3}$-lattice $\mathcal{L}_0$. By Corollary 10.2.10 L is a special explosive logic, i.e., a logic of the form $\eta^3(L')$.

Now, we assume $\neg(p \wedge \sim p) \notin L$ and $\neg\neg(p \vee \sim p) \in L$ and prove that $L = \eta^n \sigma(L)$. In this case, the Lindenbaum twist-structure $\mathcal{L}_L^{\bowtie}$ is normal. Let us consider the filter

$$F_1 := \{[\varphi]_L \mid (p_0 \wedge \sim p_0) \to \varphi \in L\}$$

and the quotient $\mathcal{L}_1 := \mathcal{L}_L^{\bowtie}/F_1^{\bowtie}$. The element $b = ([p_0]_{F_1}, [\sim p_0]_{F_1})$ is such that

$$b \wedge \sim b = ([p_0 \wedge \sim p_0]_{F_1}, [p_0 \vee \sim p_0]_{F_1}).$$

Since $[p_0 \wedge \sim p_0]_L \in F_1$, we obtain $[p_0 \wedge \sim p_0]_{F_1} = 1$ and $\Delta(\mathcal{L}_1) = \mathcal{L}_1$. Thus, $\mathcal{L}_1$ is a special normal $\mathbf{N4}^{\perp}$-lattice. Arguing as above, we check that $L = L\mathcal{L}_1$. Thus, L is determined by a special normal $\mathbf{N4}^{\perp}$-lattice. By Corollary 10.2.10 L is a special normal logic, i.e., a logic of the form $\eta^n(L')$.

It remains to consider the case when $\neg(p \wedge \sim p) \notin L$ and $\neg\neg(p \vee \sim p) \notin L$. We have $\neg((p \wedge \sim p) \wedge \neg(q \vee \sim q)) \notin L$. Indeed, this is equivalent to $(p \wedge \sim p) \to \neg\neg(q \vee \sim q) \notin L$. If $(p \wedge \sim p) \to \neg\neg(q \vee \sim q)$ belongs to L, by CIP we have

$$(p \wedge \sim p) \to \perp \in L \text{ or } \top \to \neg\neg(q \vee \sim q) \in L.$$

Both alternatives conflict with our assumption. Let us consider the filter

$$F_2 := \{[\varphi]_L \mid (p_0 \wedge \sim p_0) \wedge \neg(p_1 \vee \sim p_1) \to \varphi \in L\}$$

on $\mathcal{L}_L$. Arguing as above, we can prove that the quotient $\mathcal{L}_2 := \mathcal{L}_L^{\bowtie}/F_2^{\bowtie}$ is a characteristic model for L and that $\nabla(\mathcal{L}_2) = \Delta(\mathcal{L}_2) = \mathcal{L}_{F_2}$, i.e., $\mathcal{L}_2$ is a special $\mathbf{N4}^{\perp}$-lattice. Since L is determined by a special $\mathbf{N4}^{\perp}$-lattice, it is a special logic, i.e., $L = \eta\sigma(L)$.

□

The logics $\eta(L)$, $\eta^3(L)$, $\eta^n(L)$, and $\eta^\circ(L)$ are different for any L in Int. This fact and the above propositions imply

Theorem 10.5.15 *In $\mathcal{E}\mathbf{N4}^{\perp}$, there are exactly 28 non-trivial logics possessing CIP: $\eta(L_i)$, $\eta^3(L_i)$, $\eta^n(L_i)$, and $\eta^\circ(L_i)$, $i = 1, \ldots, 7$, where $L_1, \ldots, L_7$ are logics from Theorem 10.5.11.*

□

Chapter 11

Conclusion

Discussing the question "Why is paraconsistency worthy?" J.-Y. Béziau [11] emphasized that paraconsistent logic is an important contribution to the theory of negation and to modern logic in general. The distinction between triviality and inconsistency made in paraconsistent logic is similar to the distinction between implication and inference relation and allows one to elucidate new features of traditional logical notions. Suppose that the investigations presented in this book also contribute to the general theory of logical systems. For two explosive logics with the same positive fragment and with essentially different kinds of negation, we investigated how the lattice of extensions of a logic changes when the explosion axiom is deleted, i.e., if we pass from a logic to its paraconsistent analog. It turns out that in both cases the lattices of extensions extend in a rather regular manner. In the class of extensions of a paraconsistent logic, one can distinguish the subclass of explosive logics, i.e., the class of extensions of the original explosive logic; the subclass of logics, which can be used to represent the structures of contradictions in all extensions of the considered paraconsistent logic (see Remark after Proposition 10.2.5). Finally, all other logics can be obtained via a combination of logics from the above two subclasses. The manner of combination can be explicated via a suitable representation theory for algebras modelling the logics under consideration. Thus, admitting non-trivial inconsistent theories leads to quite natural and interesting mathematical structures arising. Moreover, the above results allow one to consider paraconsistent logic not as an alternative to traditional explosive logic, but as a more general setting for considering explosive systems. It would be natural to continue the study of lattices of extensions of paraconsistent logics for some other concepts of negation. According to the author's opinion the most natural candidate for such investigation is the subminimal negation by D. Vakarelov [111, 112]. The subminimal logic has the same positive fragment as intuitionistic logic. Its negation is a natural generalization of minimal negation and can be defined via two constants $\bot$ and $\top$ such that $\bot \to \top$ holds by the formula

$$\neg p \leftrightarrow (p \to \bot) \wedge \top.$$

In this case, the constants are definable through negation as $\bot := \neg(p \to p)$ and $\top := \neg\neg(p \to p)$. To obtain minimal logic from the subminimal one we have to add the axiom $\top$. As other candidates for such investigations one could consider the negation as impossibility by K. Došen [24, 25], different kinds of negations considered by M. Dunn in [26] and finally, the regular and co-regular negations by D. Vakarelov.

At the same time, one can continue the study of the classes of extensions of minimal logic and of Nelson's logic $\mathbf{N4}^{\bot}$ using the coordinate system depicted in this work. It should be mentioned that there are interesting results on minimal logic extensions that are not reflected in this book. L.L. Maksimova [57, 58] studied interpolation and definability properties in extensions of positive and minimal logic. She constructed several series of logics with Craig interpolation property in the class of minimal logic extensions. However, it is still unknown whether the total number of logics with Craig interpolation property in this class of logics is finite. M.V. Stukacheva studied the disjunction property in the class of minimal logic extensions [103, 104] and transferred the technique of canonical formulas by M. Zakharyaschev to this class of logics [105].

And the very final remark concerns the technique of twist-structures. In the second part of the book we used twist-structures to solve algebraization problems for Nelson's logics $\mathbf{N4}$ and $\mathbf{N4}^{\bot}$. The main difficulty here was connected with the fact that Nelson's logics are not closed under the replacement rule. But they are closed under weak forms of replacement rules:

$$\frac{\varphi \leftrightarrow \psi}{\chi^{+}(\varphi) \leftrightarrow \chi^{+}(\psi)} \quad \text{and} \quad \frac{\varphi \leftrightarrow \psi,\ \sim\varphi \leftrightarrow \sim\psi}{\chi(\varphi) \leftrightarrow \chi(\psi)},$$

where χ^{+} is a $\sim$-free formula. Namely, these rules allow one to define the semantics for Nelson's logics in terms of twist-structures. The first of these rules and the fact that $\sim$-free fragment of $\mathbf{N4}$ ($\mathbf{N4}^{\bot}$) is equal to positive (intuitionistic) logic imply that implicative lattices (Heyting algebras) should be taken as basic algebras of twist-structures for $\mathbf{N4}$ ($\mathbf{N4}^{\bot}$). The number of premises in the second rule corresponds to the number of components in elements of twist-structures. Finally, the strong negation axioms determine the twist-operations. A more general situation, where a technique of this kind could be applied would be as follows. Assume that we have a logic L in a language $\mathcal{L}$, $\mathcal{L}_0$ is a sublanguage of $\mathcal{L}$ and L_0 is an $\mathcal{L}_0$-fragment of L. Further, let the logic L_0 be extensional, the class of algebras $\mathcal{K}_0$ provides an algebraic semantics for L_0, and the logic L is closed under the rule:

$$\frac{\varphi \leftrightarrow \psi}{\chi^0(\varphi) \leftrightarrow \chi^0(\psi)}, \quad \text{where } \chi^0 \text{ is an } \mathcal{L}_0\text{-formula.}$$

Finally, assume that there are formulas $\alpha_1(p)$, ..., $\alpha_n(p)$ such that L is closed under the following form of replacement rule:

$$\frac{\alpha_1(\varphi) \leftrightarrow \alpha_1(\psi), \ldots, \alpha_n(\varphi) \leftrightarrow \alpha_n(\psi)}{\chi(\varphi) \leftrightarrow \chi(\psi)},$$

where χ is an arbitrary $\mathcal{L}$-formula. Note that in case of **N4** and **N4**$^\perp$ we have $n = 2$, $\alpha_1(p) = p$ and $\alpha_2(p) = \sim p$.

In this case, one can try to construct a semantics for the logic L in terms twist-structures over algebras from $\mathcal{K}_0$ with the number of components equalling n. Naturally, to define twist-operations some reduction axioms for connectives from $\mathcal{L} \setminus \mathcal{L}_0$ are needed. Possibly, some additional conditions should be imposed on the form of formulas α_1, ..., α_n. It is interesting to look for natural examples of logics, which are closed under rules of the form described above and to define semantics for such logics in terms of twist-structures. It may also be interesting to develop a general theory of twist-structures.

Bibliography

[1] A. Almukdad and D. Nelson. *Constructible falsity and inexact predicates.* Journal of Symbolic Logic, **49**, No. 1, 1984, 231–233.

[2] E.H. Alves, A.M. Sette. *On the equivalence between some systems of non-classical logic.* Bulletin of the Section of Logic, **25**, No. 2, 1996, 68–72.

[3] A. Avron. *On an implicative connective of RM.* Notre Dame Journal of Formal Logic, **27**, No. 2, 1986, 201–209.

[4] A. Avron. *Natural 3-valued logics: characterization and proof theory.* Journal of Symbolic Logic, **56**, No. 1, 1991, 276–294.

[5] D. Batens. *A completeness proof method for extensions of the implicational fragment of the propositional calculus.* Notre Dame Journal of Formal Logic, **21**, No. 3, 1980, 509–517.

[6] D. Batens. *Paraconsistent extensional propositional logics.* Logique et Analyse, Nouv. Ser., **23**, 1980, 195–234.

[7] D. Batens, K. De Clercq, N. Kurtonina. *Embedding and interpolation for some paralogics. The propositional case.* Reports on Mathematical Logic, **33**, 1999, 29–44.

[8] N.D. Belnap. *A useful four-valued logic.* In: G. Epstein and M.J. Dunn (eds.). Modern Uses of Multiple-Valued Logic, Oriel Press, 1977, 7–37.

[9] N.D. Belnap. *How computers should think.* In: G. Ryle (ed.). Contemporary Aspects of Philosophy, Oriel Press, 1977, 30–56.

[10] P. Bernays. *Review of [20] and [21].* Journal of Symbolic Logic, **18**, No. 1, 1953, 266–268.

[11] J.-Y. Béziau. *What is paraconsistent logic?* in: D. Batens et al. (eds.) Frontiers of Paraconsistent Logic, Research Studies Press, Baldock, 2000, 95–111.

[12] F. Borceux. *Handbook of categorical algebra 1. Basic category theory.* Cambridge University Press, Cambridge, 1994.

[13] M. Božić and K. Dośen. *Models for Normal Intuitionistic Modal Logics.* Studia Logica, **43**, 1984, 217–245.

[14] S. Burris, H.P. Sankappanavar. *A course in universal algebra.* Graduate Texts in Math., **78**, New York, Springer, 1981.

[15] A. Chagrov, M. Zakharyaschev. *The undecidability of disjunction property of propositional logics and other related problems.* Journal of Symbolic Logic, **58**, No. 3, 1999, 967–1003.

[16] A. Chagrov, M. Zakharyaschev. *Modal logic.* Clarendon Press, Oxford, 1997.

[17] R. Cignoli. *The class of Kleene algebras satisfying interpolation property and Nelson algebras.* Algebra Universalis, **23**, No. 3, 1986, 262–292.

[18] J.P. Cleave. *The notion of logical consequence in the logic of inexact predicates.* Zeitschrift für Mathematische Logic und Grundlagen der Mathematik, **20**, No. 3, 1974, 307–324.

[19] N.C.A. da Costa, J.-Y. Béziau. *Carnot's logic.* Bulletin of the Section of Logic, **23**, No. 2, 1994, 98–105.

[20] H.B. Curry. *On the definition of negation by a fixed proposition in the inferential calculus.* Journal of Symbolic Logic, **17**, No. 1, 1952, 98–104.

[21] H.B. Curry. *The system LD.* Journal of Symbolic Logic, **17**, No. 1, 1952, 35–42.

[22] H.B. Curry, *Foundations of mathematical logic,* McGraw-Hill Book Company, New York, 1963.

[23] K. Dosen. *Negative modal operators in intuitionistic logic.* Publications of the Institute of Mathematics (Beograd), Nouv. Ser. **35(49)**, 1984, 15–20.

[24] K. Dosen. *Negation as a modal operator.* Reports on Mathematical Logic, **20**, 1986, 15–28.

[25] K. Dosen. *Negation in the light of modal logic.* in D. Gabbay, H. Wansing (eds.) What is Negation?, Kluwer, Dordrecht, 1999, 77–86.

[26] J.M. Dunn. *Partiality and its Dual.* Studia Logica, **66**, No. 1, 1980, 5–40.

[27] J.M. Dunn, R.K. Meyer. *Algebraic completeness results for Dummett's LC and its extensions.* Zeitschrift für Mathematische Logic und Grundlagen der Mathematik, **17**, No. 2, 1971, 225–230.

[28] M.M. Fidel. *An algebraic study of a propositional system of Nelson.* Mathematical Logic, Proc. of the First Brasilian Conf. on Math. Logic, Campinas 1977, Lect. Notes Pure Appl. Math. **39**, 1978, 99–117.

[29] M.M. Fidel. *An algebraic study of logic with constructive negation.* Proc. of the Third Brazilian Conf. on Math. Logic, Recife 1979, 1980, 119–129.

[30] K. Fine. *Logics containing K4. I.*, Journal of Symbolic Logic, **39**, No. 1, 1974, 31–42.

[31] K. Fine. *Logics containing K4. II.*, Journal of Symbolic Logic, **50**, No. 3, 1985, 619–651.

[32] J.M. Font, P. Hajek. *On Lukasiewicz's four-valued modal logic.* Studia Logica, **70**, No. 2, 2002, 157–182.

[33] V. Goranko. *The Craig interpolation theorem for propositional logics with strong negation.* Studia Logica, **44**, No. 3, 1985, 291–317.

[34] J. van Heijenoort. *From Frege to Gödel. A Source Book in Mathematical Logic, 1879–1931.* Harvard University Press, Cambridge, 1967.

[35] A. Heyting. *Die formalen Regeln der intuitionistischen Logik.* Sitzungsber. preuss. Akad. Wiss. Berlin, 1930, 42–56.

[36] A. Heyting. *Intuitionism. An introduction.* North-Holland, Amsterdam, 1971.

[37] V.A. Jankov. *The relationship between deducibility in the intuitionistic propositional calculus and finite implicational structures.* Soviet Math. Dokl., **4**, 1963, 1203–1204.

[38] V.A. Jankov. *Constructing a sequence of strongly independent superintuitionistic propositional calculi.* Soviet Math. Dokl., **9**, 1968, 806–807.

[39] S. Jaśkowski. *Propositional calculus for contradictory deductive systems.* Studia Logica, **24**, 1969, 143–157.

[40] J. Jaspars. *Calculi for constructive communication.* ILLC Dissertation Series 1994-4, ILLC, 1994.

[41] I. Johansson. *Der Minimalkalkül, ein reduzierter intuitionistischer Formalismus.* Compositio Mathematika, **4**, 1937, 119–136.

[42] S. Kanger. *A note on partial postulate sets for propositional logic.* Theoria, **21**, No. 1, 1955, 99–104.

[43] A.S. Karpenko. *Two three-valued isomorphs of classical propositional logic and their combinations.* First World Congress on Paraconsistency, Abstracts, Ghent, 1997, 92–94.

[44] A.N. Kolmogorov. *On the principle of excluded middle.* Matematicheskij sbornik, **32**, No. 4, 1925, 646–667 [*in Russian*]. English translation [34], 414–437.

[45] A. Kolmogoroff. *Zur Deutung der intuitionistischen Logik.* Mathematische Zeitschrift, **35**, 1932, 58–65.

[46] S. Körner. *Experience and Theory.* Kegan Paul, London, 1966.

[47] M. Kracht. *On extensions of intermediate logics by strong negation.* Journal of Philosophical Logic, **27**, No. 1, 1998, 49–73.

[48] S.A. Kripke. *Distinguished constituents (abstract).* J. Symb. Logic, **24**, 1959, 323.

[49] F. von Kutschera. *Ein verallgemeinerter Widerlegungsbegriff für Gentzenkalküle.* Archiv für Mathematische Logik und Grundlagenforschung, **12**, No. 2, 1969, 104–118.

[50] A. Loparić, N.C.A. da Costa. *Paraconsistency, paracompleteness and induction.* Logique et Analyse, Nouv. Ser., **113**, 1986, 73–80.

[51] E.G.K. Lopez-Escobar. *Refutability and Elementary Number Theory.* Indagationes Mathematicae, **34**, 1972, 362–374.

[52] J. Łukasiewicz. *A system of modal logic.* The Journal of Computing Systems, **1**, 1953, 111–149. Reprinted in [54], 352–390.

[53] J. Łukasiewicz. *Aristotle's syllogistic from the standpoint of modern formal logic (2nd enlarged edition).* Clarendon Press, Oxford, 1957.

[54] J. Łukasiewicz. *Selected works, edited by L. Borkowski.* Stud. in Logic and the Found. of Math. North-Holland, Amsterdam, 1970.

[55] L.L. Maksimova. *Pretabular superintuitionist logic.* Algebra and Logic, **11**, No. 5, 1972, 308–314.

[56] L.L. Maksimova. *Craig's theorem in superintuitionistic logics and amalgamable varieties of pseudo-Boolean algebras.* Algebra and Logic, **16**, No. 6, 1977, 427–455.

[57] L.L. Maksimova. *Implicit Definability and Positive Logics.* Algebra and Logic, **42**, No. 1, 2003, 37–53.

[58] L.L. Maksimova. *Interpolation and Definability in Extensions of the Minimal Logic.* Algebra and Logic, **44**, No. 6, 2005, 407–421.

[59] A.A. Markov. *Constructive logic.* Usp. Mat. Nauk, **5**, No. 3, 1950, 187–188 (in Russian).

[60] A.A. Markov. *On constructive mathematics.* Proceedings of Steklov Mathematical Institute, **67**, 1962, 8–14 (in Russian).

[61] E. Mendelson. *Introduction to mathematical logic, 2nd edition.* D. Van Nostrand Company, New York, 1979.

[62] C.M. McKay. *On finite logics.* Indag. Math., **29**, No. 3, 1967, 363–365.

[63] S. Miura. *A remark on the intersection of two logics.* Nagoja Math. J., **26**, No. 2, 1966, 167–171.

[64] D. Nelson. *Constructible falsity.* Journal of Symbolic Logic, **14**, No. 1, 1949, 16–26.

[65] D. Nelson. *Negation and separation of concepts.* In: Constructivity in Mathematics, North-Holland, Amsterdam, 1959, 208–225.

[66] S.P. Odintsov. *Propositional relatively constructive systems.* Vychislitel'nye sistemy, **158**, Novosibirsk, 1997, 110–126 (in Russian).

[67] S.P. Odintsov. *Isomorphs of the logic of classical refutability and their generalizations.* Proceedings of the seminar of logical center, Inst. of Philosophy of RAS, Moscow, 1998, 48–61.

[68] S.P. Odintsov. *Maximal paraconsistent extension of Johansson logic.* Logique et Analyse, Nouv. Ser, **161/163**, 1998, 107–120.

[69] S.P. Odintsov. *Representation of j-algebras and Segerberg's logics.* Log. Anal., Nouv. Ser., **165/166**, 1999, 81–106.

[70] S.P. Odintsov. *Logic of classical refutability and class of extensions of minimal logic.* Logic and Logical Philosophy, **9**, 2002, 91–107.

[71] S.P. Odintsov. *On the embedding of Nelson's logics.* Bull. Sect. Log., Univ. Lodz, Dep. Log., **31**, No. 4, 2002, 241–250.

[72] S.P. Odintsov. *Algebraic semantics for paraconsistent Nelson's Logic.* Journal of Logic and Computation, **13**, No. 4, 2003, 453–468.

[73] S.P. Odintsov. *"Reductio ad Absurdum" and Łukasiewicz's Modalities.* Logic and Logical Philosophy, **11**, 2003, 149–166.

[74] S.P. Odintsov. *On representation of* **N4**-*lattices.* Studia Logica, **76**, No. 3, 2004, 385–405.

[75] S.P. Odintsov. *Negative Equivalence of Extensions of Minimal Logic.* Studia Logica, **78**, No. 3, 2004, 417–442.

[76] S.P. Odintsov. *On the Structure of Paraconsistent Extensions of Johansson's Logic.* Journal of Applied Logic, **3**, No. 1, 2005, 43–65.

[77] S.P. Odintsov. *The Class of Extensions of Nelson Paraconsistent Logic.* Studia Logica, **80**, No. 2-3, 2005, 291–320.

[78] S.P. Odintsov. *Transfer Theorems for Extensions of the Paraconsistent Nelson Logic.* Algebra and Logic, **45**, No. 4, 2006, 232–247.

[79] S.P. Odintsov. *On Extensions of Nelson's Logic Satisfying Dummett's Axiom.* Siberian Mathematical Journal, **48**, No. 1, 2007, 112–125.

[80] S.P. Odintsov. *The Lattice of Extensions of the Minimal Logic.* Siberian Advances in Mathematics, **17**, No. 2, 2007, 112–143.

[81] S.P. Odintsov. *Absurdity as Unary Operator.* in: J. Malinowski, A. Pietruszczak (eds.) Essays in Logic and Ontology, Poznań Studies in the Philosophy of Sciences and the Humanities 91, Rodopi, NY, 2006, 225–242.

[82] S.P. Odintsov, D. Pearce. *Routley semantics for answer sets.* in: G.Ch. Baral et al. (eds.) Logic Programming and Nonmonotonic Reasoning, 8th International Conference, LPNMR 2005, Diamante, Italy, September 5-8, 2005, Proceedings (LNCS 3662), Springer, 2005, 343–355.

[83] H. Ono. *Kripke models and intermediate logics.* Publ. of Research Inst. for Math. Sci., Kyoto Univ., **6**, 1970, 461–476.

[84] J. Porte. *The Ω-system and the L-system of modal logic.* Notre Dame J. Formal Logic, **20**, No. 4, 1970, 915–920.

[85] J. Porte. *Lukasiewicz's L-modal system and classical refutability.* Logique et Analyse, Now. Ser., **27**, 87–92 (1984).

[86] G. Priest. *Paraconsistent logic.* in: D. Gabbay, H. Guenter (eds.) Handbook of Philosophical Logic, Vol. 6, 2nd edition, Dordrecht: Kluwer, 2002, 287–393.

[87] G. Priest, R. Routley, J. Norman (eds.) *Paraconsistent logic. Essays on the inconsistent.* München, Philosophia Verlag, 1989.

[88] A.P. Pynko. *Algebraic study of Sette's maximal paraconsistent logic.* Studia Logica, **54**, No. 1, 1995, 89–128.

[89] A.P. Pynko. *Functional completeness and axiomatizability within Belnap's four-valued logic and its expansions.* Journal of Applied Non-classical Logics, **9**, No. 1, 1999, 61–105.

[90] H. Rasiowa. *N-lattices and constructive logic with strong negation.* Fundamenta Mathematicae, **46**, No. 1, 1968, 61–80.

[91] H. Rasiowa. *Algebraische Charakterisierung der intuitionistischen Logik mit starker Negation.* in: A. Heyting (ed.) Constructivity in Mathematics, Amsterdam, North-Holland, 1959, 234–240.

[92] H. Rasiowa. *An algebraic approach to non-classical logics.* Amsterdam, North-Holland, 1974.

[93] W. Rautenberg. *Klassische und nichtclassische Aussagenlogik.* Braunschweig, Vieweg, 1979.

[94] N. Rescher. *Many-valued Logic.* N.Y., 1969.

[95] T.T. Robinson. *Independence of two nice sets of axioms for the propositional calculus.* Journal of Symbolic Logic, **33**, No. 2, 1968, 265–270.

[96] R. Routley. *Semantical analyses of propositional systems of Fitch and Nelson.* Studia Logica, **33**, No. 3, 1974, 283–298.

[97] K.F. Samokhvalov. *Relatively constructive systems.* Vychislitel'nye sistemy, **124**, Nowosibirsk, 1988, 99–113 [in Russian].

[98] K. Segerberg. *Propositional logics related to Heyting's and Johansson's.* Theoria, **34**, No. 1, 1968, 26–61.

[99] A. Sendlewski. *Some investigations of varieties of N-lattices.* Studia Logica, **43**, No. 3, 1984, 257–280.

[100] A. Sendlewski. *Nelson algebras through Heyting ones.* Studia Logica, **49**, No. 1, 1990, 106-126.

[101] A. Sendlewski. *Axiomatic extensions of the constructive logic with strong negation and disjunction property.* Studia Logica, **55**, No. 3, 1995, 377–388.

[102] A.M. Sette. *On the propositional calculus* P^1. Mathematica Japonicae, **18**, No. 3, 1973, 173–180.

[103] M.V. Stukacheva. *On disjunction property in the class of extensions of minimal logic.* Algebra and Logic, **43**, No. 2, 2004, 132–141.

[104] M.V. Stukacheva. *Some Remarks on Constrctive Extensions of Minimal Logic.* Vestnik NGU, Ser.: Mathematics, Mechanics and Informatics, **5**, No. 3, 2005, 3–16 (in Russian).

[105] M.V. Stukacheva. *On canonical formulas for the extensions of minimal logic.* Siberian Electronic Mathematical Reports, **3**, 2006, 312–334.

[106] N.-Y. Suzuki. *Constructing a continuum of predicate extensions of each intermediate propositional logics.* Studia Logica, **54**, No. 2, 173–198 (1995).

[107] R. Thomason. *A semantical study of constructive falsity.* Zeitschrift für mathematische Logik und Grundlagen der Mathematik, **15**, No. 3, 1969, 247–257.

[108] I. Urbas. *A note on "Carnot's logic".* Bulletin of the Section of Logic, **23**, No. 3, 1994, 118–125.

[109] D. Vakarelov. *Theory of negation in certain logical systems. Algebraic and semantical approach.* Ph.D. dissertation, University of Warsaw, 1976.

[110] D. Vakarelov. *Notes on N-lattices and constructive logic with strong negation.* Studia Logica, **36**, No. 1–2, 1977, 109–125.

[111] D. Vakarelov. *Consistency, completeness and negation.* In: [87], 328–363.

[112] D. Vakarelov. *Nelson's negation on the base of weaker versions of intuitionistic negation.* Studia Logica, **80**, 2005, 393–430.

[113] D. Vakarelov. *Non-classical negation in the works by Helena Rasiowa and their impact on the theory of negation.* Studia Logica, **84**, 2006, 105–127.

[114] N.N. Vorob'ev. *A constructive propositional calculus with strong negation.* Doklady Akademii Nauk SSSR, **85**, 1952, 465–468 (in Russian).

[115] N.N. Vorob'ev. *The problem of deducibility in constructive propositional calculus with strong negation.* Doklady Akademii Nauk SSSR, **85**, 1952, 689–692 (in Russian).

[116] N.N. Vorob'ev. *Constructive propositional calculus with strong negation.* Transactions of Steklov's Institute, **72**, 1964, 195–227 (in Russian).

[117] G. Wagner. *Vivid logic. Knowledge-based reasoning with two kinds of negation.* Springer LNAI 764, 1994.

[118] H. Wansing. *The logic of information structures.* Springer LNAI 681, 1993.

[119] H. Wansing. *Semantics-based nonmonotonic inference.* Notre Dame Journal of Formal Logic, **36**, No. 1, 1995, 44–54.

[120] H. Wansing. *Negation.* In: L. Goble (ed.), The Blackwell Guide to Philosophical Logic, Basil Blackwell Publishers, Cambridge/MA, 2001, 415–436.

[121] H. Wansing. *Diamonds are a philosopher's best friends.* Journal of Philosophical Logic, **31**, No. 6, 2002, 591–612.

[122] P. Wojtylak. *Mutual interpretability of sentential logic. I.* Reports on Mathematical Logic, **11**, 1981, 69–89.

[123] P. Wojtylak. *Mutual interpretability of sentential logic. II.* Reports on Mathematical Logic, **12**, 1981, 51–66.

[124] P.W. Woodruff. *A note on JP'.* Theoria, **36**, No. 2, 1970, 183–184.

[125] A. Wroński. *The degree of completeness of some fragments of the intuitionistic propositional logic.* Reports on Mathematical Logic, **2**, 1974, 55–62.

[126] A. Wroński. *On the cardinalities of matrices strongly adequate for the intuitionistic propositional logic.* Reports on Mathematical Logic, **3**, 1974, 67–72.

Index

TRENDS IN LOGIC

1. G. Schurz: *The Is-Ought Problem.* An Investigation in Philosophical Logic. 1997 ISBN 0-7923-4410-3
2. E. Ejerhed and S. Lindström (eds.): *Logic, Action and Cognition.* Essays in Philosophical Logic. 1997 ISBN 0-7923-4560-6
3. H. Wansing: *Displaying Modal Logic.* 1998 ISBN 0-7923-5205-X
4. P. Hájek: *Metamathematics of Fuzzy Logic.* 1998 ISBN 0-7923-5238-6
5. H.J. Ohlbach and U. Reyle (eds.): *Logic, Language and Reasoning.* Essays in Honour of Dov Gabbay. 1999 ISBN 0-7923-5687-X
6. K. Došen: *Cut Elimination in Categories.* 2000 ISBN 0-7923-5720-5
7. R.L.O. Cignoli, I.M.L. D'Ottaviano and D. Mundici: *Algebraic Foundations of many-valued Reasoning.* 2000 ISBN 0-7923-6009-5
8. E.P. Klement, R. Mesiar and E. Pap: *Triangular Norms.* 2000 ISBN 0-7923-6416-3
9. V.F. Hendricks: *The Convergence of Scientific Knowledge.* A View From the Limit. 2001 ISBN 0-7923-6929-7
10. J. Czelakowski: *Protoalgebraic Logics.* 2001 ISBN 0-7923-6940-8
11. G. Gerla: *Fuzzy Logic.* Mathematical Tools for Approximate Reasoning. 2001 ISBN 0-7923-6941-6
12. M. Fitting: *Types, Tableaus, and Gödel's God.* 2002 ISBN 1-4020-0604-7
13. F. Paoli: *Substructural Logics: A Primer.* 2002 ISBN 1-4020-0605-5
14. S. Ghilardi and M. Zawadowki: *Sheaves, Games, and Model Completions.* A Categorical Approach to Nonclassical Propositional Logics. 2002 ISBN 1-4020-0660-8
15. G. Coletti and R. Scozzafava: *Probabilistic Logic in a Coherent Setting.* 2002 ISBN 1-4020-0917-8; Pb: 1-4020-0970-4
16. P. Kawalec: *Structural Reliabilism.* Inductive Logic as a Theory of Justification. 2002 ISBN 1-4020-1013-3
17. B. Löwe, W. Malzkorn and T. Räsch (eds.): *Foundations of the Formal Sciences II.* Applications of Mathematical Logic in Philosophy and Linguistics, Papers of a conference held in Bonn, November 10-13, 2000. 2003 ISBN 1-4020-1154-7
18. R.J.G.B. de Queiroz (ed.): *Logic for Concurrency and Synchronisation.* 2003 ISBN 1-4020-1270-5
19. A. Marcja and C. Toffalori: *A Guide to Classical and Modern Model Theory.* 2003 ISBN 1-4020-1330-2; Pb 1-4020-1331-0
20. S.E. Rodabaugh and E.P. Klement (eds.): *Topological and Algebraic Structures in Fuzzy Sets.* A Handbook of Recent Developments in the Mathematics of Fuzzy Sets. 2003 ISBN 1-4020-1515-1; Pb 1-4020-1516-X

21. V.F. Hendricks and J. Malinowski: *Trends in Logic*. 50 Years Studia Logica. 2003
ISBN 1-4020-1601-8

22. M. Dalla Chiara, R. Giuntini and R. Greechie: *Reasoning in Quantum Theory*. Sharp and Unsharp Quantum Logics. 2004 ISBN 1-4020-1978-5

23. B. Löwe, B. Piwinger and T. Räsch (eds.): *Classical and New Paradigms of Computation and their Complexity Hierarchies*. Papers of the conference "Foundations of the Formal Sciences III" held in Vienna, September 21...24, 2001. 2004
ISBN 1-4020-2775-3

24. G. Jäger: *Anaphora and Type Logical Grammar*. 2005 ISBN 978-1-4020-3904-1

25. M. Winter: *Goguen Categories*. A Categorical Approach to L-fuzzy Relations. 2007
ISBN 978-1-4020-6163-9

26. S.P. Odintsov: *Constructive Negations and Paraconsistency*. 2008
ISBN 978-1-4020-6866-9

Printed in the United Kingdom
by Lightning Source UK Ltd.
127958UK00002B/121/P